Performing Transversally

Performing Transversally:
Reimagining Shakespeare and the Critical Future

BY
BRYAN REYNOLDS

PERFORMING TRANSVERSALLY
© Bryan Reynolds, 2003.

Softcover reprint of the hardcover 1st edition 2003 978-0-312-29331-4

All rights reserved. No part of this book may be used or reproduced in any manner whatsoever without written permission except in the case of brief quotations embodied in critical articles or reviews.

First published 2003 by
PALGRAVE MACMILLAN™
175 Fifth Avenue, New York, N.Y. 10010 and
Houndmills, Basingstoke, Hampshire, England RG21 6XS
Companies and representatives throughout the world

PALGRAVE MACMILLAN is the global academic imprint of the Palgrave Macmillan division of St. Martin's Press, LLC and of Palgrave Macmillan Ltd. Macmillan® is a registered trademark in the United States, United Kingdom and other countries. Palgrave is a registered trademark in the European Union and other countries.

ISBN 978-1-349-63395-1   ISBN 978-1-137-10764-0 (eBook)
DOI 10.1007/978-1-137-10764-0

Library of Congress Cataloging-in-Publication Data
Reynolds, Bryan (Bryan Randolph)
   Performing transversally : reimagining Shakespeare and the critical future / by Bryan Reynolds.
   p. cm.
   Includes bibliographical references and index.

   1. Shakespeare, William, 1564–1616—Dramatic production. 2. Shakespeare, William, 1564–1616—Adaptations—History and criticism. 3. Shakespeare, William, 1564–1616—Film and video adaptations. 4. Shakespeare, William, 1564–1616—Stage history—1950– I. Title.

PR3091.R477 2003
822.3'3—dc21   2003051798

A catalogue record for this book is available from the British Library.
Design by Newgen Imaging Systems (P) Ltd., Chennai, India.

First edition: September, 2003
10 9 8 7 6 5 4 3 2

Transferred to Digital Printing 2008

For Kris

Contents

List of Illustrations	ix
Acknowledgments	xi
Foreword: Seeing Across Shakespeare	xiii
Janelle Reinelt	

1. Transversal Performance: Shakespace, the September 11
   Attacks, and the Critical Future   1
   Bryan Reynolds

2. The Making of Authorships: Transversal Navigation in the
   Wake of *Hamlet*, Robert Wilson,
   Wolfgang Wiens, and Shakespace   29
   D.J. Hopkins & Bryan Reynolds

3. Venetian Ideology or Transversal Power? Iago's
   Motives and the Means by which Othello Falls   53
   Joseph Fitzpatrick & Bryan Reynolds
   (with additional dialogue by Bryan Reynolds & Janna Segal)

4. "What is the city but the people?" Transversal
   Performance and Radical Politics in Shakespeare's
   Coriolanus and Brecht's *Coriolan*   85
   Bryan Reynolds

5. Untimely Ripped: Mediating Witchcraft in
   Polanski and Shakespeare   111
   Bryan Reynolds

6. Nudge, Nudge, Wink, Wink, Know What I Mean,
   Know What I Mean? A Theoretical Approach to
   Performance for a Post-Cinema Shakespeare   137
   D.J. Hopkins, Catherine Ingman, & Bryan Reynolds

viii \ CONTENTS

7. "A little touch of Harry in the night": Translucency and
   Projective Transversality in the Sexual and
   National Politics of *Henry V* — 171
   Donald Hedrick & Bryan Reynolds

8. Inspriteful Ariels: Transversal Tempests — 189
   Bryan Reynolds & Ayanna Thompson

9. "For such a sight will blind a father's eye":
   The Spectacle of Suffering in Taymor's *Titus* — 215
   Courtney Lehmann, Bryan Reynolds, & Lisa Starks

10. Friend or Fo, Shakespeare's Ends is the Means:
    Revising Early Modern English Iconography,
    Elisabetta Points Toward the Critical Future — 245
    Bryan Reynolds & Janna Segal

Afterword: Walk Like an Egyptian — 271
Jonathan Gil Harris

Appendix: Transversal Poetics—I.E. Mode — 287
zooz

Notes on Collaborators — 303

Index — 307

Illustrations

5.1	Witches' lair	128
5.2	Macbeth's severed head	130
8.1	Aimé Césaire's *Une Tempête* at The Gate in London (photo by Pau Ros)	197
8.2	William Hogarth, *Scene from "The Tempest,"* ca. 1735–1740 (Nostell Priory)	202
8.3	Henry Fuseli's *Ariel*, ca. 1800–1810 (Folger Shakespeare Library)	203
8.4	F. Miller's *Ariel*, ca. 1850 (*The Art Journal* 12[1850]: 278)	204
8.5	Charles Buchel's charcoal drawing of Sir Herbert Beerbohm as Caliban, 1904 (Folger Shakespeare Library)	205
8.6	Canada Lee as Caliban in Margaret Webster's 1945 production of *The Tempest* in New York (photo by Eileen Darby in *Theatre Arts Magazine*, February 1945: 88)	206
9.1	Titus and Tamora	216
9.2	Lavinia	230
9.3	Final boy	236

Acknowledgments

Collaboration can be a wonderful experience, challenging and gratifying on many levels. The experience of writing this book, however, far exceeded my expectations of how wonderful the experience could be. When I conceived of this project I was not at all apprehensive about the prospect of collaborating with so many people. I was excited by it. But my excitement was driven partially by not the healthiest of reasons. I looked forward to the conflict, to the ideological and methodological struggles, to the battles of egos and wills, as well as to engagement with alternative thoughts and feelings, unanticipated stimulations, surprise accomplishments, developing friendships, and personal growth. I imagined the project being quintessentially transversal, exemplifying transversality by both inspiring and propelling the collaborators into uncharted conceptual and emotional territories. What I got, as it turns out, was all good. In retrospect, I do not lament the lack of conflict. Of course, there were ideological, methodological, and interpretive disagreements, but these were always resolved through enjoyable rhetorical exchanges, often involving much humor ("Are we being transversal yet?"), as well as a little intoxication. From my perspective, this book is a homage to my collaborators, to those whose names are credited in the following pages, and to those unnamed, the many people who had ideas bounced off their heads—often with staggering repetition—by all of us credited folks. It goes without saying that the book could not have been written without my collaborators, and so I am most indebted to them, but also this book could not have been written without the support of many people whom it is my honor to thank here.

In addition to being grateful for all the support given to me by the university as a whole and by the School of the Arts in particular, I am grateful to my colleagues at the University of California at Irvine and in the joint UCI/UCSD Ph.D. Program in Drama and Theater. Of the many UCI and UCSD colleagues who have supported me in various ways during the writing of this book, I am especially indebted to Jill

Beck, Eli Simon, Dudley Knight, Stephen Barker, Robert Weimann, John Rouse, Jim Carmody, David McDonald, Barbara Thibodeau, Cameron Harvey, Ann Pellegrini, Janelle Reinelt, and Robert Cohen. Other colleagues, friends, and family whom I am grateful to include Monty Hom, Rodney Byrd, Julie Kiernan, Todd Wright, Greg Ungar, Karen Weber, Richard Burt, Bill Worthen, Tony Kubiak, Greg Reynolds, Donna Reynolds, Don Reynolds, Marge Garber, David Podley, Mary Dryer, Ian Munro, Matt Grant, Anke Ortlepp, Kirsten Lammich, Margo Crespin, Jennifer Gregory, Barry Crittenden, Jenn Colella, Tom Augst, Scott Albertson, and Kim Savelson. I want to give special thanks to two of my collaborators: Don Hedrick, who made me smile when I needed to and, when I did not need to, made me dance to delirium before entertaining a new idea of mine; and Janna Segal, whose unrelenting upbeat verve, wit, and diligence worked uncannily to keep me focused when my transversal affinities were inclined to take me elsewhere. I am also especially indebted to my editor, Kristi Long, who first listened politely to a sleepless me ramble on frantically over a hardy lunch about the book-in-progress at the Shakespeare Association of America Conference in Montreal, then came to believe in its potential when I showed her several chapters and some fragments, and has now shepherded it—with the help of Palgrave's Farideh Koohi-Kamali and Roee Raz—to the masses. Lastly and certainly most of all, I am grateful to Kris Lang, my partner in all things transversal and not, the best friend a person could ever have, and the love of my life.

Portions of this book have appeared in *Shakespeare After Mass Media*, ed. Richard Burt (New York: Palgrave, 2002), *Critical Essays on Othello*, ed. Philip Kolin (New York: Routledge, 2002), *Shakespeare Without Class: Misappropriations of Cultural Capital*, ed. Bryan Reynolds and Donald Hedrick (New York: Palgrave, 2000), *Social Semiotics: A Transdisciplinary Journal in Functional Linguistics, Semiotics and Critical Theory* 7 : 2 (August 1997), *The Upstart Crow: A Shakespeare Journal* 8 (1993); and are here reprinted by kind permission.

Foreword: Seeing Across Shakespeare

Bryan Reynolds has written a new book about Shakespeare—except that it defies categories, even that of being "about Shakespeare." In many ways, Shakespeare is the pretext for an intervention into postmodern theory that is intent on linking disparate, usually isolated fields. Traversing high and low culture, several media, four centuries, and multiple scholarly schools of thought, this book practices the transversal theory it preaches. Beginning with the events of September 11 and ending with a comic mock dialogue between "ooz" and "zoo" on transversal poetics, Reynolds steers a course among multiple and complex subjects of interest to theater, film, and literary scholars, and more generally, those with interests in cultural theory. A great strength of the book lies in something Reynolds calls the "principle of translucency": it provides a multicoded, multiperspectival discourse that produces for the reader the ability to "see one or two or more things through others while at the same time seeing the others themselves."

Over the past fifteen years, the fledgling discipline of performance studies has developed its relationships to cultural studies, ethnography, theater studies, and media studies while insisting on its own proper object of study—performance—taken in its broadest sense to mean an "embodied enactment of cultural forces" (Jon McKenzie, *Perform Or Else; From Discipline to Performance* [London and New York: Routledge, 2001], 8). Shakespeare scholars have often remained the most literary of scholars, but a significant number of them have broken open new areas of Shakespeare studies—I'm thinking particularly of Jonathan Dollimore, Marjorie Garber, Stephen Greenblatt, Jean Howard, Phyllis Rackin, Alan Sinfield, and others whose approaches have provided a radical realignment of Shakespeare studies with new historicism, cultural studies, feminism, queer theory, and poststructuralism generally. Yet all of these fields have their own identifying marks, their own calling cards, so to speak. Often, academic practices of power and status have seemed to construct exclusions or hierarchies where other more productive structural relations might have existed. Thus drama (meaning the literary study of texts) has been opposed to theater (meaning the concrete stagings of live performance); live performance has been opposed to film and other media; "performativity" has been debated in terms of whether or not it includes actual performances in its Austinian scope; Shakespeare has been claimed by traditionalists and innovators, and many academics in several fields still struggle with the high/low culture binary.

Refreshingly, Reynolds brushes aside these old grievances to construct a cohabitation of these various terms while not sacrificing the singularity of any. In his chapter on Robert Wilson's *Hamlet*, Reynolds couples this avant-garde theater artist's performance with the rock star Lou Reed's attendance at the performance to begin a multifaceted discussion of issues of textuality (as in Barthes) and of textual authority (as in the scholarly controversy over Shakespeare's quartos). This trajectory links to another strand of the chapter concerned with the nature of theatrical modes of production after mass media. The Wilson/Reed connection supports an insistence that live presence matters—that Wilson and Reed become the authors of their performance events, the media for themselves, for characters, and for the audiences who view them. The only adequate "quotation" of a theater or concert event would be "one that includes the performance of a portion of the event and the actual presence of Robert Wilson [or Lou Reed]." Reynolds desires the reader to be able to think about avant-garde performances and rock concerts in the same moment; similarly, the discussion of what kind of relationship—or not—Wilson's *Hamlet* has to Shakespeare's text entails philosophical questions about the status of authorship as well as the scholarly debates about textual authority that have defined one area of Shakespeare studies. The desire to have his readers think about these issues simultaneously without collapsing them into each other or subordinating them to an orderly whole creates the translucent effect of his scholarship.

Reynolds is primarily interested in developing a theory—"transversal theory"—that offers an optimistic, almost a utopian sense of possibility to cultural criticism, while posing a conception of how human subjectivity changes and morphs. In this sense, it is an attempt to account for novelty in the philosophical sense of that term: how is it that anything new ever actually enters into representation? What do we understand as the constraints but also the unpredictable novel possibilities of representation, cognition, aesthetic experience? Artistic experience can foster empathetic identification, which in turn, can lead to breaking free from "subjective territory," that space within which individuals are "subjected conceptually and emotionally, that is, developed into a subject by the state machinery of any hegemonic society or subsociety (such as the university, criminal organizations, or religious societies)." In other words, Reynolds claims for art the power to push individuals beyond their limits into "transversal territory" where multiple forms of subjectivity, contradictory feelings and cognitions, and transformative possibilities may be grasped, at least in the subjunctive.

These ideas, then, call on performance as an art form that can most easily produce transversal events, because the embodied enactment of cultural forces produces and elicits empathetic identification that can propel the border crossings beyond subjective space. "Empathy enables people to venture beyond their own conceptual and emotional boundaries, to think and feel as others do or might, and thus expand or transcend their own 'subjective territories.' . . . Transversal movements occur when one entertains alternative perspectives and breaches the parameters of their subjectification." Thus Reynolds focuses on films and live performances that offer maximum provocations for transversal movement. Julie Taymor's *Titus* is a model film, and she develops a transversal aesthetic because she creates her works by combining and proliferating techniques, cultures, and media. Combining masks and puppets with human actors, for instance, gives her representational universe a widely attractive, seductive multiplicity. Blurring the distinctions between cinematic and theatrical conventions, she still keeps both discreetly present, capable of being enjoyed and apprehended simultaneously and yet distinctly.

Among the key terms in transversal theory is the "investigative-expansive mode." In spite of its seemingly clunky terminology, it is crucial to the theoretical enterprise, and is surprisingly flexible. Reynolds and his collaborator James Intriligator recommend this methodology to those approaching either artistry or scholarship; indeed, it seems to operate as a kind of epistemological principle. In brief, the idea is that usually the emphasis in critical thinking is on definition and containment; that is, the subject matter under study is broken down into various parts for analysis, then interpreted or "read" as a coherent and unified whole. Reynolds thinks this coherence is often achieved at the expense of an exploration of the subject's relations to other aspects of its context, relations that may lead toward a more dispersed but imaginative and productive account of the original subject. So by deliberately introducing seemingly tangential ideas, objects, theories and so on into a study of, say, *Othello*, a more complex knowledge of the text or performance may emerge. Reynolds is an advocate for creating scholarship that evokes the same transversal territory he suggests artworks can create for their audiences. The value of this approach is an increase in complexity of understanding, maximizing the possibilities inherent in the subject, and expanding the horizons of disciplinary thinking.

In his chapter on *Othello*, Reynolds and his collaborators trace the critical tradition of interpretations of Othello as they vacillate between racist affirmations of his barbarous nature (the "true self" shows through in the murder of Desdemona) and attempts to avoid such a racist reading by seeing virtuous Othello (virtuous in Christian, Venetian/Elizabethan terms) ruined by the demonic Iago. Concluding that none of the existing interpretations overcomes fundamental contradictions that remain, Reynolds turns to an essay by Deleuze on sadomasochism to challenge his own analysis that has been largely ideological, turning on conceptions of Christian virtue and Venetian government by considering instead, or rather in addition, a psychoanalytic view of the sadistic and masochistic aspects of the behaviors of Iago, Othello, and Desdemona. His conclusion is that the binary oppositions the play seems to encourage lead to contradictions, and that the point is not to resolve them but to recognize the primary experience of moving through them: "The play is not built on categorizations such as that of Othello as the black man with the white soul and Iago as the white man with the black soul. Rather, it is built on such actions as the audience's realization that their expectations have been contradicted.... By contrast, our analysis is concerned primarily with this movement, the process of change, rather than the initial and final states."

Other chapters employ this investigative-expansive mode in various guises. Italian playwright/performer Dario Fo and his wife Franca Rame are praised because their mode of "dramaturgy does not seek a cohesive conclusion or holistic, unified meaning, but rather a dynamic, fluctuating collection of variables that reveal the variability and permeability of dichotomous sociocultural demarcations." In chapter 5, on Polanski's *Macbeth*, Reynolds combines a contextual comparison of the cultural environments concerning witchcraft involved in the original Renaissance performances and 1960s "hippy culture" (also containing elements of witchcraft) that formed the background for the Tate-LaBianca murders and Polanski's film. In a chapter on film acting in film versions of and about Shakespeare, Reynolds calls for a "post-cinematic" film technique, one that would take into account audiences' knowledge of cinematic contrivance and codes, make cocreators of the audiences, and would put performers at the center of the creative process by emphasizing the actor's "own self-aware textual analysis [and] simultaneous interpretation and execution." This would result, of course, in a dispersed responsibility for making meanings among audiences and artists, and would foreclose the typical director's *tour de force* (one thinks of Kenneth Branagh, although Reynolds doesn't mention him). An investigative-expansive mode of film making such as this better suits an age following that of the dominance of Hollywood cinema, one in which interactive video games, Internet surfing, and self-governed home video has begun to change the nature and desires of viewing, of being an audience.

Throughout the book, Reynolds seeks to confound the tidy limits of just about everything he touches. It is not surprising that the theorists to whom he has the most affinity are Deleuze and Guattari, although he criticizes Deleuze for the stability of his definitions. The emphasis on change, open-ended meanings, and expanding the parameters of the possible applies to Shakespeare performances, historical and contemporary public events, and essays written by scholars on any of these.

Reynolds confounds the usual conception of scholarship in yet another area: that precinct known as the ownership of ideas. This book is written with other people. Most of the chapters are joint chapters. Although Reynolds is the only author in every chapter, D.J. Hopkins and Janna Segal each coauthor two chapters, and five other people are involved in coauthoring chapters—nine if Jonathan Gil Harris and I are counted as well. Reynolds is producing a volume of transversal scholarship that refuses to be contained within the usual author-function of American scholarship in the humanities. This book isn't an edited collection of essays, and it isn't a single-author monograph either. Rather, it has a dominant author and a series of affiliated authors with more limited responsibility for the ideas therein. So how to think about this book—as written by Bryan Reynolds and friends? I like that re-vision of academic creativity. It reminds me of Brecht's work with his collaborators—or what some scholars refer to as the "Brecht Collective"—except that Reynolds credits his cocreaters while Brecht often neglected to do so. To Reynolds's invitation to readers to move into transversal territory as they journey through these pages, let me add my own utopian wish for a co-production of artists, critics, and readers who encounter these ideas in Shakespace. May we make even more out of them than they offer.

Janelle Reinelt
Irvine, California

CHAPTER 1

TRANSVERSAL PERFORMANCE: SHAKESPACE,
THE SEPTEMBER 11 ATTACKS, AND
THE CRITICAL FUTURE

Bryan Reynolds

The United States of America's immediate response to the September 11, 2001 attacks on the World Trade Center and the Pentagon was multifarious, but mainly it was horrific astonishment. The people of this country wanted to know who was responsible and why the attacks were perpetrated. They also wanted to know what the damage was and how to fix it. We know now that approximately 3000 people were murdered, yet the answers to the rest of these questions will remain uncertain and inadequate. The attacks were products of a vastly complicated history for which there can be no unmediated access, no singular or absolute truth, and therefore no totalizing resolution. Despite the strong desire for retribution and redemption, and the subsequent invasion of Afghanistan, nothing can undo what has happened. All attempts at restoration can only ever be adaptation, mimicry, and representation. Remains, organic and otherwise, can be processed chemically and/or altered imaginatively to fuel new life through assimilation, fabrication, and/or imitation. In some cases, a life's remains can achieve a powerful symbolic meaning that significantly influences not only the present, but also how we perceive the past, and where we see ourselves in the future.

Moving into the future, forever changed by the September 11 attacks, this introductory chapter considers the positive possibilities not for restoration, but for different kinds of learning and evolution for which the attacks have already become gateways. By analyzing the terrorist attacks as sociopolitical acts framed and executed in ways theoretically and affectively relatable to mass media, theater, music, dance, writing, and other modes of verbal and nonverbal social performance, I want to address issues crucial to the future of critical inquiry and the particular Shakespeare-influenced "spaces"—past, present, future, hypothetical, theatrical, social, cultural, political, historical, theatrical, textual, and critical discourses and coordinates—through which this book ventures and resounds. It is in such discursive, multidimensional "articulatory spaces," Shakespeare effected and/or otherwise, that critical discourses and coordinates interface, social performances are imbued respectively and relationally with meaning, subsequent communication transpires, and learning is achieved.

Negotiated diachronically and synchronically, articulatory spaces are always constrained by what I call "sociopolitical conductors." These are mental and physical movers, orchestrators, and transmitters, such as educational, juridical, and religious structures, multimedia broadcasting and information sources, and the institutions of marriage and family, all of which promote or oppose partially or predominantly, and often contradictorily, the dominant ideology of the society in which they function.[1] The aggregate of a society's sociopolitical conductors that support the dominant ideology I refer to as "state machinery,"[2] a concept that accounts for the singular and plural, human and technological influences that work tirelessly but ultimately futilely to manufacture societal coherence and symbiosis. An absolute state for both individuals (meaning humans individuated from other humans) and/or the society they comprise is never a real prospect as long as physical movement and change are constant realties, even though the quest for stable states and the fear of achieving them will nevertheless always stimulate solidarities and antagonisms among sociopolitical conductors with different views on what the ideal society and state should be. However manipulated by sociopolitical conductors with diverse views of the ideal world, discourse on the September 11 attacks, occupying certain articulatory spaces, will continue to influence the conductors themselves as it continues to impact all areas of critical inquiry, including the interdisciplinary fields of cultural anthropology, literary criticism, performance theory, and Shakespeare studies in which my own research concentrates.

Transversal theory

To more fully explain the way I would like us to think about the September 11 attacks, I will employ more concepts, in addition to "sociopolitical conductors" and "state machinery," that were originally developed from my research on criminality and theater in early modern

England. These concepts are linked to my own theoretical approach, what I call "transversal theory," that I introduced in my 1997 *Theatre Journal* article, "The Devil's House, 'or worse': Transversal Power and Antitheatrical Discourse of Early Modern England," and have developed in a number of publications, most recently my 2002 book, *Becoming Criminal: Transversal Performance and Cultural Dissidence in Early Modern England*.[3] Transversal theory guides the analyses of this book, and this book is an expansion of transversal theory and the concepts and powers that drive it. When I introduce transversal theory to my students, I frequently begin with the example, a hypothetical case, of an explosion unexpectedly occurring in the university classroom. I want to begin here with this example because of its relevant structural similarities, although obviously on a much smaller scale, to the events of September 11. The comparison, I believe, will enable us to examine crucial variables free from many of the immediate, personal biases that discussion of the attacks commonly invokes.

If an explosion unexpectedly occurred in the university classroom, and the teacher and other students were horribly injured, the once familiar and safe space of the classroom and, by extension, the university, would become, in an instant, radically transformed. If the cause of the explosion was ambiguous, though many students suspected foul play, it might render the students more damaged psychologically than if the source were obvious. The unidentifiable and mysterious is usually more terrifying than the readily discernable; an unknown enemy is always more difficult to comprehend, defend against, and fight. Whether the source of the explosion is uncertain or evident, the explosion itself would produce a drastic shift from familiar and safe to unfamiliar and dangerous, and this sudden transformation would have a disorienting effect on the students for some duration, at the very least. According to transversal theory, what I call the students' "subjective territory" would have been altered.

Subjective territory refers to the conceptual and emotional spatial range from which a given subject perceives and experiences the world; this applies to all individuals living in a society who are, consciously and/or not, self-governing in accordance with their state-sanctioned conceptuality and emotionality.[4] Put differently, subjective territory accounts for the individual human who has been subjugated conceptually and emotionally, that is, developed into a subject by the state machinery of any hegemonic society or subsociety (such as the university, criminal organizations, or religious societies); and thus the individual's subjective territory, and corresponding self-government, reflects his or her socioeconomic positioning. Subjective territory reinforces the society's sociopolitical conductors that work to inculcate individuals with the appropriate ideology, the ideology that in turn confirms for them their prescribed addresses, their subjective territory, within the society's hierarchical geography. The resulting determination is physical as well as conceptual and emotional; physical constraints (such as traffic laws, ghettoization, and regionalization) influence the conceptual and emotional aspects of subjectivity, just as they are symptoms and extensions of these aspects.

Consider the constructed life experiences, the subjectivity, of people occupying specific social and class identities (male or female; rich or poor; Catholic, Jewish, or Muslim; and so on). Because of commonalities in experience and education, there is typically much interaction, overlap, and imbrication among subjective territories, hence creating a righteous feeling of homogeneity and universality that links a society's members. The system operates self-consciously, indeed symptomatically, with great efficacy: each individual's interiority is networked with triggers that set off feelings of guilt, shame, and anxiety whenever their subjective territory is threatened. Threats result from inadvertently dissident wanderings and slippages, willful conceptual and/or emotional border crossings, or challenges and disruptions imposed by outside forces, such as the destabilizing effects of a sudden explosion. When one moves outside of their subjective territory, conceptually or emotionally, and crosses into the subjective territory of another and/or into unidentifiable territory, and/or finds oneself disconnected from familiarity, one has engaged in what I call "transversal movements."[5]

Transversal movements are feelings, thoughts, and actions alternative to those that work to circumscribe and maintain a particular subjective territory. Most people engage in them to some degree, in one form or another, everyday. People most often move transversally when they empathize or imagine they are empathizing with others. They may actually have no idea of what someone else is thinking or feeling, but they are nonetheless still thinking and feeling atypically in their attempt to empathize, "as if" they are someone else, which pushes them transversally. By occupying, if only imaginatively or ephemerally, the subjective territory of another one's own subjective territory expands and reconfigures. Empathy is also probably the most common way by which people venture into what I call "subjunctive space," the hypothetical space of both "as if" and "what if," which is an in-between space operative between subjective territory and what I have termed "transversal territory."[6]

In subjunctive space, unlike in transversal territory, the subject necessarily retains agency and can self-consciously hypothesize scenarios and experiences, thereby self-activating her/his own transversal movements. These can work to stabilize, empower, and/or disempower the subject, which may lead to either further subjectification or further occupation of transversal territory. Chaotic and boundless, transversal territory is a mysterious, challenging, and transformative space through which people journey when they defy or surpass the conceptual and/or emotional boundaries of their prescribed subjective addresses, in effect subverting the hierarchicalizing and homogenizing state machinery of the governing organizational structure. People usually engage in and occupy transversal territory only transitionally and temporally. To reside permanently in transversal territory, rather than pass through it, would be to subsist without any kind of cognitive stability or control.

People can generate their own transversal movements or the transversal movements of others through emotional, conceptual, and bodily performances that provoke strong reactions, such as empathy, terror, fury, alienation, ecstasy, love, abjection, trauma, or psychosis. Unconventional ways of thinking and behaving are both spurred by and generate what I refer to as "transversal power," which fuels transversal movements. Transversal power, a fluid and discursive phenomenon that can be found in anything from terrorist acts to natural catastrophes to heart-wrenching poetry to philosophical inquiry, may not come from a recognizable source or with obvious purpose. It is unclear whether an individual, for instance, can generate transversal power internally, or if the individual instead harnesses transversal power already existent and in circulation: consciously and/or not, the individual taps into the power, either drawing it out or unleashing it. Whatever the circumstances of its activation, transversal power induces people to transverse and permeate the organized space of subjectivity, of all subjective territory, and venture into the inestimable space of transversality, where learning and metamorphosis can be dynamic and limitless.

With the example of the unexpected explosion in the university classroom, we can add to the equation, for instance, that it is likely that some of the students would suffer from traumatic neurosis, a debilitating mental disorder characterized by insecurity, confusion, anxiety, hysteria, or phobias. This is where the society's monitoring mechanisms come into play. The clinical psychologists would be prepared to match the traumatized students with the appropriate template (they might refer to the American Psychiatric Association's *Diagnostic and Statistical Manual*) in hopes of curing them (stopping their pain) and thus restoring them to their prescribed subjective territories. The clinician seeks to relativize and normalize each student's experience, strategically trying to manipulate him/her with psychology, behavioral therapy, and/or medication into snapping back emotionally and conceptually, and, finally, physically, which would be demonstrated through the student's return to school and other "normal" activities. But reversal and restoration are never real possibilities. Once exposed to alternative thoughts and feelings, one can never return to the way one used to think and feel. This is the case for all experiences, but for those that inspire transversality, the effects are greater, and hence more likely to have significant lasting impact. Should any of the students exposed to the explosion develop a serious neurosis or psychosis, the parameters of the student's subjective territory would have been significantly breached, if not shattered. As a result, the student's subjectivity would most likely remain to a considerable extent in the transitional space of transversal territory.

Transversal methodology

When analyzing any thing, idea, or event, and especially something like an unexpected explosion in a university classroom, which necessitates transversal movement and increased potential for discordance and chaos, transversal theory calls for what cognitive neuroscientist James Intriligator and I have termed the "investigative-expansive mode" of analysis (also referred to as the "i.e. mode").[7] This is a methodology opposed, for instance (to give a common example), to what we call the "dissective-cohesive mode" ("d.c. mode"), the interpretative strategy, often referred to as the "scientific method," which characterizes most dialectical argumentation. The investigative-expansive approach breaks the subject matter under investigation into variables that are then partitioned and examined in relation to other influences, both abstract and empirical, beyond the immediate vicinity. A chief objective of this expansive methodology is to contextualize historically, ideologically, and critically both the subject matter and the analysis itself within local and greater milieu. Mobile and vine-like, the investigative-expansive mode resists anything resembling predetermination or circumscription and requires continuous maneuverings and reparameterizations in response to unexpected, even sudden, emergences of glitches, quagmires, and new information as it deduces, trail blazes, follows off-beat leads, takes tangential excursions, while rigorously sprawling analytically.

Like the investigative-expansive mode, the dissective-cohesive mode also breaks down the subject matter under investigation into variables and then partitions them as a means by which to examine them discreetly. Conversely, the dissective-cohesive approach does this in order to later reassemble the variables into a unified and accountable whole, such as a human subject, a social totality, or a deterministic history. An example is psychoanalysis' attempts to locate formulaically the subject firmly within its explanatory geography. Such preoccupation with discerning patterns and conforming them to a preexisting and inflexible model ineluctably "affirms" established parameters and finally demarcates investigation. Determination of this kind leaves little room and tolerance for innovation and/or deviation, since any variance may work to undermine the credibility and value of the analytic system. The d.c. mode may situate subject matter relationally within an environment and consider what I call articulations in "absent-spaces" as significant factors, that is, historical, subterranean, and/or apparently sourceless expressions, receptions, and experiences. Yet the overall goal of the d.c. mode is to gain a better understanding of what is believed to have been the subject matter's original whole or coherent presence and what that totality "should" be. It often does this at the expense of exploring or acknowledging the subject matter's teleological, diachronic, and/or synchronic interconnectedness to other objects and/or forces, in such forms as expressive, receptive, and experiential articulations, in both "present-spaces" and "future-present-spaces," the latter of which is subsequent to the former and thus less immediately occupied and observable.

Investigative-expansive analysis is a part of transversal poetics; it's transversal theory in practice. While in the throes of critical inquiry, the investigative-expansive thinker ("i.e. thinker") demonstrates the transversal slogans, "Become what you aren't" and "Meanings are vehicles," as she/he ventures processually through sundry recipes often written in incomprehensible languages, remarkable osmoses with unidentifiable substances, and hitherto unimagined becomings-others. Embodying the i.e. mode, i.e. thinkers are ready to abruptly shift courses and gears as terrains change across dimensionalities and temporalities, perhaps while contending with the daunting recognition that what was once there is gone, and what was least expected now looms large. Within the i.e. thinker's field of consciousness, notwithstanding the free-range of opportunities, certain variables are tracked and focused on as they move through space–time, although the scope, lenses, and emphases of observation and cognizance fluctuate. Among these are: (1) expectations, contexts, and coordinates of the analytical process itself; (2) degrees of agency and prospects of the researcher, probable audiences, and subject matters of the investigation performed; (3) varying affinities among, more than differences between, elements within both territories transversed and bypassed; and (4) future potentialities as compelling impetuses, rather than just the past as reasoned motives, for scholarship in the present. Because articulatory spaces are vital to contextualizing and understanding how values assigned to subject matters in present-spaces have been fabricated through the traces of absent-spaces as potential avenues to future experiences, i.e thinkers stress hermeneutic amplification in "future-present-spaces" ("The future is now, now, now . . .") and vigilantly grasp the capricious waves between past and future on which they inquisitively surf. Such concentration on research expansion in future-present-spaces is consistent with transversal theory's privileging of space over time, presupposing, as evidenced by the preponderance of spatial metaphors used in all vernaculars (such as, "I don't want to go there," "there" being an emotional rather than a physical place), that we exist first in space then in time, two-into-one, biunivocally, in almost instantaneous succession ("I am in my study; it's 12:14 a.m"). It is the space–time progression that accounts for why it is easier to imagine and experience timeless space than spaceless time. (When was the last time you forgot what space you were in?)

Before moving on, I want to make clear that the i.e. mode's simultaneous emphasis on present-spaces and future-present-spaces does not exclude, but rather promotes, investigation into origins, causes, and intentions since these are variables that contribute substantially to the "affective presence" of articulations (social, cultural, historical, ideological, fantastical, and so on) through absent-spaces that strongly influence the distribution of powers within a society.[8] Nevertheless, the investigative-expansive approach does give primary attention to the relational positionings of most of the immediate expressions, receptions, and experiences, discerned by them and their environments' presences and absences, as a means by which to situate the analysis and subject matter in space–time. Then, consideration is given to the specific critical enterprise's implications for future-present-spaces rather than the more fuzzy and distant "future-absent-spaces," which refers to all that is contained in the "now, now, nows" that cannot be uttered, and thereby observed, fast enough. (Like with the nighttime howls of an Irvine coyote, the "nows" echo away, and where the coyote goes nobody knows.) Future-absent-spaces are also what is conjectured, predicted, or anticipated, but with less probability of actualization than, say, vicissitudes in future-present-spaces. Although a somewhat arbitrary threshold measured in the wake of our struggling and failing senses, focus on future-present-spaces is more dependable, responsible, and practical than pursuing an absent truth that will always appear obfuscated by historical process, and consequently can only ever rear its recondite head phantasmagorically. Please welcome the ghost of William Shakespeare.

Shakespace

In this book, my collaborators and I elucidate and employ the i.e. mode in our analyses of Shakespeare's influence on and participation in sociopolitically informed readings, productions, and adaptations of Shakespeare's plays and the historical–cultural contexts in which Shakespeare functions as an icon and avatar. With the chapters exploring different theories, texts, performances, themes, and media, the i.e. mode operates respectively in each chapter and altogether by expansive explorations into interconnected subject matters and multivariate parameters of investigative inquiry. For instance, in chapter 4 on *Coriolanus*, I explore representations of authority in both Shakespeare's play and Bertolt Brecht's adaptation, *Coriolan*, in light of their respective sociopolitical contexts. Rather than follow one line of argument and come to a totalized conclusion, such as that sovereignty in both plays resides ultimately with the commoners, I entertain a variety of leads, oscillating between the sixteenth and twentieth centuries, and wind up making a number of claims subsidiary to an overarching argument, expounding on the purposefulness of ideological criticism of Shakespeare. Similarly, in chapter 8, Ayanna Thompson and I analyze representations of the character Ariel, moving across space–time via adaptations of *The Tempest* from early modern England through the English restoration to Cuba, Martinique, and the United States in the twentieth century. In our travels, we consider a mixture of political implications of Ariel's implementation in diverse productions, and suggest alternative employments for future performances. In these chapters, as in all the others, my collaborators and I engage with many manifestations of Shakespeare in often-conflicting transhistorical, sociocultural spheres, which I refer to as articulatory spaces.

Informed by transversal theory, "Shakespace"—Shakes (as in Shakespeare) space—is a term coined by Donald Hedrick and myself for the particular articulatory space through which discourses, adaptations, and uses of Shakespeare have suffused the cosmopolitan landscape transhistorically.[9] Unlike most writers and other figures in history with mighty spectral presence, the spaces through which Shakespeare has passed from the sixteenth to the twentieth and into the twenty-first century as an icon have been extraordinarily diverse and numerous. Of all authors in Western history, only Karl Marx and Sigmund Freud could be seen as comparable to Shakespeare, given that they too have stimulated, occupied, and affected countless commercial, political, social, and cultural spaces; there is "Marxspace" and "Freudspace." Yet, only Shakespeare has been influential for four-hundred years (he had a three-hundred year head start on Marx and Freud); only Shakespeare is consumed by people of most ages, especially nowadays, since he is mandatory reading at many levels of education in the West; and, because of Shakespeare's phenomenal status in Hollywood (with over twenty cinematic adaptations of his plays having been released between 1980 and 2002, not to mention one Academy-Award winning film about him, *Shakespeare In Love*), his popularity transverses class divisions. Furthermore, because Marx and Freud themselves often employ Shakespeare to support their arguments, they too are part of the imposing Shakespeare machine that in turn must always also inhabit their respective territories.

The Shakespeare-influenced spaces, however conventional, alternative, or sometimes both, is Shakespace, a term that accounts for these particular spaces and the time or speed at which they move from generation to generation and from era to era. Shakespace's unique functions as a resistor, generator, and conductor of transversal power are resultant from phenomena that within and passing through Shakespace are the epochal forces and transformations wrought by a multiplicity of influences: by early capitalism; by the great experiment of the new public entertainment industry in early modern England; by the interrogation of socially prescribed gender roles; by aristocratic and legitimation crises; by the desacralization of absolutist sovereignty; by cross-cultural collisions and relativizations deriving from exploration and colonization; by the scientific revolution and its confounding of official knowledge; and—to take this space into our own time—by the recursive force of the Western canonical tradition itself on Shakespeare's work.

The accumulation of these interacting forces gives additional transformative power to Shakespace, whether radical or conservative, and hence produces what is often read as the ideological complexity of any object of critical and political scrutiny. Both despite and because of the contradictory nature of Shakespace throughout history—from reactionary to complacent to radical—Shakespace has continued to expand into the twenty-first century as simultaneously the manifestation of and the inspiration for transversal movements across conventions' borders and outside of dominant sociopolitical parameters. It is through these transversal movements that Shakespace emerges as a culturally imaginable space where Shakespeare may function without class in both the socioeconomic and aesthetic senses of the word. This is because Shakespace's cultural power, or Shakespeare-effect, resists and sometimes transcends all classification that is either reductive or totalizing. Shakespace becomes, as it were, a socially and historically contingent playground on which class differentiation and class conflict sometimes slip transversally into an ambiguous space that makes possible, and in fact encourages, alternative opportunities for thought, expression, and development. Whereas Shakespace has certainly been, and frequently still manifests itself as, what I call an "official territory" (the ruling proprieties and legalities within a social body), and thus has worked to promote various organizational social structures that are discriminatory, hierarchical, or repressive (including structures in academia itself), it is most powerfully a transversal territory. It has radiated territorially as a transportive, transformative space in which dreams have been made, pursued, virtualized, and realized.

It is Shakespace's transversal territories that this book most diligently pursues. Focusing on Shakespearean and spatially occupied/conceived of-as-Shakespearean manifestations of transversality, my collaborators and I offer new readings of plays and the performances of plays in various media while at the same space–time employing transversal poetics as a progressive means by which to comprehend our collaboration with Shakespeare in the generation and unleashing of Shakespace's transversal power, whether that be through the process of reading or performing Shakespeare. On Othello, in chapter 3, Joseph Fitzpatrick and I use the i.e. mode to explore numerous possible explanations for Iago's actions and Othello's response, including, with the help of Janna Segal, considering Desdemona as a willful agent rather than a passive victim in her own destruction; Iago emerges as a sociopolitical conductor for an alternative ideology that works in the context of the play to activate transversal power that ultimately serves rather than problematizes the Venetian state. In chapter 5 on Roman Polanski's cinematic adaptation of *Macbeth*, I investigate expansively various agents of sedition, terrorism, transversality, and state power, from Macbeth to the witches to Charles Manson to the U.S. judicial system, while contextualizing comparatively the topicality of both the film at the end of 1960s radicalism and Shakespeare's play at the height of early modern England's witch craze. Venturing into yet unexplored Shakespace while simultaneously producing, expanding, and complicating already existent Shakespace, the chapters in this book present not only fresh analyses of Shakespeare's plays, adaptations of them, and their critical contexts, but also theoretical models for further adventures in Shakespace.

Discourse on tragedy

At this point, in light of Shakespace, I want to consider what I call "Osamaspace." Of course, it is probably many years too soon to measure the extent of bin Laden's influence, but this is not my purpose. Like Shakespace, Osamaspace is an articulatory space marked by a dynamic of transversality, but unlike Marxspace and Freudspace, which are also transversally marked, the trajectory of Osamaspace is yet to be plotted, and can therefore be more easily manipulated. Already Osama bin Laden has prompted what is just the beginning of innumerable textual remains in the forms of articles and audiovisual recordings by him and about him, as well as posters, buttons, and t-shirts devoted to him and his terrorist network Al Qaeda. Countless children have been named after him, and "Osama" may become as popular a name as Mohammed. However, at this historical moment, these remains have minimal impact when compared with the atrocity he helped orchestrate on September 11, 2001. The attacks' most prominent vestiges cannot be represented: they are the pain, loss, grief, and suffering experienced by all of the victims, from those directly injured to the families and friends of victims to the rest of the United States and to those sympathetic across the world. Thus, above all else, for much of the world's population, bin Laden is associated with tragedy and horrendous crimes.

Bin Laden's terrorist performance has been evaluated in tragic terms: the more tragedy he produced and continues to produce, the greater the terrorist (performer) he is. I want to provide an understanding of tragedy before discussing in detail my concerns with Osamaspace's transversality, a space which in many respects operates in spite of its namesake's own declared purpose: to combat U.S. foreign policy and to institute, through terrorism, and later by totalitarian means, a totalized Islamic state composed of all Muslim countries, stratified with overlapping subjective territories from the top down (dictator, upper echelon of clerics, men, male children, all females). Rather than seek to explain tragedy in the traditional Aristotelian sense, which is to say, how tragedy often functions in theater to arouse pity and fear in order to accomplish the catharsis of such emotions on the part of the audience, I want to consider tragedy, what I believe is the most heinous reality of human experiences, in relation to authorship, intentionality, and the link between ethics and celebrity status, as seen in the public's desire for celebrities to do good. After all, it is precisely the link between ethics and celebrity status that works to produce mobile spaces of social power under the auspices of prominent cultural icons like Socrates, King Tut,

Jesus Christ, Mohammed, Shakespeare, Marx, Freud, Hitler, and Osama bin Laden.

Several factors determine the extent to which an action or event is a tragedy. First of all, as with any performance (defined here as a self-consciously presented expression for an intended audience), how the event is framed needs to be taken into consideration. Framing refers to the aesthetic, social, cultural, political, ideological, and historical context established for the performance, whether it takes place in a stage-play or during a sports game or at a particular venue, like in a courtroom, university classroom, chapel, theater, or a pub. The sociopolitical conductors that have worked to instill our biases and predilections shape our relationship to that context and thus to what is expected to happen there. In most cases, when the framing is less apparent and less understood, the potential for a tragedy-producing performance is greater. In other words, the more sudden and unexpected the injuring action, the more tragic potential it has. But framing also has to do with investment. The degree to which the audience invests itself in the damaging event and its victims emotionally, conceptually, physically, financially—in terms of space–time, energy, and emotion—directly affects the potential for tragedy; the greater the investment, the greater the potential for loss, the greater the potential for tragedy.

The investment formula of tragedy can also apply to the victim. The more the victim cares about her/himself and others, the more others care about her/him, the greater the potential for personal tragedy. Hence, tragedy depends largely on people's expectations. Expectations relate immediately to framing and, often at a distance, to the sociopolitical conductors that predispose us and negotiate our relationship to the frame. Already contextualized within a society's dominant perspective, that is, in relation to its official territory, framing provides contextualization, which occurs through layers of references, from the immediate to the remote (for instance, what would you expect from a production of *Romeo and Juliet* by students at a Southern Methodist high school in Houston?). The frame is the discernible, referential contextualization, providing a measure for evaluation, which makes explanation both possible and acceptable. The most powerful tragedies, however, like the most powerful performances, are those that shatter expectations, and consequently transverse established frames. Whereas an unexplainable performance might make it less powerful because it would be less identifiable and thus disinclined to arouse empathy, the less explainable the tragedy, the deeper we experience it and the more severe the impact.

For tragedy to happen at all, someone has to care, and the greater the social value the tragic victim possesses within a given society, the greater the tragedy for the society's members. According to life insurance companies, the value of a human life is determined by one's social responsibilities; for example, a parent is worth more than a childless adult. For the general population, nevertheless, social value is usually based on the potential to do good things, as well as on previously noted accomplishments. This is why it is often the case that people feel worse when the victim is a small child rather than an older person. The child has more potential, more time and chances in life in which to perform goodness. Should the child have already been acknowledged as precocious and likely to do much good, the tragedy is perceived as worse. But is such assessment multipliable or exponential?

Both quantity and quality need to be taken into consideration. People typically speak of casualties, and, in most cases, individuals within the collective pool of casualties that possess friends and family are seen as more valuable than those without.[10] The death of one famous person, a celebrity who has made important contributions to society, and is thus known by more people (often in an imaginary relationship with them), is generally valued more than many unknown or less important people, regardless of whether they have friends and family; if there are many casualties, with a famous person among them, the masses have at least one victim they can identify and mourn for. The need to identify a victim, as a way of being included, can be seen as the inversion of the need for an author.

The notion of innocence, as directly related to sympathy and value, comes into play as well. The younger the person, the more innocent people ordinarily assume them to be. In addition to having more potential than an older person, injury done to younger people is deemed more tragic the less "worldly" they are perceived to be. So, while the innocent are not seen—because of their innocence—as more socially valuable than people who have already demonstrated their wisdom and made noteworthy contributions to society, the innocent are seen as society's responsibility. They are the ones whom the socially valuable need to care for and protect. After all, for many purposes, the more innocent people are, the more easily socialized and subjectified, and therefore the greater their potential to serve the official culture and state machinery. In chapter 9, on Julie Taymor's film *Titus*, Courtney Lehmann, Lisa Starks, and I show how the audience is implicated into taking responsibility through an act of bearing witness to the unbearable that renders the audience complicit in the horror, inducing abjection, paradoxically, as a transversal passage to a new life symbolized ultimately at the film's end by the fact that Young Lucius survives the carnage and has ensured the survival of the more innocent, potentially more valuable victims, himself and Aaron and Tamora's infant son.

The unexpectedness or suddenness of an injuring event can relate to framing in another way. It is more tragic for someone to die suddenly from an accident than to die slowly over time. Sudden death resists the framing for both audience and victim that develops as both parties become accustomed to the prospect of death. The more prepared to die, the less tragic the experience or event of death. Prevention is also crucial to understanding tragedy. If an accident could have been easily prevented, but was not, then the tragedy is worse. Victims produced through ignorance are more tragic than victims caused by carelessness. Moreover, self-victimization through an accident happening while the individual is in the process of perpetrating a morally reprehensible act is less tragic than accidental self-injury occurring during normative affairs. Still it is more tragic if the accidental self-injury occurs while the victim is pursuing a noble cause. In cases of accidental self-injury, according to Aristotle's theory of tragedy, the more noble the victim, the more tragic the incident. But, for Aristotle, whose theory pertains specifically to tragedy in ancient Greek theater, this scenario would only work if the noble person were someone like ourselves; in other words, this person would have to be someone with whom we can identify. Nevertheless, Aristotle's theory does not apply to "real life." In fact, unless people have an actual (interactive social, not imaginary) personal relationship to the victim, the tragedy is perceived as less severe. However, when the victim is famous (someone with symbolic value), and therefore not someone with whom most people can relate on an interactive level, the tragedy worsens. Celebrities are privileged in this regard: their deaths have more meaning than the "average person's"; however, an "average person" can momentarily achieve celebrity status through a tragic death that is sensationalized in the media.

Oftentimes it is unclear who the most tragic figure is. Is the accidental death of a small child worse for the child or for the mother? From a different perspective, consider the lack of satisfaction the public experiences when a murderer commits suicide. Explanation is more elusive without an author, without available and specific accountability. In this case, attention to the victim, and any other members of the event's audience, as would-be sources of elucidation or causality frequently increases (as discussed in chapter 9, on Julie Taymor's cinematic adaptation of *Titus*). This brings us back to the self-value of the injured party.

The more we recognize that the individual values him- or herself, the more value we place on his or her life. Yet, the extent of the tragedy is often also determined by the degree of the damage done.

Entering into a subjunctive space to elucidate the argument, I want to consider a series of scenarios based on the "iffy" factors informing a tragedy. Should a girl slip on the street and only bruise her elbow, her tragedy is less severe than if she had broken her hand. If the girl had slipped on a banana peel accidentally left by someone else, then the tragedy is worse: it was both preventable and the result of negligence or carelessness, as well as sudden and unexpected. If the victim were a pianist, then the tragedy would be even worse. If the pianist were on her way to an important audition, then the tragedy would be still worse. If the banana peel were strategically placed by a competitor, then the tragedy would be substantially worse: there would be an author, an identifiable enemy testifying to a failure in "humanity" (a tragedy in itself), compounded with the tragic damage. If the pianist were an eight-year-old who was so skilled that she was going to audition for Julliard, then the tragedy would be worse. If the pianist were going to audition for the opportunity to make money for her family, to save them because they are poor and starving, then the tragedy would be worse still. As this example shows, there are endless possible variables that can be combined to increase the social value of even the most seemingly accidental tragic event. Authorship, as we have seen, is nevertheless always crucial to the assessment.

If the audience were to later find out that the pianist only got as far as she did because she had sabotaged the pianos of her peers in a preliminary competition, then the tragedy would be much less severe. It would be up to the audience to determine whether the tragedy was fortuitous and an appropriate punishment: whether or not it was a tragedy at all. But how can it be a punishment if it was not self-consciously administered by someone or some people, perhaps representing an institution (like the U.S. government), who have evaluated the situation and passed judgment? Who would be the author of the punishment? Pursuing the issue of authorship, in this case the authors of both the tragedy and the punishment, I want to return to the attacks of September 11, 2001 and the radical expansion of Osamaspace that they have generated, an expansion that has tested the limits and increased the scope of countless subjective territories.

Responses to the September 11 attacks have been, and will continue to be, profuse and contradictory, and there is no end to what can be done, especially since "doing" can take many forms. The attacks, as well as everything related that has occurred in their aftermath, invite discourse that can work to commandeer Osamaspace, transposing it into a relativistic terrain of learning and tolerance rather than of terror and hatred, converting it into a conductor through which transversal power can fuel movement outside of the subjective territories that resist making the acknowledgment and respect of difference common ground. Osamaspace can work to disseminate an understanding of social identity and cultural difference that is not dependent on negation or opposition, but rather on the positive recognition of intersections and disparities.

To see both this negation and positive recognition happening, one needed only to watch *Oprah* and *Nightline* during the weeks after the attacks. In my view, because of the diverse perspectives they give voice to, these shows are among the most conscientious and purposeful on network television. To see an opposing model, where Osamaspace is conflated with JesusChristspace, one need only turn to the Christian Broadcasting Network's *700 Club*. After the attacks, two of its leading spokesmen, Jerry Falwell and Pat Robertson, shared their views. Falwell lamented, "God continues to lift the curtain and allow the enemies of America to give us probably what we deserve," to which Robertson declared, "Jerry, that's my feeling." Falwell continued: "I really believe that the pagans and the abortionists and the feminists and the gays and the lesbians who are actively trying to make that an alternative lifestyle, the A.C.L.U., People for the American Way . . . I point the finger in their face and say, 'You helped make this happen'" (Dunne 180). As offensive and absurd as this sounds to me, I remind myself that Falwell's views are shared by many people in the United States, and that even President Bush invokes the ridiculous notion of "pure evil" as the driving force behind the attacks, rather than encourage exploration into why people gave up their lives to injure America, and why millions of people support bin Laden and Al Qaeda.

In terms of both national and personal tragedy, for most United States citizens the attacks are among the worst events in history. They occurred on an unusually beautiful September morning. In New York, people were moving about their business with the quintessentially upbeat verve that has long distinguished that city—my hometown—from all others. Nationwide, Americans felt relatively safe compared to people in other parts of the world. A terrorist attack, much less an "accidental" airplane crash into the World Trade Center, was among the furthest things from their minds. Also in this category of distant thoughts was what life might be like under Taliban rule in Afghanistan.

In fact, because Shakespeare is taught regularly in schools across the country, it is more likely that on September 11 Americans were thinking about life in Elizabethan England than in contemporary Southwest Asia. Basically, for most Americans, the attacks were framed such that they were unexpected and came suddenly.

The circumstances of the September 11 attacks provide evidence illustrating the complex relationship between the sociopolitical function of authorship and the subjective and subjectifying elements of tragedy. The attacks were logistically elaborate, and therefore preventable. Most of the people killed were directly involved—because of their white-collar occupations—in the capitalist socioeconomic system that works to sustain all the freedoms and luxuries the United States has to offer and is accustomed to. (People would have responded very differently if the planes had crashed into San Quentin Prison.) With subjective territories radically different and opposed to those of most Americans, the terrorists immediately responsible for hijacking the planes and crashing them into the World Trade Center, Pentagon, and onto the ground in Pennsylvania valued their mission more than their own lives. This willingness to suicide was especially shocking for Americans; a self-consciously executed suicide mission has never been an acceptable option for the Christian-influenced American military. The American people and everyone supporting them throughout the world had no opportunity to learn from the terrorists themselves why the attacks were executed; they received no statement from a terrorist organization claiming responsibility; they had little terrorist-connected material, in the form of personal evidence, from which to glean meaning. With no obvious enemy, or author, against whom to retaliate, the United States had to put a name to the faceless enemy.

While these authorship-related lacks are not in the same category of damage as the human losses suffered, the overall tragedy was exacerbated by their absence. Without access to an author's intentions, say in the form of a separate recorded statement (written or spoken), for interpretive resolution we must look to the immediate text itself (in this case, the available material substantiation that the attacks occurred), the contexts of its reception (in this case, people's responses to the attacks), and the environments from which it sprung (in this case, the political situation in the United States vis-à-vis the rest of the world). The subject matter needs to be synthesized, analyzed, and presented. It is only through discourse that such a resolution can be achieved; and only through performance can resolution be demonstrated. (If people don't act happy, then why should we believe that they are?)

Critical inquiry

The trend in literary criticism over the last twenty years has been to approach the author problem much like I just described. Critics first contextualize a given literary text both historically and culturally, and then analyze it as an expression of its moment, rather than of an individual author's intentions. In most cases, we have no unmediated access to authorial intention, and even if the author had stated her intentions, we have no way of knowing if what was stated is true, and thus the author is not considered a reliable source for deducing meaning.

The higher the stakes, the more pressing this authorship issue becomes. Consider cases of slander, copyright, plagiarism, and intellectual property. To take the issue further, what if someone commits a violent crime, but claims that it was not her/his intention, then are we to believe the person or look to the damage done in order to determine intention, and thus responsibility? Of course, this has been a major concern in legal studies for a long time, and I raise it here only to give perspective, from a related area of inquiry, to our comparison of more quotidian issues of authorship ("What did Shakespeare mean?") with more serious ("Why did bin Laden author—if he in fact did author—the murder of thousands of innocent people?"). Moreover, just as one of the goals of literary criticism is to determine which works of literature are good and which are not, we must ask what comprises a successful act of terrorism, how successful must it be to be terrorism, and does this have anything to do with authorial intention or our understanding of the terrorist act?

If a woman claims that a male coworker regularly harasses her sexually, and this terrorizes her, this is her experience. Regardless of either the coworker's intentions or his awareness of her terror, the woman is the victim of sexual harassment, and the coworker the terrorist. The coworker's actions reinforce a patriarchal ideology by which women and men learn that sexual objectification of women in the workplace is acceptable and should be tolerated, even if the governing legal system stipulates otherwise. The coworker's subjective territory is the product of the same patriarchal ideology for which it is a conductor, working to promulgate terror among a vast population of women across the world. In return, the terror promotes the ideology, working to further objectify and disempower women. Who, ultimately, is the author of the woman's terror?

Actions are symbols of ideology, just as symbols are the objects of actions. In the lexicon of critical theory, the September 11 attacks can be seen in deconstructionist terms. The World Trade Center has long been a symbol of capitalism and America's power, but despite the word "Center" in its name, it is by no means the center of the capitalist socioeconomic system or U.S. power structure. While this building has been an important symbol, it is one among many, and it will soon be replaced by another symbol. As explained by Jacques Derrida's deconstructionist philosophy, through a process of "supplementarity," each sign used to replace the center of any signifying system is in fact a supplement to an always already absent center.[11] In other words, the capitalist socioeconomic system has never had a center, only symbols of centeredness, like the World Trade Center, and therefore, as long as the system continues to function, there can always be more symbolic centers. When the system fails to produce these, it will fritter away.

Long after September 11, Osamaspace continues to occupy conceptual centrality in the capitalist West, still saturating dominant media with discourse on biological terrorism, supplanting commentary on, say, American sports. Shortly after the attacks, anthrax was mailed anonymously to key governmental and media locations, infecting and killing many people. The search for the author of those mailings, and the letters they also contained, is still nonstop, insofar as the cases remain unresolved, even though their headliner status has long since dwindled. The mailings' authorial specters dissipated and have been suffused by other specters, such as the authors of the suicide bombings in Israel, Bali, Riyadh, Casablanca, and elsewhere. Again, we have terrorist performances without identifiable authors. Through an ongoing analytical process, we try to situate the terrorism, like most subject matter of critical inquiry, by locating its individual and communal sources in history in hopes of constructing a narrative that makes sense; and what makes narrative sensible, is agency, responsibility, intentionality—in short, authorship. Events often occur happenstantially, such as slippages spurred by surges of transversal power, and therefore as symptoms without motives. Afterwards, in their wake, people construct a narrative explanation, extrapolated from apparent evidence of the event's source. The narrative is then, retroactively, imposed on our present-space perspective on the past-present, now absent-space. The story generated helps people come to terms with the event, however tragic or joyous, and these terms depend on onus: someone or something must be responsible, and the more totalized and singular those sources, the more comfortable people's relationships to them.

American society's compulsion to attribute responsibility to individuals rather than to groups (a social clustering of otherwise individuated humans) reflects its desire for celebrity personalities, which is a byproduct of each member's own wishful aspirations, which are, in turn, byproducts of the capitalist socioeconomic system's dependence on the concept that the best individual will prevail. The focus on individual authors that is the mainstay of modern thinking about human performance reflects the predominance of monotheism. But bin Laden will not suffice as our all-purpose author of terrorism, operating as both center and supplement; he cannot be in all places at once, and he is not solely responsible for the September 11 attacks and other terrorist acts committed by Al Qaeda. If another author is not discovered or manufactured (such as Saddam Hussein), then, according to the dominant paradigm, there can be no sense, no resolution.

On the horizon

In my view, the methodological paradigm needs to shift transversally, and the quest for individual authors needs to be displaced by a quest for a more comprehensive account of both causality and responsibility. I have tried to show, among other things, that in the aftermath of the tragedy of the September 11 attacks, the future provides an especially powerful opportunity for learning and evolution. History has led to a critical juncture, but it does not necessarily determine the future. The irony of the ethics/celebrity nexus reveals that as we struggle to understand Osamaspace, the subjective territories it supports, and the issues of authorship, tragedy, intentionality, and social value that manipulate it, we must confront the parameters of our own subjective territories and the ideologies that inform them. We are moving transversally, and this can be productive if, rather than attempt to snap back, we conduct the power and perform transversally by trying, for instance, to understand the Taliban's ideology and goals, their treatment of women, and their support of Al Qaeda. By empathizing with our attackers and/or asking ourselves subjunctively "what if" we were them, we can see our own country's destiny as linked to a past that we have tried to ignore, a past of imperialism and self-centeredness, which is no longer viable on practical or moral grounds. In accounting for bin Laden as a coauthor of moral activity, we become ourselves open to examining not only our nation, but ourselves as coauthors and as sociopolitcal conductors; we can come to understand our preferred convictions with greater awareness and responsibility: the role of coauthor is our own, and the narrative continues. There is no guarantee of positive reconfiguration, but there is the transversal power of recent events that can be harnessed to effect change.

Taking the idea of coauthorship further, I want to stress the fact of collaboration always present but seldom acknowledged in all modes of expression, especially in my own profession as an academic researcher and critical theorist. Ideas are everywhere around us, meandering with the ebb and flow of the historical moment, often synthesized and conveyed to us through mass media, colleagues in other fields, our students, friends, and loved ones. Trends in literary-cultural criticism mutually reflect theoretical trends in other disciplines as well as social, cultural, and political trends at large (consider similarities between the "uncertainty principle" and the "affective fallacy" or chaos theory and poststructualism). Transversal theory in particular encourages the getting outside of one's subjective territory synchronically, diachronically, and fantastically in order to interact and merge with the subjective territories of others, pushing the limits of emotional and conceptual ranges. This is why transversal thinkers often pursue collaborations with others, wanting to foster transversal movements, dynamic evolution, and increased intimacy and community.

Transversal poetics works investigative-expansively on as many levels as possible, from artistic expression to focused research to casual communication with peers. Of course, there are practical and personal factors: there are only so many people with whom one can comfortably and effectively work. Transversal methodology, like all critical approaches, is collaboratively pursued and affected. All scholarly work, whether single authored or collaborative, contributes to the greater field of research in a given area, as well as to other fields, in effect becoming a part of the cumulative research and culture it affects—a never-ending process (a consequence today becomes tomorrow's cause). Nevertheless, as deconstructionist theory has emphasized, under scrutiny each single-authored or collaborative work has an indeterminate source, ultimately lacking a single author or seed of conception, and, as such, testifies to the decenteredness and multi-voiced, processual nature of creativity and critical thought. Transversal methodology acknowledges and capitalizes on this situation, yet it does so without forgetting that when the deconstructionists' rhetorical dust settles discursively or continues discursively to cloud the air, no less than a phenomenon remains: human beings experiencing idiosyncratically fields of consciousness and possessing degrees of agency and potential peer on.

This book, the bulk of which I wrote in collaboration with other human beings, seeks to exemplify transversal poetics. It develops transversal theory in a number of ways, refining and expanding on theories and terms already disseminated, thereby demonstrating the value of collaboration and the transversal approach to critical inquiry. The repetition of explanations of theories and terms in many of the chapters allows each chapter to stand alone as an independent essay, while at the same time working to make the book cohesive. The intentional repetition has far-reaching potential: the chapters can be used individually, such as in college courses, by scholars, and by film and theater practitioners, making the book more purposeful, usable, and accessible to diverse audiences. As such, this book is a model for future approaches not only to Shakespeare and the "spaces" through which Shakespeare travels transhistorically, but also for academic approaches to communication within and outside academic communities. By facilitating discourse among authors, and generating a conception of author-as-simultaneous-audience, this book expands the conception of authorship beyond the confines of single-handed scholarship, beyond the concept of self, and into the conceptual and emotional, non-delineated transversal territories that I envision as a new poetics and aesthetics for the critical future.

Notes

1. I coined the term "sociopolitical conductors" as a corrective to Louis Althusser's ahistorical concepts of "Ideological State apparatuses" and "Repressive State Apparatuses" that have informed much scholarship, particularly of cultural materialists and new historicists, in the field of early modern studies. The term "sociopolitical conductors" accounts for the historical specificity of sociopolitics and conditional states of both dominant and counter-hegemonic machinations for prescribing ideologies, whereas Althusser reduces these conductors to a binary system of an ideological-repressive monolithic state, and simultaneously does not acknowledge the similarly functioning, often multiple and contradictory ideological machinations of non-state promoting entities (see Louis Althusser, *Lenin and Philosophy and Other Essays*, trans. Ben Brewster [New York: Monthly Review Press, 1971]: 127–88). From a more classical Marxist standpoint of theorists like Nicos Poulantzas, Althusser's division of the state is problematic not only because of its ahistorical qualities, but also because it "diminishes the specificity of the *economic state apparatus* by dissolving it into the various repressive and ideological apparatuses; it thus prevents us from locating the state network in which the power of the hegemonic fraction of the bourgeoisie is essentially centered" (*State, Power, Socialism* 33). For a detailed discussion on sociopolitical conductors, see Bryan Reynolds, "The Devil's House, 'or worse': Transversal Power and Antitheatrical Discourse in Early Modern England," *Theatre Journal* 49.2 (1997): 143–67, and Bryan Reynolds, *Becoming Criminal: Transversal Performance and Cultural Dissidence in Early Modern England* (Baltimore: Johns Hopkins University Press, 2002).
2. I also coined the term "state machinery" as a corrective to Althusser's conceptual apparatuses. State machinery is in many ways a departure from as well

as a fusion of his "Repressive State Apparatus," which includes governmental mechanisms, such as the military and the police, that work to control our bodies, and his subsidiary "Ideological State Apparatuses," the inculcating mechanisms that work to control our thoughts and emotions. "State machinery," a term indicating both singularity and plurality, makes explicit the multifarious and discursive nature of "state power," which are forces of coherence, whether acted consciously and/or not by or upon "individuals" (meaning humans individuated from other humans) or "groups" (a social clustering of otherwise individuated humans) that prevents the misconception of the sociopower dynamic as resultant from a conspiracy, historically determined or not, led by a fully organized or monolithic state. This is not to say, however, that conspiracies do not occur and take the form of state factions. On the contrary, this must be the case for the more complex machinery to run. Incidentally, it should also be noted that my conception of "state power" as any force from any source that works to consolidate social entities (working toward the creation of a society) differs from that of more classical Marxists like Nicos Poulantzas, who maintains, "When we speak for example of *state power*, we cannot mean by it the mode of the state's articulation and intervention at the other levels of the structure; *we can only mean the power of a determinate class* to whose interests (rather than to those of other classes) the state corresponds" (*Political Power and Social Classes* 100). For a detailed discussion on "state machinery" and "state power," see Reynolds, "The Devil's House," 143–67, and Reynolds, *Becoming Criminal*.
3. For more on "transversal theory," see, in addition to the texts mentioned in the foregoing notes, Bryan Reynolds and Joseph Fitzpatrick, "The Transversality of Michel de Certeau: Foucault's Panoptic Discourse and the Cartographic Impulse," *Diacritics* 29.3 (1999): 63–80. I am indebted to Félix Guattari for the terms "transversal" and "transversality," which I have adapted for my purposes. In the section on "Institutional Psychotherapy" in his collection of essays entitled, *Molecular Revolution: Psychiatry and Politics* (1971–1977) Guattari uses these terms to discuss the phenomenon of group desire: "Transversality in the group is a dimension opposite and complementary to the structures that generate pyramidal hierarchization and sterile ways of transmitting messages. Transversality is the unconscious source of action in the group, going beyond the objective laws on which it is based, carrying the group's desire" (22). While Guattari's use of the term is incorporated in my own, I extend his definition of transversality to conceptuality and its territories, and allow it to apply to the existential processes of individuals as well as of groups. Incidentally, Guattari borrowed the term from Jean-Paul Sartre, who speaks of "consciousness" as being composed of " 'transversal' intentionalities, which are concrete and real retentions of past consciousnesses" (*Transcendence of the Ego* [New York: Noonday Press, 1957]: 39).
4. I would like to take the opportunity this note affords me to differentiate briefly my concept of subjective territory from concepts of the subject and social identity that have dominated much critical theory over the last twenty years, particularly those formulated by Chantal Mouffe and Ernesto Laclau, who emerged influentially on the academic scene in the 1980s with the

publication of their *Hegemony and Socialist Strategy* (1985), and continue to maintain, as they state in their preface to the 2001 edition, their same views in the twenty-first century (vii).

For Laclau and Mouffe, whose theory of subject formation comes predominantly from Lacanian psychoanalysis, the subject, like desire, is predicated on lack and can only establish itself as a subjectivity through a common grounding and knotting together of equivalential identifications (subject positions with a recognizable and/or persuasive degree of sameness) into a singular "nodal point" (derived from Jacques Lacan's *point de capiton*—in English literarily, "quilting point" [Lacan 268–69]), which is an empty signifier capable of partially fixing and centering the substance of a range of floating signifiers, and thus creates and endeavors to sustain a fully achieved social identity (112). The nodal point, as Slavoj Žižek explains Lacan's formulation, "*as a word*, on the level of the signifier itself, unifies a given field, constitutes its identity: it is, so to speak, the word to which 'things' themselves refer to recognize themselves in their unity" (95–96). By extension, through linked equivalential identifications, society is also pursued by hegemonic forces. However, according to Laclau and Mouffe, such "suturing" together of seeming equivalences can ultimately never produce either a subject, an individual, or a society because there are always antagonisms from without that work hermeneutically to "overdetermine" (a Laclau-Mouffeian borrowing from Freud) the presence of some of the sutured social identities and social positions vis-à-vis others. This causes either their condensation or displacement, and thereby transfers the interpretive focus onto another set of equivalencies so as to fixate another nodal point, and so on. Laclau and Mouffe's sense of stepping-stone subject formation is much like Jacques Derrida's theory of *défèrance*, which accounts for the phenomenon that a signifier has significance given that it is a product of difference, but that its significance is simultaneously unstabilizable and always deferred because it can never achieve absolute meaning as a transcendental signified (Derrida 289–93).

Unlike Michel Foucault, who sees antagonism as internal and "never in a position of exteriority" to power relations of a particular society (93–96), Laclau and Mouffe argue that "antagonisms are not *internal* but *external* to society; or rather, they constitute the limits of society, the latter's impossibility of fully constituting itself" (125). As Žižek eloquently summates:

> The thesis of Laclau and Mouffe that "Society doesn't exist," that the Social is always an inconsistent field structured around a constituted impossibility, traversed by a central "antagonism"—this thesis implies that every process of identification conferring on us a fixed sociosymbolic identity is ultimately doomed to fail. The function of ideological fantasy is to mask this inconsistency, the fact that "Society doesn't exist," and thus to compensate for us for the failed identification. (127)

But for Laclau and Mouffe it is not only society that is negated, but also the identifiable human agents who might have sociopolitical power should it be possible for them to be members of a society. Laclau expresses this concept succinctly:

> The question of *who* or *what* transform social relations is not pertinent. It's not a question of "someone" or "something" producing an effect of transformation or articulation, as if its identity was somehow previous to this effect. [. . .] It is because the lack is constitutive that the production of an effect constructs the identity of the agent generating it. (Laclau 210–11)

Articulated in the space of "everyday life," should you want a mechanic to work on your car, you might as well solicit anyone to assist you, because it is impossible for you to know who is or is not a mechanic until a person demonstrates his/her mechanical skills, and even then, we might ask: How skilled must he/she be to be a mechanic, and who is qualified to make this evaluation? Of course, we might respond to such queries with: Would seeking a state-certified mechanic at a state-licensed automobile repair shop increase the odds of her being skilled enough? Empirically, if what appeared to need fixing appears to have been fixed by the mechanic, then chances are it was. Within the relevant, local interpretive community, agency was identified and meaning was achieved, as predicted by the preestablished parameters of the quest for car repair. Transposing this example to one more in line with Laclau and Mouffe's political agenda, we might ask: Doesn't the revolutionary potential of individuals, perhaps measured by past expressions of dissidence, matter when pursuing alliances for the revolution?

Alternatively, as will be further explained in this introduction, transversal theory and methodology are all about potential. They work discursively to empower social identities and groups recognizably striving—conceptually, emotionally, and/or physically—to transcend their subjective territories. The idea of totalized or absolute subjectivities, identities, societies, or states is an obvious impossibility insofar as movement and change are fundamental to the world's existence. Subjective territory, very different from the always already lacking and failed subject for Laclau and Mouffe, is not an ahistorical phenomenon continuously manifested, expanded, and discombobulated through just antagonism, but a space socially, culturally, and politically generated and maintained through antagonisms as well as other conceptualizations and experiences, including growth, evolution, curiosity, imagination, kindness, empathy, desire, passion, and love. Subjective territory—unfixed, mutative, and permeable—is a conceptually and emotionally constrained, multidimensional space through which individuals, that is, actual humans with social identities and varying degrees and kinds of agency, navigate their consciousness and presence within specific social, cultural, and historical parameters of space–time. For more detailed discussions of the differences between my concept of subjective territory and other theories of identity formation, including the work of Laclau, Mouffe, Žižek, Guattari, Althusser, Bourdieu, Greenblatt, Lacan, Deleuze, Jameson, and Paul Smith, see Reynolds, *Becoming Criminal*; and Bryan Reynolds and Janna Segal, "The Reckoning of Moll Cutpurse: A Transversal Enterprise," *Rogues in Early*

Modern English Literary Culture, ed. Craig Dione and Steve Mentz (University of Michigan Press, forthcoming 2004).
5. For more on "transversal movement," see the texts mentioned in notes 1–3.
6. For further discussion of "subjunctive space," and in relation to the "Mary/Mollspace" of Thomas Middleton and Thomas Dekker's *The Roaring Girl*, see Bryan Reynolds and Segal, "The Reckoning of Moll Cutpurse."
7. Conceived of by Bryan Reynolds and James Intriligator, the "investigative-expansive mode" was first introduced in print by Bryan Reynolds and Joseph Fitzpatrick in their article, "The Transversality of Michel de Certeau: Foucault's Panoptic Discourse and the Cartographic Impulse," *Diacritics* 29.3 (1999): 63–80. Bryan Reynolds thanks Janna Segal for coining the catchy abridgement, "i.e. mode."
8. For discussion of "affective presence," beyond the coverage in this book, see Reynolds, *Becoming Criminal*.
9. See Donald Hedrick and Bryan Reynolds, "Shakespace and Transversal Power," *Shakespeare Without Class: Misappropriations of Cultural Capital*, ed. Donald Hedrick and Bryan Reynolds (New York: Palgrave/St. Martin's Press, 2000), 3–47.
10. While recently visiting a friend in her hospital room, the hospital pastor paid us a visit. Urging us to have our friends pray for my ill friend, the pastor explained that studies have proven that the more people who pray for an individual, regardless of whether or not the individual is aware of the people praying for her, the more likely God is going to cure her. In fact, he explained, it does not matter if the people know her because, for God, they need only to know of her and see her as worthy of their prayers. Apparently, according to the studies cited by this pastor, God values people more who are valued by others. Although I am not a believer in a Judeo-Christian god, or of any god(s), I did find this account fascinating. But most of all I found it disturbing, especially as I watched the pastor visit room after room each day as I visited my friend, and wondered how he made people feel when he told them that God's verdict on their life is commensurate with how many friends and fans they have. To read about case studies on the power of prayer, see, Larry Dossey, *Healing Words: The Power of Prayer and the Practice of Medicine* (San Francisco: Harper Collins, 1993) and *Prayer is Good Medicine* (San Francisco: Harper Collins, 1996); Bill Banks, *Three Kinds of Faith for Healing* (St. Louis: Impact Christian Books, 1989); and Gordon Lindsey, *Prayer & Fasting: The Master Key to the Impossible* (Dallas: Christ for the Nation, 1971).
11. For more on Derrida's concept of "supplementarity," see the chapter entitled, "Structure, Sign, and Play in the Discourse of the Human Sciences" in *Writing and Difference*, trans. Alan Bass (Chicago: University of Chicago Press, 1978), 278–94.

Works cited

Althusser, Louis. *Lenin and Philosophy and Other Essays*. Trans. Ben Brewster. New York: Monthly Review Press, 1971.

Derrida, Jacques. *Writing and Difference.* Trans. Alan Bass. Chicago: University of Chicago Press, 1978.
Dunne, Dominick. "Mourning in New York." *Vanity Fair.* Vol. 495. November 2001: 178–83.
Foucault, Michel. *The History of Sexuality: An Introduction.* Vol. 1. New York: Random House, 1990.
Guattari, Félix. *Molecular Revolution: Psychiatry and Politics.* New York: Penguin Books, 1984.
Hedrick, Donald and Bryan Reynolds. "Shakespace and Transversal Power." *Shakespeare Without Class: Misappropriations of Cultural Capital.* Ed. Donald Hedrick and Bryan Reynolds. New York: St. Martin's Press, 2000. 3–47.
Lacan, Jacques. *The Psychoses.* Trans. Russell Grigg. New York: W.W. Norton, 1993.
Laclau, Ernesto. *New Reflections on the Revolution of Our Time.* London and New York: Verso, 1990.
Laclau, Ernesto and Chantal Mouffe. *Hegemony and Socialistic Strategy: Towards a Radical Democratic Politics.* London and New York: Verso, 1985.
Poulantzas, Nicos. *State, Power, Socialism.* London and New York: Verso, 2000.
———. *Political Power and Social Classes.* London: NLB and S & W, 1973.
Reynolds, Bryan. *Becoming Criminal: Transversal Performance and Cultural Dissidence in Early Modern England.* Baltimore: Johns Hopkins University Press, 2002.
———. "The Devil's House, 'or worse': Transversal Power and Antitheatrical Discourse in Early Modern England." *Theatre Journal* 49.2 (1997): 143–67.
Reynolds, Bryan and Joseph Fitzpatrick. "The Transversality of Michel de Certeau: Foucault's Panoptic Discourse and the Cartographic Impulse." *Diacritics* 29.3 (1999): 63–80.
Sartre, Jean-Paul. *Transcendence of the Ego.* New York: Noonday Press, 1957.
Žižek, Slavoj. *The Sublime Object of Ideology.* London and New York: Verso, 1989.

CHAPTER 2

THE MAKING OF AUTHORSHIPS:
TRANSVERSAL NAVIGATION IN THE
WAKE OF *HAMLET*, ROBERT WILSON,
WOLFGANG WIENS, AND SHAKESPACE

D.J. Hopkins & Bryan Reynolds

Whence we depart

Manhattan, July 7, 1995: On this hot, muggy evening, an audience rapidly filled Lincoln Center's air-conditioned Alice Tully Hall. Whereas the majority probably hailed from New York City and its nearby environs, one member of the audience had driven down from Cambridge, and another had flown in from San Diego, just to see this particular performance of William Shakespeare's *Hamlet*. At the time, these two, the authors of this paper, did not know each other.

While the Cambridgian (Reynolds) and his date were taking their seats somewhere in the back of the auditorium (they were tardy), the San Diegan (Hopkins) was sitting comfortably, engrossed in his press package, when a cigarette-roughened voice politely interjected: "Excuse me." Hopkins retrieved his legs to allow the fashionably late audience member and his companion to slide past to their seats beside him. Looking up as they passed, Hopkins recognized that the man, wearing a purple crushed velvet jacket, was rock star Lou Reed, and that the woman with him, sporting her signature dandelion-gone-to-seed hairdo, was performance artist Laurie Anderson. They too had come to see this production of *Hamlet*, presumably for the same reasons as Hopkins and Reynolds: not out of sheer devotion to Shakespeare or *Hamlet*, but because of the individual to whom ownership of this particular production was assigned—a "star," not unlike Reed and Anderson themselves.[1] With a horizon of expectations informed by, among other related forces, our preconceptions about the theatrical magic of Robert Wilson, all of us had willfully sought an experience in what Donald Hedrick and Bryan Reynolds call "Shakespace."[2]

Shakespeare has inspired and inhabited innumerable commercial, political, social, and cultural spaces, both ideational and physical, like academic discourse and Lincoln Center's Alice Tully Hall, respectively. These Shakespearean or Shakespeare-affected spaces, however conventional, radical, or something in between, Hedrick and Reynolds refer to as Shakespace, a term that accounts for these specific spaces and the timespeed, the "pace," at which they pass from generation to generation and from era to era. Shakespace is a socially and historically determined playground on which cultural and class differences sometimes slip "transversally" into an equivocal space that allows for, and in fact fosters, alternative opportunities for thought, emotion, expression, and development.

The Shakespearean production we discuss here offers just such a space of alternative opportunities, and its mode is one with which Shakespeare himself was engaged: theater. Although we are living in a culture that exists *after* mass media, our culture has not (necessarily) done away with those modes of production that existed *before* mass media. In other words, people still produce Shakespeare's plays on the stage. What we will be discussing here is the nature of the stage and theatrical modes of production after mass media. Though the mode of production we consider is not itself a mass medium: theater is surprisingly marginal. While Lincoln Center may seem like a privileged site of production, particularly within New York City, its cultural importance and influence pale in comparison to the relative importance and influence of other modes of production in our mass media age. The film, television, and publishing industries dwarf theater production in terms of financial clout and cultural saturation; tellingly, one cannot speak of a "theater industry."

Increasingly marginalized in academic contexts, as well, theater has evolved to meet many of the criteria Richard Burt offers for his definition of "schlock." Theater is often regarded as an obsolete mode of production, which lies, as Burt puts it, "beyond the usual academic purview" (7), perhaps specifically because it is not one of the reproducible mass media. But theater is not schlock per se, nor can we see the abstract and abstruse theater works of Robert Wilson as adhering to Burt's other descriptions of schlock as popular and trivial. For better or worse, the subject matter under consideration here—theater, Wilson, *Hamlet*—is anchored in "high culture," elevated (or relegated) to the realm of the official canon.

According to Reynolds's "transversal theory,"[3] which accounts for the critical approach of the present analysis, Shakespace often manifests as "official territory," and has therefore promoted various organizational social and political structures that are discriminatory, hierarchical, and repressive. Yet it has most powerfully manifested as "transversal territory."

Transversal and official territories function in relation to "subjective territory." As Reynolds describes elsewhere,[4] subjective territory is the range of the conceptual and emotional experience of those subjectified by the "state machinery," which includes all indoctrinating mechanisms of a society's "official territory," such as the educational, juridical, religious, and familial structures that support the ideology of the dominant classes and the government. In agreement with its investment in cultural, social, political, and economic fluctuations and aspirations, the organizing state machinery operates over time and space, sometimes consciously and sometimes unintentionally, to consolidate social and state powers in order to construct a society of subjects and thus "the state": the totalized state machinery. Subjective territory is therefore delineated by the conceptual and emotional boundaries that are defined by the prevailing science, morality, and ideology of official territory. These boundaries create common ground for an aggregate of individuals, and work to maintain the coexistence of the social body that this aggregate becomes. In short, subjective territory accounts for the existential and experiential realm in and from which a given individual who has been subjugated conceptually and emotionally, that is, developed into a subject by the state machinery of any hegemonic society or subsociety (such as the university, criminal organizations, or religious societies), perceives and relates to the universe and her/his place in it; thus the individual's subjective territory, and corresponding self-government, reflects his/her socioeconomic positioning.

Whereas the infinite possibilities of conceptual and emotional experience are powerfully constrained by the state machinery's subjectification of the individual, transversal theory insists that subjective territory is not capable of fully containing the subject. Beyond the limits of subjective territory exist non-subjectified regions of one's conceptual and emotional range, the open space of transversal territory. This uncertain, unpredictable space houses, and is the gateway to, all of the possibilities that official territory wants to preclude and exclude. Transversal territory can be reached by transgressing the boundaries that the state machinery enforces. This transgression of subjective territory's boundaries is made possible by "transversal power," and through "transversal movements." Transversal power, which can be found in anything from criminal acts to natural catastrophes to the daily practice of walking in a city, is a dynamic of change.[5] It is what spurs people to move transversally, consciously or not, out of their subjective territory, into transversal territory, and possibly into the subjective territory of others. When people expose themselves to alternative ideas, emotions, and experiences, such as in the theater or classroom, and engage with those alternative energies, such as through empathy or trauma, they move transversally, thereby altering and expanding their own subjective territory. Their subjective territory becomes existentially and experientially bigger and different, incorporating new associations, sensibilities, and perspectives. In the present analysis, we move transversally, navigating in the wake of Wilson's *Hamlet: A Monologue*. In doing so, we engage with several important implications to the transversality we encountered in the remarkable Shakespace that both encompasses Wilson's production and informs the rest of this book.

Our aim here is to analyze the Shakespace in which *Hamlet: A Monologue* operates, and to analyze, from within this Shakespace, the discourse on authorship, media, and performance that this Shakespace is so importantly informed by and informs. In the context of this discourse, it may be that this production becomes a kind of "antischlock," troubling the relationship between elitist and popular culture from the top down. As we will demonstrate, *Hamlet: A Monologue* relies in no small measure on the popular knowledge of the play *Hamlet*. Wilson's transversal performance works to exploit the audience's familiarity with this Shakespearean source text to create a theater product that cannot be reduced to a popular (schlocky) commodity.

Our transversal method of analysis, which we hope to promote with this application, is what Bryan Reynolds and James Intriligator call the "investigative-expansive mode," as opposed to what they call "the dissective-cohesive mode," which characterizes most research.[6] Like the dissective-cohesive mode, the investigative-expansive mode requires that the subject matter of interest be divided up and partitioned into variables in accordance with the selected parameters of the investigation. But unlike the dissective-cohesive mode, whose goal is ultimately to reconstruct the variables as a coherent whole, the investigative-expansive mode examines both the internal connectedness of the variables (among themselves) and the external connectedness of the variables (to other forces) in coordination with an ongoing readiness to reparameterizing as the analysis progresses. The investigative-expansive mode welcomes the emergence of unanticipated problems, information, and ideas. Thus, the purpose is not to assemble a fully accounted for totality, but to better our comprehension of the subject matter's changing relationships to its own parts as well as to the environments of which it is a part. The subject matters of the present analysis are *Hamlet: A Monologue* and its critical relationship to issues of authorship, media, and performance. The chosen variables include Wilson's and his dramaturg Wolfgang Wiens's respective contributions to the performance of *Hamlet: A Monologue*, the current "conceptual crisis in drama studies" (as W.B. Worthen describes it), the textual history of Shakespeare's *Hamlet*, and competing theories of authorship. And the particular greater environments of concern are both the Shakespace and the hermeneutic space in and through which *Hamlet: A Monologue* and the related discourses on authorship, media, and performance will continue to journey.

A matter of presence

The set for the July 7, 1995 performance of *Hamlet: A Monologue* consisted of a bare stage, except for the bold presence of a tiered stack of dark-gray rock slabs, a sort of stylized "sterile promontory," a lit backscreen that changed colors (from blue to yellow to many others), and a single actor, who was usually clad in black. The play began with a single prone actor, Robert Wilson, lying on top of the rock stack, where he then struck a hieratic pose and spoke:

—Had I but time . . .
—O, I could tell you—but let it be.
—Wretched queen, adieu!
—I follow thee . . .
—Follow my mother! Drink of this potion!
—Venom, to thy work!
—I can no more—the king, the king's to blame.
—Lo, here I lie, never to rise again.
—O villainy! ho! let the door be locked—
—The drink, the drink!
—No, no, the drink, the drink—
—How does the queen?
—Look to the queen there, ho!
—Nay, come again.
—Come on.
—Come for the third, Laertes.
—I dare not drink yet, madam—by and by.
—Good madam!
—A touch, a touch!

—Another hit!
—Give him the cup.
—A hit, a very palpable hit.
—No!
—One!
—Come, my lord.
—Come on, sir.
—Come, begin! (Wilson/Wiens, *Hamlet, A Monologue* 1)

This is the first speech of Scene One of *Hamlet: A Monologue*, a one-person show consisting of fifteen scenes. In performance, this introduction was followed by the lone actor coming to his feet, and punctuating his fragmentary, backwards-tending narrative by executing a series of complex gestures calculated to abstractly express death throes, collapse, pain, combat, and similar correlatives to the spoken text. The actor playing the character of Hamlet spoke all the words quoted above, though they are recognizable as lines attributed to various characters in the last scene of Shakespeare's play. Indeed, all of the words in the performance, and in the published script, are attributed to that character: Hamlet himself is the only character we see, though he in turn performs the personae of the Ghost, Claudius, Gertrude, Ophelia, and Polonius, as necessary. All of the "unnatural acts, accidental judgments, casual slaughters" that constitute the action of a conventional staging of *Hamlet* were here distilled and funneled into a solo performance, *Hamlet: A Monologue*. In addition to being the only actor, Robert Wilson was also the director, scenographer, and co-adaptor responsible for this marathon condensation of Shakespeare's text.

Our introduction to this chapter, with an emphasis on personal experience and a tourist's eye for rock stars and artists, is not an indulgent and anecdotal foray into pop culture. We introduce the individual personalities of Reed, Anderson, Wilson, and ourselves to stress that at a theatrical performance, as at a rock concert, *presence* is essential to the event. We want to show that this particular production responded to, and exemplified, a culturally activated shift in the textual authority of Shakespeare's plays. Under the present-tense conditions that governed what we experienced as a "transversal theater performance," *Hamlet: A Monologue* demonstrated a textual authority over the written text that would seem to be its requisite, citational predecessor: the written text of Shakespeare's *Hamlet*.

Wilson's performance of *Hamlet* is predicated on a single premise: that the recent events of Hamlet's life "flash before his eyes" at the moment of his death. *Hamlet: A Monologue* begins at this moment of death and moves backwards through the events that would have occurred already in a conventional, "textually obedient" version of *Hamlet*. Although this *precis* of *Hamlet: A Monologue* is factually true, its usefulness as a summation of the impact of the performance text is challenged on two points. First, it runs the risk of giving the impression that the performance was reductive and superficial, which it most definitely was not. The script of *Hamlet: A Monologue*, conceived and adapted by dramaturg Wolfgang Wiens, recreates Hamlet's world as though Denmark were indeed bound within the subjective territory of the Prince. As the quoted synthetic fragments show, this adaptation is not simply the scenes in reverse order and characters walking backwards. Every character and event in Shakespeare's *Hamlet* are portrayed in Wiens's written text and Wilson's performance text through the expressive medium of one character: the idiosyncratic half-mad, half-dead Hamlet. The second, more significant reason that our "reverse order" description of *Hamlet: A Monologue* is ultimately invalid, despite its relative validity for literary criticism, is that it gives the false impression that the "events" of Hamlet's "story," as told in Wilson's "play," are of principal importance to the performance text of *Hamlet: A Monologue*. They are not.

His own public Elsinore

In performing the role of Hamlet, Wilson became multiple, plural; he became a *many*, yet so did his Hamlet. Wilson's Hamlet defined both what we call a "subjective territory" and a "transversal territory." His Hamlet moved transversally through the subjective territories of the other characters depicted in Shakespeare's play, synthesizing their conceptional and emotional experiences, and reflecting them in his own burgeoning subjective territory. Hamlet's subjective territory became the transversal territory from which multiple characters emerged, and across which they traveled throughout the course of the performance. In other words, Hamlet articulated the dynamic, permeable subjective territory that delineated the other characters that appeared in this version of Shakespeare's play. From within this multivalent subjective territory, the other characters moved transversally and reparameterized their own conditions of existence despite the seemingly univocal and singularly present character/medium who/which made their own identities possible. In effect, Wilson transformed himself into the media for the transversal movements of a field of characters. He functioned as the medium for the character Hamlet and, as Hamlet, as the medium for the others.

As a complex signifier for multiple referents, including the actor Wilson himself, Wilson's performance asserted fantastic presence. The history of Shakespeare's *Hamlet*—its multifarious presence in Shakespace (its entrails, tentacles, and gases)—was transposed and disseminated by *Hamlet: A Monologue*.

To recapitulate, with an added philosophical rendering: In the move from a five-act, multicharacter play to a fifteen-scene monologue, Wilson created the terms for a subjective territory; a subset of all conceptual and emotional space in which individual subjects are constructed and constrained by the shared paradigms of a community, be those social structures, government regulations, or, as in the case of Wilson's Prince of Denmark, the (*dramatis*) *personae* of a play, all of whom are assembled within a single character's shattered sense of self. Shakespeare's *Hamlet* became Wilson's *Hamlet: A Monologue* in the subjective territory of a Hamlet dancing transversally at his mortality's vanishing point. In this theatrical Shakespace where Hamlet's life flashed before his eyes, Wilson and Wiens evolved a philosophical discourse on death. The life that flashed before Hamlet's eyes in the moment of his death was a life that simultaneously included presences and traces of the subjective territories of other characters that influenced his life. Death, then, requires either the hallucinated or real participation of others; a fracturing, transversal self (or selves) or a culminating, communal expression that fuels life's departure. Wilson's character Hamlet was in and of itself a transversal theory of subjectivity that begins with the miracle of death. Hamlet's death moment was infused with the consciousnesses of Shakespeare's other characters, of Wilson, of Wiens, of Shakespeare. From our own experience, and from our conversations with other audience members, we discovered that this infusion inspired the audience either to relate to multiple characters, and thus multiple perspectives, or to become alienated from Hamlet, making identification impossible. Or, it inspired both. Insofar as we anticipate identification with Hamlet, inasmuch as we have been taught that we should identify with Shakespeare's Hamlet (because such famous critics as Coleridge, Lamb, and Hazlitt have), Wilson's Hamlet's identification or alienation effects stimulated transversal movements: the performance was an irresistible invocation. Recall that at a séance, the "medium" is the controlling, conjuring figure through whom others—no longer living—enter to be among those still alive and in attendance. In the Wilson/Wiens's discourse on death, Wilson performed the medium Hamlet; but it was not the dead who passed through the living. In *Hamlet: A Monologue*, the character on death's threshold became the transversal medium for those who are still alive.

Becoming investigative-expansive

At this point, we want to shift directions to a subject that our analysis points toward, but has yet to address: *Hamlet: A Monologue* is a theater text with particular relevance to the current "conceptual crisis in drama studies" notably articulated by Shakespeare scholar W.B. Worthen (1093). In a tremendous study of the factional disputes among different academic disciplines, Worthen meticulously outlines the means by which "performance came to be defined in opposition to theater structures and conventions," as well as the means by which, in an academic context, the "different disciplinary investments of performance studies and literary studies" often bring those fields into conflict with the disciplinary investments of theater studies (1093). Worthen casts doubt on those who would claim a " 'ministerial' or 'derivative' relation to the dramatic text," arguing that "although the sense of dramatic performance as a performance *of* a play is widespread, as John Rouse remarks, just 'what the word *of* means' in this context is 'far from clear' " (1094–95).

In the essay to which Worthen refers, Rouse explores the issues of "Textuality and Authority in Theater and Drama," using as his chief example a very different *Hamlet* adaptation: Heiner Müller's *Hamletmachine*, directed by Robert Wilson for a 1986 production at the Thalia Theater in Hamburg.

> It isn't hard to understand why audiences took nearly a decade to learn to "make sense" out of Wilson's performance texts. In the first place, he doesn't subordinate the visual score to the acoustic; all of his productions subvert narrative organization, even when they make use of traditional dramatic texts. Wilson also subverts normal rules for the internal structuring of each of his scores. He came to theater from art and architecture and tends to privilege codes from these disciplines when combining the signs of his visual scores. (152)

Here, Rouse describes Wilson's production method as one that works in opposition to a conventional, illustrative mode of theater. Rather, within the defined performance space, "the production ostends two texts and gives the spectator control over writing the intertext" (Rouse 154). Wilson himself seconds this description of his theater: "We make theatre so that the spectator can put it together in [her or] his own head" (qtd. in Rouse 154).

Worthen stops short of prescribing the redefinition of "drama studies," which might be necessary for the solution of this contentious issue, perhaps even for the preservation of the field itself. The problem of definition may begin with the conventional nomenclature that Worthen has inherited: Worthen begins his essay by referring to the field in dispute as "drama studies" (1093). The persistent use of the word "drama" to describe an academic discipline vexed by issues of "citationality" is more than merely ironic, it is the very site of the interdisciplinary impasse that Worthen seeks to negotiate. If the text is, as Worthen rightly suggests, merely "material for labor, for the work of [theatrical] production," then the retention of the word "drama" in the name of this field is a counterproductive vestige of theater's fealty to a preexisting literary work (1098). To extend Worthen's argument, any performance that calls itself "dramatic" is by definition "a hollow, even etiolated, species of the literary" (1098). As a result of this conditioning of the word "dramatic," when Worthen argues that "one of the ways both literary studies and performance studies have misconceived *dramatic* performance is by taking it merely as a reiteration of texts, a citation that imports literary or textual authority into performance" (1098), he is harboring nomenclature within his own rebuttal that preserves the key concept of the opposition's position: the literary citationality of "drama."

Perhaps by way of a corrective, one might take a cue from Terry Eagleton, who concludes his book-length survey *Literary Theory* by rejecting altogether his title phrase, and the limited concept it embodies:

> My own view is that it is most useful to see "literature" as a name which people give from time to time for different reasons to certain kinds of "discursive practices," and that if anything is to be an object of study it is this whole field of practices rather than just those sometimes rather obscurely labeled "literature." I am countering the theories set out in this book not with a *literary* theory, but with a different kind of discourse—whether one calls it of "culture", "signifying practices" or whatever is not of first importance—which would include the objects ("literature") with which these other theories deal, but which would transform them by setting them in a wider context. (205)

Eagleton prohibits the word "literature" from its associations with "theory" in order to broaden the possibilities open to theoretical discourse. A similar prohibition in nomenclature might lead to a resolution to the "crisis in drama studies," a shift in language that would bring about a corresponding conceptual shift in the definition of the field of study. Academic disciplines that focus on a study of "drama," whether that drama be read or performed, are rightly the province of Literature Studies. It is not useful to challenge claims of literary citationality by calling live theater "drama," or the study of live theater "drama studies." The study of theater is "Theater Studies," not "Drama

Studies." A live theater performance is "theater," not "drama." Many Theater Studies programs emphasize the dramatic component of their study, and many works of theater art emphasize a dramatic component in performance. But not all do this. One could argue that the issue could be clarified by clarifying the division between the disciplines under discussion. If all "Drama Departments" would just change their names to "Theatre Departments," then the problem would be solved. But the whole question of an interdisciplinary turf war is blown apart by Worthen's essay itself. How does one categorize an essay like his? It was published in *PMLA*, so it could be called "Literature Studies." But Worthen is on the faculty of a theater department: does that make his essay "Theater Studies"? And, despite the fact that Worthen's goal in this essay is to rehabilitate a notion of theater as a valid field of study, the essay relies to a large degree on Worthen's extensive understanding of performance studies. In its persistent interdisciplinarity, Worthen's essay is a work of cultural studies: an examination of conflicts among three different academic disciplines and the disciplines' differing notions of what should be valued as a subject of study. One can even imagine a performance of a canonical work of dramatic literature that, in performance, as Wilson's appropriation of Shakespeare's *Hamlet* demonstrates, does not emphasize a dramatic component. Though taking as its point of departure no less grand a literary masterpiece than *Hamlet*, *Hamlet: A Monologue* shows the possibilities for theatrical texts to defy a ministerial relationship to a written text. Wilson's theater text works against premises of literary citationality, not by doing violence to the written text of *Hamlet*, but by acknowledging that the conventionally "read" text of *Hamlet* is only one unstable part of the theater text.

We have been referring to the "written text" and the "theater text," respectively "Shakespeare's text" and "Wiens/Wilson's theatrical text." These appellations are not arbitrary, but rather incited by the complex signifying conditions that surround *Hamlet: A Monologue*. These elements must be distinguished from each other in order to pursue effectively questions concerning citationality, authority, and the nature of Rouse's performative "of." Equally important is the recognition that the history of "textual" issues is closely linked to issues of textuality, which have haunted Shakespeare's *Hamlet* for centuries.

The eternal "retext"

The "text" of *Hamlet* has been central to controversy over textual authority for hundreds of years. Moreover, the word "text" has evolved, and now has two very different meanings of importance for us here, both of which are meanings very much applicable to a consideration of *Hamlet*. Many of Shakespeare's plays, as is well known, exist in multiple forms. The traditional definition of "text" is used in Shakespeare studies to refer to specific different versions of Shakespeare's plays, the contents of which may differ, from one degree to another, although the different versions of the play may have the same, or roughly the same, title. "Textual scholarship" is the field of study that considers the history of these different versions: how different versions may have come into being, how editors over the years have responded to the versions and their differences, and even added to those differences. Yet the vast majority of Shakespeare scholars focus their attention on a different Shakespearean "text" altogether. Studies of the "textuality" of Shakespeare's works consider the relationships between the plays and the complex network of social, cultural, and political forces in which those plays were originally situated, as well as the impact of those same forces over the hundreds of years since.

This latter meaning of "text" has evolved more recently, and is defined by Roland Barthes as "a methodological field" (157). But Barthes's understated *precis* of "text" fails to capture the sense of far-reaching interconnectedness and conceptual referentiality that has come to be associated with his use of the word. The earlier use of the word text is closer in meaning to Barthes's term "work," which he uses to refer to the "concrete" item or object that might be found "occupying a portion of book-space (in a library for example)" (157–58). The object of textual scholarship are those items that occupy a portion of book-space, which could be called in Barthesian terms, Shakespeare's "works." To make matters more complicated, these works have had value judgements applied to them over the years: some are "good" and some are "bad."

Writing at the beginning of this century, A.W. Pollard (1909) initiated the popularly accepted distinction between "good" and "bad" quarto versions of Shakespeare's plays. Pollard initially identified five texts that seemed particularly deficient in their first Quarto (Q1) versions, and particularly at variance from their Folio versions: *Romeo & Juliet, Henry 5, Merry Wives of Windsor, Hamlet,* and *Pericles*.

Pollard sought to justify his identification of these plays as "bad" by referring to the introduction to the first Folio, written by the Folio publishers Heminge & Condell, in which they denounce "frauds and stealths of injurious impostors" who had printed and sold "stolen and surreptitious copies" of Shakespeare's plays. Pollard concluded that this accusation was leveled at these Q1 versions. To this list of five were subsequently added *The Taming of A Shrew, 1 Contention,* and *Richard*

Duke of York, better known as *The Taming of The Shrew, 2 Henry 6*, and *3 Henry 6*, respectively. The list was brought to ten with Q1 *Richard 3* and Q1 *King Lear*.

In 1910, W.W. Greg performed a close analysis of the "bad" quarto version of *The Merry Wives of Windsor*. He concluded that the lapses and deficiencies in that text could be explained if one were to presume that the actor playing the role of "Mine Host" had written down the text from memory. This theory inaugurated the "memorial reconstruction" hypothesis, which accounts for the origin of the "bad" quartos by suggesting that an actor or shorthand thief in the audience had written out (from memory or by shorthand) a version of the script and would subsequently sell the script to a printer. Based on Greg's tentative hypothesis, "memorial reconstruction was adopted as blanket classification for these ten Shakespearean texts, ossifying Greg's original tentative suggestion into textual orthodoxy" (Maguire 5).

The late 1970s and 1980s brought a sustained challenge to the memorial reconstruction hypothesis, introducing the question of revision. Michael Warren argued in 1976 that both versions of *King Lear* are "good," the later version of the Folio being Shakespeare's own revision of that text. This approach to Lear resulted in a collection of essays edited by Warren and Gary Taylor, *The Division of the Kingdomes, Shakespeare's Two Versions of King Lear*, as well as prompting Taylor and Stanley Wells to include both versions of *King Lear* in the *Oxford Complete Shakespeare*. Yet the textual complexity of *Lear* is slight when compared to the situation surrounding the texts of *Hamlet*.

The text of *Hamlet* exists for us in three versions: Q1 of 1603, Q2 of 1604/1605 and the first Folio of 1623. After the discovery of Q1 in 1823, most scholars regarded it as a rough draft, which was subsequently revised in Q2. Early in the twentieth century, the memorial reconstruction hypothesis took hold and became the norm of opinion whereby the existence of Q1 was explained. Since the first Quarto version has been perennially relegated to second-class status, Q2 has most often served as the basis for contemporary printed editions. However, Wells and Taylor have adopted the Folio as the principal text for their Oxford editions on the grounds that the Folio version is Shakespeare's own revision.

In her assessment of the status of Q1, Leah Marcus makes the observation that Q1—however understood as revised, memorized, or otherwise—is "bad" by definition, "and will continue to be bad so long as we rank the early texts of the play on the basis of their adherence to culturally predetermined standards of literary excellence" (Marcus 135). Marcus argues that advocates of memorial reconstruction find memory (and the oral tradition implied by the use of memory in theater) to be inherently untrustworthy. She concludes:

> Over and over again in Shakespeare's plays, but particularly in *Hamlet*, bad taste is associated with an outmoded oral theatrical culture. Similarly for twentieth century adherents of the theory of memorial reconstruction, "bad" Shakespeare is the product of defective memory and insufficient literacy.... [G]ood taste is associated with writing as opposed to orality; and "good" Shakespeare with the creation of a theater that is specifically literary. (137)

In her chapter, "Bad Taste And Bad Hamlet," Marcus presents a theatrical situation that illustrates the cultural conflict between "good" and "bad": "In a small theater, Hamlet nears his most famous soliloquy" (132). Marcus sets the stage for the audience's much-anticipated reception of this famous character's most famous set of words. The physical tension is heightened, theatrical elements focused, the history of acting invoked, and then: "something goes radically *wrong*" (132). Instead of the words that the audience knows and are culturally conditioned to expect, they hear a very different Hamlet:

> To be or not to be; aye, there's the point.
> To die, to sleep; is that all? Aye, all.
> No; to sleep, to dream; aye, marry, there it goes. (Q1 58)

The sense of dissonance here lies in the antipathy for the mode of oral performance and the trend toward a literary measure of a theatrical "good." The blunt staccato language of Q1, quoted here, seems far removed from the blank verse and carefully articulated philosophical probing of Q2:

> To be, or not to be, that is the question,
> Whether tis nobler in the minde to suffer
> The slings and arrowes of outragious fortune,
> And by opposing, end them, to die to sleepe
> No more, and by a sleepe, to say we end
> The hart-ake, and the thousand naturall shocks
> That flesh is heire to; tis a consumation
> Devoutly to be wisht to die to sleepe,
> To sleepe, perchance to dreame, I there's the rub ... (Q2 G2)

By comparison to Q2, the Q1 version sounds like an actor struggling to remember a complicated and incompletely memorized part. Or, to give some measure of credit to the text of *Hamlet* most often dismissed, the Q1 passage can be read as a character struggling to comprehend a complicated set of issues, and exploring these ideas out loud in what appears as an (intentionally) inarticulate oral expression of his inner turmoil. This is the question of Q1: intentional or unintentional? Author or pirate? The answer to these questions—however fascinating and cleverly formulated, as is Marcus's "hypothetical reconstruction"— is irrelevant (Marcus 133). The sole verifiable conclusion that we can make relates more to our contemporary assessment of these texts than it does to the texts themselves. Q1 represents a threat to the purity of Shakespeare's *literary* legacy. Simultaneously, however, it represents one of the clearest links to Shakespeare's legacy as a practitioner of live theater, to a performance tradition widely regarded as "more honour in the breach, then the observance," but a tradition that nevertheless extends continuously from Shakespeare's time to ours (Q2 D). Consider our own description of the opening of *Hamlet: A Monologue*. A Lincoln Center patron expecting Shakespeare but getting Wilson would feel much the same "culture shock" as that which Marcus suggests a playgoer would feel if, when culturally conditioned to expect Q2, the playgoer got Q1.

This brief review of the history of the different versions of Shakespeare's scripts is intended to provide a glimpse of the general suspicion of any suggestion that a theatrical or oral medium might be given a privileged position over a literary one. With this sense of the "badness" attached to performance media, we would like to return to a consideration of Wilson's version(s) of *Hamlet*.

A new medium for the text of *Hamlet*

Wilson says of the monologue, "It's all in a sense in his mind," Hamlet's own "autobiographical" recounting of events (*Making* 1:28). But, as we suggested earlier, the visual and theatrical means by which Wilson and his collaborators (re)present the events of the play on stage receives the focus of Wilson's efforts, not the events themselves. And these means correspondingly become the focus of the audience's attention.

Earlier we provided a quote from the opening of *Hamlet: A Monologue*. Now we need to qualify that quotation by saying that it is a quote from the *written* text of the adaptation. Wilson's name appears on the title page of the unpublished "text," where Wilson also gets billing ahead of Wiens in the byline.[7] But it is possible to identify the script principally with the work of the dramaturg, Wolfgang Wiens, and not with Wilson—largely because all the other artists on the project do. In an interview about the origins of the project, the German assistant director, Ann-Christin Rommen (an unusually active contributor, given the conventional limitations of the position of AD), had this to say: "We come together—Wolfgang Wiens, who wrote this adaptation, and Bob and I—and we were reading the text, and Wolfgang had made the decision already that it would be fifteen scenes" (*Making* 12:04). It is clear from Rommen's unguarded statement that Wiens is regarded as the individual behind the script, responsible for both its structure and its content. However, while clarification of Wiens's significant role in this project may be interesting to dramaturgs and other theater professionals, it would be inappropriate to assign to Wiens definitive authorship of *Hamlet: A Monologue*. Wiens's written text, citational in its relationship to Shakespeare's written text(s), was ultimately only a navigational aid for Wilson's final project, a compass that assisted the progress toward the performance text, but did not determine that text.

What happened on stage in *Hamlet: A Monologue* happened there alone, though the theatrical performance was supported by several (physical) texts. The second text, an unusual medium for a "dramatic text," is equally applicable to *Hamlet: A Monologue*, and is, in fact, more readily identifiable with Wilson and his body of work. The following pictorial quote from this script corresponds to the portion of the written adaptation that we previously quoted:

the sleep of death
(*Hamlet*, Program 15)

This, too, is Scene One of *Hamlet: A Monologue*. A "script" of sorts, that is separate from the written adaptation that bears the same title, this visual text version of the fifteen-scene monologue is composed of seventeen charcoal sketches drawn by Wilson (two of the scenes are rendered in two panels).[8] In a production by another director, a series of sketches would not constitute a second script, certainly not one with an impact on the performance equal to that of the lengthy and complex English-language script composed by one of Germany's foremost dramaturgs. And for what production of *Hamlet* could a series of sketches—or a work in any media—be said to supercede Shakespeare's masterpiece as a guiding principle of the production?

THE MAKING OF AUTHORSHIPS / 45

The storyboard sketches are a kind of Quarto version, if you will, of the Wilson production. He conceived a visual script for the performance that privileges only visual elements and spatial volumes: "What the stage looks like... how dense it is or how dark it is or how light" (*Making* 8:36). Can a series of storyboard drawings graphically representing fifteen scenes be considered as a "version" of Shakespeare's *Hamlet*? The answer is: yes, but only a "bad" version. Marcus cuts the Gordian knot that ties critical discussion of Shakespeare's texts to the distinctions "good" and "bad," and her conclusions are equally applicable in a critical discussion of Wilson's texts. The only "good" version of the "To be or not to be" soliloquy is the version in which that line ends with "that is the question." In fact, one might go so far as to say that the only "good" version of *Hamlet* is a literary one that exists only on a page: any deviation from the reading experience of the play inherently constitutes one degree or another of "badness." But given that Q1 *Hamlet* is dismissed as "bad" because of its associations with an "outmoded oral theatrical culture," it is possible to accept other text versions as equally *Hamlet*, though equally "bad" because of their shared relationship with a performative medium rather than a literary one (Marcus 137). By embracing "badness" as the site of orality and theatricality, it is in fact possible to admit Wilson's series of charcoal renderings as a postmodern "Qw" of *Hamlet*.

murder most foul   mad call i it   get thee to a nunnery
                  (*Hamlet*, Program 15–16)

Transversal authorship

A series of questions surround any radical appropriation of a classic text. The main question here is: Who is the author of *Hamlet: A Monologue*? Is it Wolfgang Wiens, the dramaturg who conceived this adaptation of Shakespeare's *Hamlet*, divided it into fifteen scenes, and composed the script? Is it the distant figure of Shakespeare, more cultural construct than verifiable human? Or is it Wilson? Wilson is the clear nominee for what might be called the "auteur" of this production: director, scenographer, coadaptor, and solo performer, not to mention his status as the only living "famous" person associated with this production.

Wilson himself is the medium for this theater event, but given the nature of his work and the number of different contributions he makes to the final product, might he be the *media* of the event? Is it because of Wilson's "star" status that his "authorship" is so easily accepted? This train of thought ends with another question: "Who is the author of *Hamlet* itself?" In his essay "What Is an Author?" Michel Foucault writes,

> In a civilization like our own there are a certain number of discourses that are endowed with the "author-function," while others are deprived of it.... The author-function is therefore characteristic of the mode of existence, circulation, and functioning of certain discourses within a society. (148)

Foucault offers Freud and Marx as two exemplars of the group "founders of discursivity." These are the authors who have "produced... the possibilities and the rules for the formation of other texts" (154). Foucault posits that those who engage the rules established by the author-function are themselves authors, engaged in a valid form of discourse, one that extends and mediates the discourse introduced by one of a select few founding authors. Albeit dependent upon preexisting writing, Foucault considers this engaging mode of discourse neither "ministerial" nor "citational" in nature, and he does not condemn it as "a hollow even etiolated species of the literary" (Worthen 1098). Why then is theater considered "a mode of production fully inscribed within a discourse of textual [literary] and cultural authority" (Worthen 1094)? The medium of theater is capable of negotiating a field of discourse, such as that defined by the author-function "Shakespeare," just as is literature.[9] A performance of *Hamlet,* any performance of that play, moves into the field of discourse produced by Shakespeare and the Shakespace in which it circulates. Productions of *Hamlet* can be judged hollow or not hollow based on individual merit, but to dismiss all of theater as hollow negates all performative social semiotics that inform theater as well as theater's necessary interactions with the author-function.

Additionally, there are certain assumptions about the nature of an author that remain latent in Foucault's formulation. According to Foucault, the author-function developed "in the course of the nineteenth century" (153). Given that Foucault's definition of the author-function is associated with relatively contemporary texts—unlike his history of authorship that extends, say, from Homer through Saint Jerome to the present—it is not entirely clear why the author-function should emerge at one particular time. There must have been something about that historical moment that constituted a favorable mode of reception for these authors and the rules that they developed. In other words, it is not that Freud and Marx established revolutionary modes of discourse, but that they wrote during an era in which there existed (and exists to this day) a population of readers predisposed to enter into discourse with rule-generating authors, as perhaps the readers of Hippocrates and Saint Jerome were not. In other words, it is not with the individuals Freud and Marx that the author-function resides, but with the willingness of the population at large to enter into discourse with them, or rather, with the methodological field of their texts. Some might call this population a readership, others an audience. Ultimately, this discursive and interpretive population is the medium of authorship.

By accepting that Shakespeare can constitute an author-function, we begin to construct a theory of the theatrical text that inscribes, or *enacts*, textual authority in the present tense of the theater text. The significant difference between the author-function of Shakespeare and that of Marx or Hippocrates is that, by virtue of being not merely literary but theatrical, there is more than one way to engage the discourse of the author-function: critical writing *about* and theatrical performance *of*. But now, in investigative-expansive fashion, we have crossed back over the conceptual and textual divide drawn by Rouse: we have used the phrase "performance of" to describe the theater text's relationship to the written text. How then do we reconcile the authority of the written text of *Hamlet* with the notion of a performance text, one that lives its brief, but wholly self-authorized, existence only on the stage? The relationship between theater and writing must be scrutinized. Any quotation from the written text of *Hamlet* can only appear on the page. A purely literary reference, while convenient for the purposes of writing articles such as this one, produces the impression of a purely literary referent. While the status of *Hamlet* as a literary text cannot be denied, neither should its status as the object of past theatrical texts, nor its potential as the object of future theater texts. The text of *Hamlet: A Monologue* resists easy quotation. Although Wilson's adaptation has taken as its point of origin the written text of Shakespeare's *Hamlet*, the principal text of *Hamlet: A Monologue* is a theatrical one. A Wilson sketch should perhaps accompany any quote from the Wiens text. Wilson's storyboard sketches are as much a part of the "script" for *Hamlet: A Monologue* as Wiens's adaptation, and, as a result, Wilson's set of images are as much the performed object of Rouse's "of" as is Wiens's deconstructed assemblage.

Or perhaps one cannot quote from a theater text at all. Just as the ephemeral event itself must end, leaving behind hardly a Derridean "trace" of its performance (or, as Prospero would say, "Leaving not a rack behind"), so too after the theater event one has no easy point of reference for discussion. One can quote Shakespeare's play, the adaptation, some sketches, or some production photos, but these are hardly pieces of the media that is "theater." Simply quoting lyrics from a song by Lou Reed does not remotely constitute an adequate citation of a Lou Reed concert. Hopkins's excitement at seeing Reed take a seat nearby embodies the sense of authorship, and the concert citationality that Reed's presence implies: Reed is undeniably a major "author" of his own concert appearances, and his presence is a kind of quote from his performances. In the same way, scripts, sketches, and photos can only "refer" to *Hamlet: A Monologue* and the only adequate "quotation" of the theater event called *Hamlet: A Monologue* would be one that includes the performance of a portion of the event and the actual presence of Robert Wilson.

Had I but time

The productive excursions that comprise our analysis exemplify transversal theory's investigative-expansive mode of analysis, which has spurred our explorations down relevant lines of inquiry, all of which, as we will now show, intersect at the opening of *Hamlet: A Monologue*. Of his process of textual assemblage, Wiens has said, "Trying to find the first line, it was the heaviest thing to do" (*Making* 13:20). We quoted that first line earlier as "Had I but time—." However, we took that quote from the first page of the adaptation (Wiens's written text version) not from the actual performance (Wilson's theater text version). In so doing, we have misrepresented the actual opening of *Hamlet: A Monologue*, especially given that we couched that quote in the context of a documentary of our attendance at this theater event. Here is a transcription of the actual "first line" of the performance that Wilson delivered in Alice Tully Hall on July 7, 1995:

> Had I but time—
> Had I but time—
> Had I but time— (*Making* 13:30)

The "heaviness" of Wiens's effort clearly paid off: Wilson liked the first line so much that—in typical Wilsonian fashion—he felt that it should be said not once but three times. This repetition represents in microcosm the kind of performative "badness" active throughout

Hamlet: A Monologue. The strength of the version of *Hamlet* that Wilson performed on stage is demonstrated in the above quote. In the words of Gregory Boyd, Artistic Director of the Alley Theater, there is no preexisting text that Wilson is not willing to use as "raw material, as grist for his own mill" in the service of "setting up a conversation with the play, and standing in a contentious relationship to the play" (*Making* 15:33). Ultimately, each of the preexisting texts that inform *Hamlet: A Monologue* is extended and mediated in the theater.

In this way, the theater text of *Hamlet: A Monologue* dispels the notion of a derivative relationship between the theatrical performance and any written text or script. Wiens's adaptation itself is already at a significant interpretive remove from any text that might be attributed to Shakespeare. Wilson's staging of this mediated adaptation only extends the distance between Shakespeare's source text and the theater text. *Hamlet: A Monologue* has a relationship with *Hamlet* inasmuch as the former presupposes the latter. But Wilson's performance text cannot be said to have been a performance "of" any preexisting text, not Shakespeare's, not even Wiens's. Wilson bypassed the contemptuous critical underestimation of theater as merely the reiteration of dramatic literature: he took the stage himself in order to embody the primary text of *Hamlet* in a theatrical mode, as the principal site of authority, disjunct from any preexisting text.

The authority of the theatrical text is established in performance. But who is the author of a theatrical text? This may be the wrong question. The inquiry should be not for a "who" but for a "where." Just as we have suggested that it is the receiving population, or audience, who determines the author-function, so too could it be said that authority for a radical appropriation of a classic text, like *Hamlet: A Monologue*, rests with the audience. The written text of *Hamlet* was invoked continuously throughout Wilson's performance. It was the specter that haunted the theater text at a genetic level before that text established its conceptual and authorial independence in performance. Additionally, Shakespeare's *Hamlet*, a knowledge of *Hamlet*, a memory of the experience of reading or even rereading the written text of *Hamlet*, was presupposed as a condition of membership in the audience of *Hamlet: A Monologue*. Shakespeare's *Hamlet* was a discursive presence in the subjective territories of the audience of *Hamlet: A Monologue*, and it was within the space of this audience that the discursive circuit closed. Inevitably, the transversal power of the performance extended beyond the audience's effected subjective territories, beyond the Shakespeare of which the audience was a part, and into all conceptual and emotional spaces that the audience henceforth inhabits.

Notes

1. In *Liveness*, Philip Auslander argues that live performance and mass media are inherently "rivals" in our "cultural economy" (1). The "televisual" has superceded other modes of discourse to become "the determining element of our current cultural formation" (2). Within the discursive framework of the televisual, theater may in fact be unable to effectively convey the "liveness" that would seem to be its hallmark and the trait that would distinguish live performance from the reproducible mass medias. One culturally demarcated site where even Auslander suggests that live performance may retain unmediatized liveness is the rock concert. We find Reed's attendance at Wilson's theater event evidence of one possible point of intersection between the rock concert and the theater; we look to Wilson's *Hamlet* in search of the possibility of theatrical liveness despite the evidence of ubiquitous moribund televisuality.
2. For a detailed discussion of Shakespace, see Don Hedrick and Bryan Reynolds, "Shakespace and Transversal Power," in Hedrick and Reynolds ed., *Shakespeare Without Class: Misappropriations of Cultural Capital* (New York: St. Martins Press, 2000).
3. Transversal theory was introduced by Bryan Reynolds in his article, "The Devil's House, 'or worse': Transversal Power and Antitheatrical Discourse in Early Modern England" (*Theatre Journal* 49, 1997: 142–67), where Reynolds discusses the subjective territory of Elizabethan London and the ways in which plays and players (including by association Shakespeare himself) pursued strategies of resistance in response to a restrictive political and social milieu. The terms "transversal" and "transversality" are adapted from Félix Guattari, and extended to include "conceptuality and its territories" (Reynolds 148). See also, Donald Hedrick and Bryan Reynolds, "Shakespace and Transversal Power"; Bryan Reynolds and Joseph Fitzpatrick, "The Transversality of Michel de Certeau: Foucault's Panoptic Discourse and the Cartographic Impulse" (*Diacritics* 29:3, 1999); and Bryan Reynolds, *Becoming Criminal: Transversal Performance and Cultural Dissidence in Early Modern England* (Baltimore: Johns Hopkins University Press: 2002).
4. See Reynolds, "The Devil's House," 145–51.
5. See Reynolds and Fitzpatrick, "The Transversality of Michel de Certeau."
6. Bryan Reynolds and James Intriligator introduced this methodology in a paper entitled "Transversal Power," given at the Manifesto Conference at Harvard University on May 9, 1998.
7. Both the official program published by the Alley Theater and the press release distributed by Lincoln Center use the phrase "adapted by Wolfgang Wiens and Robert Wilson" to attribute authorship for *Hamlet: A Monologue*. However, the cover of the script of the adaptation provided to the authors by the archive at Wilson's fund-raising organization, the Byrd Hoffman Foundation, explicitly reads "adapted by Robert Wilson and Wolfgang Wiens." It should also be noted that two of the documentary sources produced by the Alley Theater, a program and a video, refer to the title "*Hamlet*" rather than "*Hamlet: A Monologue*" for reasons that are not clear.

8. The storyboard sketches that we refer to here were reprinted in the program for *Hamlet: A Monologue* (see note 7).
9. One might conclude that other art forms are capable of founding and expanding discursive fields of this sort: in music, e.g., contemporary composers and performers must negotiate the author-functions "Mozart," "John Cage," and "Lou Reed."

Works cited

Auslander, Philip. *Liveness: Performance in a Mediatized Culture.* New York: Routledge, 1999.
Burt, Richard. "To e- or not to e-? Disposing of Schlockspeare in the Age of Digital Media." *Shakespeare After Mass Media.* Ed. Richard Burt. New York: Palgrave, 2002: 1–32.
Eagleton, Terry. *Literary Theory.* Minneapolis: University of Minnesota Press, 1983.
Foucault, Michel. "What Is An Author?" *Textual Strategies: Perspectives in Post-structuralist Criticism.* Ed., Trans. Josué V. Harari. Ithaca, NY: Cornell University Press, 1979: 141–60.
Hamlet, Alley Theatre. Program. Eds. Gregory Boyd and Valdemar Zialcita. Alley Theater, Houston, TX. May 1995.
Hamlet: A Monologue. Dir. Robert Wilson. Lincoln Center, New York City. July 6–8, 1995.
Hamlet: A Monologue. Adapted by Robert Wilson and Wolfgang Wiens. Second version. Byrd New York City Hoffman Foundation, (no date).
Hamlet, The First Quarto. William Shakespeare. Ed., Kathleen O. Irace. New York: Cambridge University Press, 1998.
Hamlet, The Second Quarto 1604–5. William Shakespeare. Facsimile reproduction. New York: Oxford University Press, 1964.
Maguire, Laurie E. *Shakespearean Suspect Texts.* New York: Cambridge University Press, 1996.
The Making of a Monologue: Robert Wilson's Hamlet. Documentary video. Dir. Marion Kessel. Arts Alive!/Alley Theater, 1995.
Marcus, Leah. *Unediting the Renaissance.* New York: Routledge, 1996.
Worthen, W.B. "Drama, Performativity, and Performance." *PMLA* 113 (October 1998): 1093–1107.

CHAPTER 3
VENETIAN IDEOLOGY OR TRANSVERSAL POWER? IAGO'S MOTIVES AND THE MEANS BY WHICH OTHELLO FALLS

Joseph Fitzpatrick & Bryan Reynolds
(with additional dialogue by
Bryan Reynolds & Janna Segal)

This chapter presents a new perspective on the means by which Othello falls from his position of high status both within Venetian society and in the eyes of the play's audience. By examining the pertinent critical history of *Othello*, especially criticism accounting for the effects on Othello of Iago's machinations, this analysis explores the social and ideological relations between Iago and Othello through the lenses of several different critical approaches, including new historicist and cultural materialist. Our aim is to reveal the contradictions and limitations in this critical history that encourage an examination of the play according to a more inclusive approach that we call "transversal theory," which will be explained as the analysis progresses.[1] In doing this, we hope to emancipate the critical history of *Othello* from the delimiting analytical parameters that have characterized it, and to show the transversal ways in which Iago paradoxically opposes and reifies Venetian state ideology.

Critical traditions

In order to situate our transversal analysis in terms of the critical discourse on *Othello*, we will begin with G.K. Hunter's "Othello and Colour Prejudice" for two important reasons: it was groundbreaking in examining how racial issues inform Othello's fall; and it was keenly aware of its own intervention in *Othello* criticism. Hunter shows "how mindlessly and how totally accepted in this period was the image of the black man as the devil" (39).[2] Yet, according to Hunter, Shakespeare's Moor defeats this prejudice of the Elizabethan audience as soon as he appears on stage in Act One, Scene Two: "This is no 'lascivious Moor,' but a great Christian gentleman, against whom Iago's insinuations break like water against granite. Not only is Othello a Christian; moreover, he is the leader of Christendom in the last and highest sense in which Christendom existed as a viable entity, crusading against the 'black pagans'" (45).

Following this passage, we may infer that the audience's acceptance of Othello, if Hunter is correct and Othello was accepted, stemmed less from his redeeming personal traits than from his association with the two dominant sociopolitical institutions of the period: Church ("Christendom") and state (the government and military).[3] Thus, the audience would have largely dismissed the paganism and demonism associated with Othello because of his race once they learned of his conversion to Christianity.[4] And much of the antipathy that the Elizabethans might still have felt toward Othello, not specifically as a black man, but more generally as an obvious non-European, is mediated by the position of authority he has earned in the Venetian military.

It is generally accepted that as the audience's potential prejudices against the black protagonist are reversed in the play's first act, when Othello is identified as worthy by both the Venetian state and by one of Venetian society's prized possessions (Desdemona), a simultaneous reversal in the opposite direction reveals the unexpected vileness of the white, Venetian antagonist.[5] Hunter describes the situation in powerful terms: "We might say that Iago is the white man with the black soul while Othello is the black man with the white soul" (46).[6] Janet Adelman advances this reading, noting, "This trope [of Iago as "monstrous progenitor" (1.3.401–02)] makes the blackness Iago would attribute to Othello—like his monstrous generativity—something already inside Iago himself, something that he must project out into the world" (130). The "blackness" that is transferred to Iago is accompanied by the designation of "devil." Iago repeatedly refers to himself as the devil, and is joined in doing so by those who recognize his villainy in like terms. Thus a chiasmic double reversal occurs within the first act: the suspicious "outsider" Othello becomes, for the audience especially, a respected member of Venetian society, an "insider," and the outwardly "honest" Iago is unveiled as a villain. The relationship between Othello and Iago is therefore reduced by these critics, to different degrees, to symmetrical antithetism: the semblance of wholeness or totality is constructed methodologically through the binary oppositions (light/dark, white/black, insider/outsider, self/Other, good/evil) either cited from or imposed onto the text to give an explanatory coherence to the characters' relationship, and to the play overall. We call this analytical strategy, which is characteristic of most dialectical argumentation, the "dissective-cohesive mode."

As an alternative, transversal theory insists on the "investigative-expansive mode" of analysis.[7] Like the dissective-cohesive mode, the investigative-expansive mode requires that the subject matter of interest be divided up and partitioned into variables in accordance with the selected parameters of the investigation. But unlike the dissective-cohesive mode, whose goal is to reconstruct the variables into a coherent whole, the investigative-expansive mode examines both the internal connectedness of the variables (among themselves) and the external connectedness of the variables (to other forces, such as the play's critical history) in coordination with a fluid openness to reparameterizing as the analysis progresses—in response to the emergence of unanticipated problems, information, and ideas. Thus, the purpose is not to assemble a fully accounted for whole, but to better our comprehension of the subject matter's changing relationships to its own parts, as well as to the environments of which it is a part. The subject matters of the present analysis are *Othello* and its critical history; the chosen variables include the characters Iago and Othello and various readings of the play, and the greater environment of concern is the hermeneutic space in which the play and its criticism resides, and across which they both will continue to travel.

To review, we have seen that critics who focus on the play's first act reveal the chiasmic double reversal, which is accentuated, as Adelman notes, by the act's concluding soliloquy in which Iago forms his fiendish scheme: "Hell and night," Iago exclaims, "Must bring this monstrous birth to the world's light" (1.3.402). Furthermore, we have shown how critics have found ample fodder in this act for dissective-cohesive formalizing. What happens in the following four acts, however, is an issue of much debate. Hunter and Adelman both claim that "the dark reality originating in Iago's soul spreads across the play, blackening whatever it overcomes and making the deeds of Othello at last fit in with the prejudice that his face at first excited" (Hunter 55).[8] Hunter contrasts this reading with what he calls "a powerful line of criticism on *Othello*, going back at least as far as A.W. Schlegel, that paints the Moor as a savage at heart, one whose veneer of Christianity and civilization cracks as the play proceeds, to reveal and liberate his basic savagery: Othello turns out to be in fact what barbarians *have* to be" (55).[9] This line of criticism might be best represented in the twentieth century by Bloom and Jaffa in *Shakespeare's Politics*: "Shakespeare's Moor . . . returns at the end to the barbarism that the audience originally expected. The first, primitive prejudice against Othello seems to find justification in the conclusion" (Bloom and Jaffa 43).[10] In terms of the earlier transfer of the role of "devil" from Othello to Iago, the two major critical traditions might be summarized thus: the Schlegel/Bloom and Jaffa line of criticism claims that Othello and Iago were *both* devils from the beginning; just as Iago hides his identity from the other characters in the play, so does Othello hide his own identity from the other characters, the audience, and himself. Hunter's and Adelman's alternative line of criticism considers the original role reversal valid, but has the devil Iago corrupting the virtuous Othello, thereby transforming him into the fiend the Elizabethans assumed him to be.

Alan Sinfield offers what amounts to a revision of Bloom and Jaffa's reading from a critical perspective that is potentially capable of avoiding the racist implications that Bloom and Jaffa see in the play. Sinfield writes, "In the last lines of the play, when he wants to reassert himself, Othello 'recognizes' himself as what Venetian culture has really believed him to be: an ignorant, barbaric outsider—like, he says, the 'base Indian' who threw away a pearl. Virtually, this is what Althusser means by 'interpellation': Venice hails Othello as a barbarian, and he acknowledges that it is he they mean" (31). Sinfield introduces the Althusserian vocabulary of ideology that informs much new historicist and cultural materialist criticism to explain Othello's transformation. Notwithstanding his high position within what Althusser would call the "Repressive State Apparatus" of the military, Othello fails to sufficiently adopt the proper Venetian ideology, falling victim instead to a contrasting "barbarian" ideology that hauntingly informs his characterization and is actualized. This reading is in effect similar to Schlegel's, but without the suggestion that Othello is somehow inherently *defective* in his otherness.[11] Yet Sinfield's approach to this line of criticism leaves us with a number of questions, including: Where and when is Othello indoctrinated with this "barbarian" ideology? Why does he recognize himself as being like the "base Indian" only long after he apparently succumbs to the "barbarian" ideology?

Even though this type of ideological interpretation jibes well with the action of the play,[12] Sinfield does not follow through with this interpretation. Instead, he leaves it stranded just a step away from the old reversion-to-savagery interpretation, separated from Bloom and Jaffa's and Schlegel's readings, one might argue, by a veneer of Althusserian jargon. It seems, though, that Sinfield has keenly pointed to the need for an understanding of how Othello's initial status in Venice and his subsequent actions throughout the play are ideologically formed.

A transversal disjunction

At this point, we will turn from considering an ideological reading of *Othello* to discussing these critical limitations. If the negating of one's subjectivity is a path to empowerment—even greatness (as Bradley would have it)—in the world of the play, Desdemona's final self-effacing declaration, "Nobody, I myself" (5.2.125), is a fluke: she dies powerless. If Iago does use empathy as a means of accessing Othello's psychology and thereby manipulating him, as Greenblatt persuasively argues, then Iago cannot possibly be outside of ideology, but rather is moving between ideologies and across their boundaries. It appears, then, that personal empowerment is obtained not through self-negation alone, however provisionally negated, but more effectively through imagining oneself outside of one's self, or through the active acknowledgment that one is not limited to a singular, constant self. In fact, Iago thrives within a conceptual framework of identity multiplicity. He reveals this to Roderigo: "our bodies are gardens, to the which our wills are gardeners, so that if we will plant nettles, or sow lettuce, set hyssop, and weed up thyme ... why, the power, and corrigible authority of this, lies in our wills" (1.3.320–26). Iago seizes the "power" and "authority" to create and shape his own identity by placing the burden of its formation entirely upon himself, his "will." As noted previously, the language Iago uses in an earlier conversation with Roderigo anticipates this revelation: "It is as sure as you are Roderigo, / Were I the Moor, I would not be Iago: / In following him, I follow but myself" (1.1.56–58). "Being Iago" denotes an ability to play many different roles, rather than signifying a static identity. His statement that following Othello is equivalent to following his own desires suggests a connection and identification with Othello that would make empathy, and therefore manipulation, possible.

Put in transversal terms, empathy enables people to venture beyond their own conceptual and emotional boundaries, to think and feel as others do or might, and thus expand or transcend their own "subjective territories." We call this process "transversal movement." "Subjective territory," as Bryan Reynolds explains, "refers to the scope of the conceptual and emotional experience of those *subjectified* by the state machinery or any hegemonic society" (146). Specifically, "subjective territory is delineated by conceptual and emotional boundaries that are normally defined by the prevailing science, morality, and ideology.... It is the existential and experiential realm in and from which a given subject of a given hierarchical society perceives and relates to the universe" (146–47). Transversal movements occur when one entertains alternative perspectives and breaches the parameters of their subjectification.

Such movements require engagement with, and, at the very least, a passing through "transversal territory," which is "the non-subjectified region of one's conceptual territory [of one's imaginative range]" (149). Thus, transversal territory is entered through the transgression of the conceptual and emotional boundaries of subjective territory.

The ease with which Iago accesses transversal territory enables him to enter freely into the subjective territories of others where he can readily comprehend their sensibilities and idiosyncrasies. He then uses this information to force them into transversal movements across the emotional and conceptual spaces of alternative ideologies. Transversal theory, however, accepts the premise that humans exist in, and are influenced by, ideology inasmuch as they are social beings. Hence, Iago can never be totally outside of ideology unless he becomes psychotic, or his consciousness is somehow shattered. Moreover, because humans are social, the "scientific objectivity" suggested by Althusser is only possible insofar as humans can experience unmediated access to social events, which is impossible.

Moving beyond the dissective-cohesive mode

We have demonstrated that Othello's acceptance into Venetian society depends on his Christianity.[13] Yet, as the Schlegel line of criticism shows, Othello's jealous actions throughout the later portions of the play are antithetical to both the Christian ideology and the broader social norms enforced by the overarching Venetian state ideology. They reflect his movement, propelled by Iago, beyond the officially sanctioned parameters for living and operating in the world. Iago's subversion of Othello's Christianity is powerfully manifested in Othello's final acts. Indeed, murdering Desdemona, revenging himself upon Iago, and then taking his own life flout the most fundamental precepts of Christian ideology including "thou shalt not kill." More subtle but no less telling of his departure from Venetian state ideology is his transgression of the essential Christian imperative to "temper justice with mercy." Othello strangles Desdemona in the marriage bed, which she allegedly defiled, because, for him, "the justice of it pleases" (4.1.205). At the moment when Othello is prepared to mete out that justice, Desdemona appeals to the Christian mercy she believes is still in him:

> Oth. Thou art to die.
> Des. Then Lord have mercy on me!
> Oth. I say, amen.
> Des. And have you mercy too! (5.2.57–60)

Operating now outside of the Christian ideology, Othello can acknowledge Christian values without believing in them. He can wish the Lord to have mercy on Desdemona without feeling any imperative to grant such mercy himself. Indeed, he responds to Desdemona's plea by renewing his accusations, and finally by carrying out his threat.

Iago's method in toying with and eventually reversing Othello's Christian beliefs can best be seen by looking closely at Act Three, Scene Three, where Iago sets to work on Othello. Iago begins by playing dumb to Othello's question as to whether Cassio is "honest," causing Othello to remark, "By heaven, he echoes me, / As if there were some monster in his thought, / Too hideous to be shown" (3.3.110–12). Iago continues to stall, bluff, and insinuate, until finally Othello bursts out, "By heaven I'll know thy thought" (3.3.166). The phrase "by heaven" is repeated, a simple, no doubt unconscious exclamation that comes naturally to the Christian Othello. In both cases, though, it suggests that heaven alone knows what Iago is thinking. Iago picks up on this usage, and quickly twists it to his own purpose, warning Othello: "I know our country disposition well; / In Venice they do let God see the pranks / They dare not show their husbands" (3.3.205–07).[14] Here Iago refers explicitly to his status as an insider, as someone who understands Venice in a way that Othello never will. His claim regarding Venetian practices rouses Othello's jealousy, but also plants certain associations in his mind. It not only suggests that the man who catches his unfaithful wife has seen what was meant for God's eyes only, but it also implicates God as a silent witness to cuckoldry, telling Othello that the God in Venice is content with letting adulterers go unpunished, cuckolds unrevenged. That Othello has picked up on these meanings is clear when he says, "If she be false, O, then heaven mocks itself" (3.3.282). Othello has already made the link between Desdemona's alleged falsehood and the injustice, or even perversity, of a God that would allow this to happen. By the end of the scene, Othello has knelt down with Iago and vowed his revenge, "Now by yond marble heaven" (3.3.467).[15] Later in the play we do not see "heaven mocking itself," but rather Othello mocking heaven: he perverts the sacrament of confession first in a dialogue with Iago, in which Cassio's confession leads not to forgiveness, but to damnation— "To confess, and be hanged for his labour" (4.1.37)—and then later with Desdemona, when he tries to make her confess so that he will feel more justified in killing her.

There is evidence, then, that Othello's relationship to Christian ideology has been subverted and altered by Iago. But what about his relationship to Venetian ideology? In Act One, Scene Three, we see

Venetian ideology at work in the pseudo-legal proceedings initiated by Brabantio against Othello. Implicit in this scene is the assumption that those involved are living within the same Venetian state ideology; there does not appear to be any dissent beyond Brabantio's initial vehemence and lingering bitterness. Hence, this scene revealingly contrasts Othello's shifting system of values in later acts. In the "temptation scene," for instance, as Iago is just beginning the processes of indoctrination, he offers Othello "imputation and strong circumstances, / Which lead directly to the door of truth" (3.3.412–13). Of course, it is "imputation" and "circumstantial" evidence that Brabantio brings before the Duke to support his case against Othello. Yet, while the Duke finds this kind of evidence against Othello insufficient in court, it is here being offered by Iago as the best proof available against Desdemona, and Othello appears willing to accept it. In the next scene, Othello questions Desdemona about the handkerchief, and we find yet another inverted parallel with the earlier court scene. Othello describes the allegedly "charmed" handkerchief and its mystical history. He claims that the handkerchief was to "subdue" his father "entirely to [his mother's] love," but would have the opposite effect if lost or given away, so much so that, "To lose, or give't away, were such perdition / As nothing else could match" (3.4.65–66). This strange fetishism recalls Brabantio's earlier argument in which he accuses Othello of witchcraft:

> I therefore vouch again,
> That with some mixtures powerful o'er the blood,
> Or with some dram conjur'd to this effect,
> He wrought upon her. (1.3.103–06)

Once again the speaker, this time Othello, masks his unreasonable expectations in a supernatural discourse, accepting this mysticism as a part of his own righteous justice. Even more blatant, however, is Othello's outright refusal to grant Desdemona the same due process that saved *him* from Brabantio's accusations. Othello was allowed to speak his case, and to send for the one witness who could resolve the problem peacefully and justly: "I do beseech you, / Send for the lady to the Sagittar, / And let her speak of me before her father"(1.3.114–16).

Under Iago's influence, Othello never confronts Desdemona with the actual accusation of adultery until he has already passed sentence on her. The exact point in the play when we can see Iago retraining Othello comes in Act Two, Scene Three. Pouring out liberal helpings of alcohol, Iago manages to orchestrate a brawl between Montano and Cassio.

But it is only after Othello has stopped the fight and demanded an explanation that we see the wicked brilliance of Iago's plan: the two combatants, drunk and wounded, are unable to present Othello with a clear narrative of what has happened, and must defer to Iago to explain everything. Whatever his affiliation with the Venetian state, whatever his belief in civil law, Othello is operating at this point in a military capacity that allows him to make unilateral decisions based on the best information he can gather. Iago uses this situation to insinuate himself further into Othello's trust, planting in Othello's mind the idea that Iago's word should count as legal proof. This idea takes hold more and more as Iago weans Othello from Christian and Venetian state ideologies. At the end of the play, as Othello prepares to kill her, Desdemona begs him to let Cassio explain the situation: "Send for the man and ask him" (5.2.50). Othello refuses her; he has already ordered Cassio killed. Proof therefore is pursued and authorized not through justice, but through rash retribution.

Iago's motives?

> I hate the Moor,
> And it is thought abroad, that 'twixt my sheets
> He's done my office; I know not if 't be true . . .
> Yet I, for mere suspicion in that kind,
> Will do, as if for surety. (1.3.384–88)

If we choose to take at face value Iago's own professed motives, he himself operates according to what one might describe as a subset of Venetian ideology, that of the cuckold, a perspective that opposes cuckoldry's threat to patriarchal hierarchy. Because Iago is an organ of Venetian state power, he condemns cuckoldry; however, Iago exceeds state-prescribed subjective territory and enters transversal territory when his condemnation becomes pathological in its violation of the Venetian state's judicial process. Iago circumvents the Venetian state's legal system when he acts from a sensitivity to what is "thought abroad" about him, and dismisses the procedure, requiring burden of proof, demonstrated in the trial scene, acting "as if with surity" without searching for proof beyond "mere suspicion." As we have seen, Iago is not *trapped* within either this cuckold ideology or a particular subjective territory. The nonsubjectivity of "I am not what I am" permits him to move transversally beyond the ideology and view it *as* an ideology. Nevertheless, this self-negating phrase, "I am not what I am," is Iago's *modus operandi* only in the actual processes of empathizing and disseminating his ideology.

But Iago's function as a representative of Venetian state power, manifest in spreading his repudiation of cuckoldry, transgresses subjective territory when he circumvents that same state ideology's legal standards. Iago's movement into transversal territory, then, challenges the Venetian state because it flouts its judicial codes. However, Iago's transversally inspired actions can be interpreted as paradoxically reinforcing Venetian state power, even while it undermines that power through the destruction of Othello.

This brings us back to Sinfield, and what he calls "the 'entrapment model' of ideology and power, whereby even, or especially, maneuvers that seem designed to challenge the system help to maintain it" (39). This model, also known as the new-historicist "subversion/containment paradigm," uses a dissective-cohesive mode of analysis in that it insists that all social phenomena work to produce a singular sociopolitical system or dominant power structure.[16] According to this model, Iago's subversion of the ideologies that have created Othello as a leader of Venice's military appears to "challenge the system" but, in actuality, would "help to maintain it." The transversal movement that Iago demonstrates by taking the Venetian state's repudiation of cuckoldry too far makes suspicion count as proof and posits the cuckold himself (and not the state) as the executioner who destroys both Iago and Othello. While the state is clearly better off without the mercilessly self-serving Iago, the benefit of destroying Othello may not be obvious until we turn to F.R. Leavis's interpretation of Othello's character.

Leavis critiques what he considers to be the sentimental idealizing of Othello's character by critics such as A.C. Bradley, arguing instead that Othello only succumbs to Iago's insinuations because his character—self-important, self-deceiving, and sentimental—is inherently prone to the jealousy that Iago suggests. For Leavis, it is "Othello in whose essential make-up the tragedy lay" (151). This reading makes Othello's downfall truly *tragic*, and not merely the lamentable result of Iago's demonic machinations. In this case, rather than undermining the order of the state's military apparatus, Iago has done the state a service, albeit inadvertently: he exposes Othello's tragic flaw in a time of peace. This is especially fortunate for the state given that the consequences of this flaw surfacing while at war would be unpredictable and potentially disastrous. Hence, Iago's vindictive measures play directly into Sinfield's entrapment model.

Yet this reading supposes that Iago really acts from the motives he cites in his concluding monologue of Act One. What are we to make, then, of the fact that Iago himself lays out a completely different set of motivations in his Act One, Scene One, dialogue with Roderigo? In lines 8–33, Iago explains that he was passed over for promotion by Othello, and that his job was instead given to Cassio, who, unlike Iago, "never set a squadron in the field, / Nor the devision of a battle knows, / More than a spinster" (1.1.22). Iago's motive is therefore revenge: revenge against the military general who did not promote him, against the captain who, lacking his qualifications, got his place, and against the system that permitted it. Put differently, Iago seeks revenge against an unjust and capricious hierarchical institution of the state by causing the destruction of the institutor of hierarchical order (symbolized by Othello) and its nepotistic beneficiary (Cassio). More than in the jealousy scenario in which Iago indirectly challenges the Venetian state, Iago is here seen as directly attacking it. But it is still easy to see the entrapment model at work. By directing his revenge primarily against the representative of state hierarchy (Othello), Iago not only causes the elimination of himself and Othello, but he also carelessly allows Cassio to live. Since Iago's machinations necessarily brought out the flaws in Othello's character, it is reasonable to assume that at the end of the play Cassio stands an excellent chance of being restored to the official position he held before he was cashiered by the Othello that Venice now sees was deluded by jealousy. Iago's attempts at revenge, it seems, have in the end done no substantial harm to his hated rival Cassio, and have even potentially *helped* him by creating an open position further up in the military hierarchy.

The problem with the subversion/containment paradigm in both cases, however, is that it does not account for the complexity of the situation. As Reynolds argues, "It is especially significant that this paradigm, as it is often applied, precludes or ignores micro-subversions, or small revolutionary changes" (148). In the two scenarios discussed here, the immediate impact on the morale and organization of the state is not considered. For instance, Iago can be seen as scoring a small victory against state order in that his vigilantism challenges the ability of the military leader to make unilateral decisions regarding his inferiors without fear of violent retribution. It is also entirely likely that Iago's subversive activity will affect how the state is likely to treat converted Moors in the future. Moreover, the parameterization behind these entrapment scenarios is dissective-cohesive and state-centered; as a result, the analysis does not consider questions such as: What happens to Iago, Cassio, and the state at the play's conclusion? Where did Othello and Iago come from in the first place? What impact will Desdemona's murder have on her family and friends, and therefore the state? And how might the audience respond to the play's action?

In order to foreground this transversal analysis it is helpful to go beyond Althusser, and look instead to Gilles Deleuze and Félix Guattari.[17] They accept an essentially Althusserian model of the state, dividing the state apparatus into "two heads: the magician-king and the jurist-priest" (Deleuze and Guattari 351). These "heads" of the state "are the principal elements of a State apparatus that proceeds by a One-Two, distributes binary distinctions, and forms a milieu of interiority" (352). In this "milieu of interiority," the state apparatus creates an ordered, "striated," hierarchical society, organized by the state-imposed system based on supposedly absolute categories and binary oppositions. Opposed to this state apparatus is the "war machine," a "pure form of exteriority" (354) that operates outside of the state apparatus. The war machine opposes the state's rigid hierarchy, its order, its "striation," its binary oppositions, its stasis: "the war machine's form of exteriority is such that it exists only in its own metamorphoses" (360). Yet, while the war machine is defined in opposition to the state, as "a form of pure exteriority," it can become involved in the state system when it is "appropriate[d] . . . in the form of a military institution" (355), and in the relation between the two "heads" of the state apparatus: "One would have to say that it is located between the two heads of the State, between the two articulations, and that it is necessary in order to pass from one to the other. But 'between' the two, in that instant, even ephemeral, if only a flash, it proclaims its own irreducibility" (355). While the war machine can surface within the state apparatus, its presence is problematic because by its nature it lies outside of the state's control.

If Iago is attacking the military order that passed him over for promotion, then he can be seen as a manifestation of Deleuze and Guattari's war machine. Iago originates in the state's military institution, appropriates the war machine, and moves insidiously into the state apparatus. He subverts in Othello both Christian ideology and Venetian state ideology as they are manifested in judicial codes; these two ideologies are aligned with the "jurist-priest" head of the state, characterized as "the clear," "the calm," "the weighty," and "the regulated" (Deleuze and Guattari 351). What Iago turns Othello into resembles more closely the "magician-king" head of the state as described by Deleuze and Guattari: "the obscure," "the violent," "the quick," and "the fearsome" (351). Thus, " 'Between' the magical-despotic State and the juridical State containing a military institution, we see the flash of the war machine, arriving from without" (353). From here we can understand Iago's relationship to the Venetian state's ideological machinery as an almost ironic reversal of Deleuze and Guattari's statement that the state supporter "can only appropriate [a war machine] in the form of a military institution, one that will continually cause it problems." Iago, the "problematic" part of the military war machine that the Venetian state has appropriated, attacks the state hierarchy by appropriating and transversalizing the methods of Venetian state ideology.

In order to further reveal Iago's transversal relationship to Venetian state ideology—that is, his movements in and out of subjective and transversal territories—it is necessary to review the two motivations that Iago gives for his actions, either as a cuckold or as an undervalued military official. Turning to the play, we see that the latter explanation is highly dubious. Bradley cautions us against believing Iago too readily: "One must constantly remember not to believe a syllable that Iago utters on any subject, including himself, until one has tested his statement by comparing it with known facts and with other statements of his own or of other people, and by considering whether he had in the particular circumstances any reason for telling a lie or for telling the truth" (198). By these well-justified standards, Iago's conversation with Roderigo must be considered suspect. The only statement he makes in that dialogue, which can be verified by recourse to facts within the play, is that Cassio is Othello's lieutenant. There is no evidence that Othello passed Iago over for the promotion, or that Iago was ever even considered for it. Iago's own statements do not support this motivation anywhere else in the play, and his subsequent monologue at the end of Act One, Scene Three, offers a completely different motivation without mention of Cassio's promotion. Iago would certainly benefit from lying to Roderigo: he has to offer some sort of motivation, and this is one that Roderigo would easily understand and believe, without the condescension he might feel toward a cuckold.

Perhaps, then, Iago really does believe he has been cuckolded. There does not seem to be any reason for him to tell lies in a monologue directed to the audience; moreover, there appears to be support for this motive in Emilia's lines, "Some such squire he was, / That turn'd your wit, the seamy side without, / And made you to suspect me with the Moor" (4.2.147–49). This support, Ridley says, "disposes of any idea that Iago's suspicions of Othello (1.3.385; 2.1.290) were figments, invented during momentary 'motive-hunting' and dismissed. They have been real, and lasting, enough for him to challenge Emilia" (159). Even if Iago's suspicions were real and lasting, however, this does not prove that they are in fact the motivating force behind his actions, a fact that becomes clear upon closer examination of Iago's language in the Act One, Scene Three, monologue. In this speech, Iago first states his motivation of cuckoldry in the following words: "I hate the Moor, / And it is thought abroad, that 'twixt my sheets / He's done my office" (1.3.384–86). This is supposedly a statement of motivation, of cause and effect. Yet the sentence is not constructed on the logical model of "A then B," but rather on the model of a different Boolean operator, on "A *and* B": "I hate the Moor, / *And* it is thought abroad..." (our emphasis). On the most basic level of language and of symbolic logic, Iago is not stating a motive; he is merely listing two separate facts that are simultaneously true. Here we can see why Coleridge labeled this monologue "the motive-hunting of motiveless malignity" (49). Iago knows only that he "hates the Moor"; he puts this together with his belief that "it is thought abroad...," and realizes that this is sufficient motive to act on his preexisting, "motiveless" hatred of Othello.

If Iago does represent this "motiveless malignity," where does that leave him in relation to Venetian state ideology? If he does not work as a normal agent of the state, fully inculcated in one of its minor ideologies (that of the cuckold), and does not work against the state, as a war machine that appropriates the methods of state ideology, then how *does* he work? In Act Three, Scene Three (right before the temptation scene in which Iago interpellates Othello as cuckold), Othello makes the prophetic statement in his apostrophe to Desdemona: "Excellent wretch, perdition catch my soul, / But I do love thee, and when I love thee not, / Chaos is come again" (3.3.91–93). Even proponents of the entrapment model would have to admit that Iago brings some degree of "chaos" into the functioning of the state; and those, like us, who dismiss the idea that he is a true mechanism of the state may see this as the primary result of his actions. Without ascribing to the Iago-as-war-machine model, then, we may still believe that his effects are very much like those of a war machine, breaking down the hierarchical order of the state apparatus, turning the violence inherent in a military man such as Othello against the state's ordered society. What is needed, however, once we have dismissed his Act One, Scene One, motive of revenge against the military system, is a model that does not rely on the aggression toward the state that Iago does not seem to possess.

Such a model can be found in our concept of "transversal power." Transversal power is any force (physical or metaphysical) that causes movements in or through transversal territory. As Reynolds puts it, "transversal power induces people to transversally cut across the striated, organized space of subjectivity—of all subjective territory—and enter the disorganized yet smoothly infinite space of transversal territory" (150). Like the war machine of Deleuze and Guattari, "Transversal territory invites people to escape or deviate from the vertical, hierarchicalizing and horizontal, homogenizing assemblages of any organizational social structure" (149). Iago enters the "nonsubjectified" (once again, "I am not what I am") transversal territory outside of the conceptual boundaries set on him by the state. Without just cause, and separate from any existing suspicion of cuckoldry, Iago hates his commanding officer and desires the "chaos" of Othello's downfall. As would the war machine, Iago appropriates the form of the Venetian state official, and recreates Othello's subjective territory in a manner that reveals—relative to the subjective territories created by the state—transversal movement; Othello's conceptual and emotional boundaries are reconfigured under the influence of Iago's transversal power. Thus, from the "motiveless" irruption of transversal power into Iago's subjective territory we see a chain of events that lead inexorably to "chaos."

There may be some danger of implying with this interpretation that *Othello* should be read as a reactionary text, condemning all subversive action because of the results of Iago's deeds. Of course, this reactionary conclusion could never actually be reached from the evidence given in the analysis: the power that Iago wields over Othello demonstrates the frightening power that the state itself can wield if its ideological mechanisms and institutions are not constantly questioned and challenged. Yet this raises an important point, which is the inadequacy of all interpretations that attribute a didactic moral to Shakespeare's works, whether reactionary or "progressive." Bloom and Jaffa summarize the moral of *Othello* with the following formula: "*Othello* appears . . . to leave us with this choice—a mean life based on a clear perception of reality or a noble life based on falsehood and ending in tragedy" (66). An Althusserian reading of the play, if it accomplishes nothing else, at least shows the error of Bloom and Jaffa's moralizing oversimplification. Othello's downfall is not simply a matter of being mistaken about the facts; it is rather a result of an entirely different way of relating to the world around him that is developed in him by Iago. Othello is indoctrinated with an ideology that does not require the same degree of proof of wrongdoing as does Venetian state ideology, and which is harsher in its punishment than the Christian faction of that ideology. Both the old ideology of the Venetian state and the new, minor ideology of the cuckold are "based on falsehood" in the sense that they are "imaginary" (in Althusser's definition), and mutually contradictory. Othello's tragedy comes from his indoctrination into an ideology that encourages him to ignore (or at least distort) the facts and act swiftly, "justly," and without mercy.

And now for something completely different[18]

In keeping with the analytical strategy of transversal theory, Bryan Reynolds and Janna Segal here expand upon Fitzpatrick and Reynolds' investigation by introducing a new variable into the discussion: Deleuze's deconstruction in "Coldness and Cruelty" of "the spurious sadomasochistic unity" (40). Looking at the relationships among Iago, Othello, and Desdemona in light of Deleuze's pedagogical distinctions between "the sadistic 'instructor'" and "the masochistic 'educator'" (19), Reynolds and Segal will offer another parameter of exploration that will further elaborate upon Iago's transversal power, and reassess the submissive status of the martyr Desdemona.

Criticism analyzing sexual and marital relations, and, in particular, male desire in *Othello* has generally ignored sadism (named after the Marquis de Sade), masochism (named after Leopold von Sacher-Masoch), and the sadomasochistic model, preferring instead to see the relationships among the characters through the lens of normalized notions of what constitutes a healthy relationship between the sexes, and then to criticize Shakespeare's characters for breaching those constructs in their behavior.[19] For example, owing to the "abnormal" violent relationship between Iago and Emilia, their marriage has been negatively described as a "part-time job" (Kállay 42), and "a weary stale alliance, entrenched in the reactionary rut of dominance/subordination and routine abuse" (Deats 248). Iago himself has been chastised for his supposed inability to love (Gronbeck-Tedesco 261–62; Kállay 42), and has been cast as "the most abysmal sexual villain in all literature" (Cummings 80). Desdemona has also suffered at the hands of critics who have dismissed her as a passive female character, frequently pointing to her final words ("Nobody. I myself. / Farewell. Commend me to my kind lord—O, farewell" [5.2.122–23])[20]—which are delivered in response to Emilia's request for the name of her murderer—as proof of her unflinching obedience to Othello and the patriarchal order.[21] While critics have shown how Iago, Desdemona, and Othello do not conform to dominant Renaissance and modern codes of "proper" behavior, we want to explore how the characters function in their "perverse" relations with one another.

In "Coldness and Cruelty," Deleuze defines sadism and masochism as two separate "perversions"[22] that operate through different means to achieve a shared, subversive end. According to Deleuze, "nothing is in fact more alien to the sadist than the wish to convince, to persuade, in short to educate" (18), while the masochist "needs to educate, persuade and conclude an alliance with the torturer in order to realize the strangest of schemes" (20). Although each uses a different pedagogical approach, both the sadist and masochist function to subvert the law: the sadist transcends the law by creating "anarchic institutions of perpetual motion and permanent revolution" (87), and the masochist demonstrates its absurdity by converting "the very law which forbids the satisfaction of a desire under threat of subsequent punishment [. . .] into one which demands the punishment first and then orders that the satisfaction of the desire should necessarily follow upon the punishment" (88–89). Deleuze further distinguishes these two forms of perversion/subversion by their relationship to the law: the sadist creates an institution of anarchy that seeks "the degradation of all laws" and establishes itself as "a superior power" to accepted authority (77), and the masochist draws a contract that "generates a law," rather than negating all laws (77).

In transversal terms, by degrading the law and thereby the official culture from which it was drafted and that reinforces it, the sadist can be said to be creating an institution that moves beyond subjective territory into a transversal terrain that threatens the foundations upon which subjective territory is constructed and maintained. The masochist's contract also similarly challenges official territory; however, the masochist does not seek to overturn the law and official culture, but to perpetuate their capacity as punishment in a different pursuit: pleasure. With the masochist, the function of the law is subverted into a means of obtaining pleasure while at the same time working to maintain, at least ostensibly, the law's structure and power of authority over individuals. With the sadist, the law is destroyed and replaced with an institution that resists maintenance and, because it remains in "perpetual motion and permanent revolution" (87), denies itself the ability to become a new form of official territory.

Based on Deleuze's division, it could be argued that Iago is a masochist: he succeeds in manipulating others into participating in his plots, and forms alliances with his enemies in order to realize his "schemes." Yet, in his attempts to orchestrate the events that lead to Othello's destruction, Iago does not so much persuade as "instruct" his victims as to how to carry out his revenge.[23] For instance, when Iago suggests to Roderigo that he kill Cassio, he does not persuade Roderigo to do it, but rather outlines how it will be done: "He sups tonight with a harlotry, and thither will I go to him. [. . .] if you will watch his going thence—which I will fashion to fall out between twelve and one—you may take him at your pleasure. I will be near to second your attempt, and he shall fall between us" (4.2.235–40). Following his lecture on the details of the murder, Iago offers to "show" Roderigo "a necessity in his death" (4.2.242). While "show" suggests that Roderigo needs to be persuaded, the scene ends before Iago "educates" Roderigo as to why Cassio must die. In Act Five, Scene One, when the deadly duo next appears, Iago is again issuing instructions to Roderigo: "Here, stand behind this bulk, straight will he come. / Wear thy good rapier bare, and put it home" (5.1.1–2). Iago's commands aside, Roderigo suggests that in the interim of the planning and the execution of the scheme he has been persuaded by Iago to kill Cassio: "I have no devotion to the deed / And yet he hath given me satisfying reasons: / 'Tis but a man gone. Forth, my sword: he dies" (5.1.8–10). The "satisfying reasons" given, however, are less than compelling, which suggests that little convincing was needed to spur Roderigo on. After repeating the reason Iago gave, Roderigo simply follows Iago's orders, which further establishes Iago as more the "instructor" than "educator" of his revenge schemes.

Iago can be described as persuading Othello to believe that Desdemona has been false by spurring suspicious thoughts and images of her sexual impropriety, indirectly leading his superior officer on a self-destructive path. But the seeds of suspicion planted by Iago are generally coupled with instructions for how Othello should proceed in light of Iago's insinuations. For example, in the first temptation scene, after acknowledging his lack of "proof," Iago tells Othello, "Look to your wife, observe her well with Cassio. / Wear your eyes thus, not jealous nor secure; / I would not have your free and noble nature / Out of self-bounty be abused: look to't" (3.3.199–204). Specifically, Iago instructs Othello not only to "look to't," but also where to cast his gaze (on Desdemona "*with* Cassio," which implies that she has been "with" him before) and how to perceive what he then sees (to be "not jealous nor secure"). In effect, Iago is able to insure that Othello sees not what is before him, but what Iago desires him to see. Iago uses a similar tactic when he instructs Othello as to how to punish his wife: "Do it not with poison, strangle her in her bed—even the bed she hath contaminated" (4.1.204–05). Again, Iago blends images of sexual misconduct—here asserting that Desdemona has been unfaithful in the bed she shares with Othello—with commands for precisely how Othello should respond to her adultery. Of course, Iago's influence is confirmed by the fact that Othello follows Iago's orders, suffocating Desdemona in "her bed."

In the final scene of the play, when the roles are reversed and Iago becomes the captive facing punishment, Iago is still able to maintain his position as the "instructor." Asked by Othello to explain, "Why he hath thus ensnared my soul and body?" (5.2.299), the imprisoned Iago flatly refuses to cooperate: "Demand me nothing. What you know, you know. / From this time forth I never will speak a word" (5.2.300–01). By taking a vow of silence, Iago retains his dominance over Othello, further torturing him by withholding the knowledge Othello so desperately seeks. Iago's refusal to "educate" his captor is also issued with an order, "Demand me nothing," which Othello obeys. With their positions switched, Iago the "instructor" preserves his authority.

While Iago is, according to Deleuze's distinctions, the sadist in Shakespeare's story, Desdemona is the masochist whose ability to derive pleasure from pain is presented as the facilitator of her love for Othello. Prior to her arrival in the story, Desdemona's fascination with cruelty is introduced by her husband, who describes how he was able to woo his wife by telling her violent stories, which she would "with a greedy ear / Devour" (1.3.150–51). According to Othello, Desdemona, upon hearing of his painful youth, often cried and "wished she had not heard it, yet she wished / That heaven had made her such a man" (1.3.163–64).[24] Due to her capacity to simultaneously feel what are typically seen as opposing emotions—pain and pleasure—Desdemona was able to relish the cruel content of Othello's stories and achieve love, being the climax of the sorrow she experienced upon hearing Othello's autobiographical narratives.

Equally engaged in his role in Desdemona's masochistic scenario, Othello admits that he was able to enjoy the suffering his stories caused her. According to Othello, he fell in love with her precisely because "she did pity" him upon hearing his tales of woe (1.3.169). The "pity" that he found so attractive in Desdemona not only signals the pleasure he derives from her pain, but the dialectical nature of their relationship. Since her pity was issued from a position of privilege—she pitied that which she did not experience as a white woman—her sympathy served to validate the pain of his enslaved youth.[25] Having decided that Othello's pain was worthy of her tears, Othello deemed her worthy of his love. Likewise, although Othello is Desdemona's torturer and she his victim, he is controlled by her, which he is keenly aware of. So afraid of her ability to sway him, Othello vows to forbid her to speak at the time of her murder: "I'll not expostulate with her, lest her body and beauty / unprovide my mind again" (4.1.201–03). Desdemona also identifies her husband as both her servant and her master. Confident in her influence over him, she promises Cassio that she will "tame" her husband and restore their friendship (3.3.23). However, wary of angering her husband, Desdemona dismisses Emilia at Othello's "bidding," answering Emilia's protests with, "We must not now displease him" (4.3.13–15). Both husband and wife share a revolving status in their marriage as the dominator and the dominated, and both are equally capable of inflicting pain on, and receiving it from, each other. Although Desdemona and Othello both love and fear each other, it is Desdemona who can be more appropriately described as the "masochistic 'educator.'" Unlike Iago, who issues orders to enact his schemes, or Othello, who so willingly follows Iago's commands, Desdemona tries to persuade her husband to do as she pleases. In describing to Cassio how she will convince Othello to renew their friendship, Desdemona reveals her role as the persistent teacher in their marriage: "My lord shall never rest, / I'll watch him tame and talk him out of patience, / His bed shall seem a school, his board a shrift, / I'll intermingle everything he does / With Cassio's suit" (3.3.22–26). Deats's argument that in this passage Desdemona "promises to play the shrew if her husband does not heed her suit" (242) implies that Desdemona intends to conform to a constructed notion of "negative" feminine behavior if her husband does not agree to her demands, and dismisses the possibility that Desdemona plans to use language to her advantage, empowering herself with her persuasive "talk." Desdemona's verbalized intentions reveal that she is not merely a passive voice in the play, but rather that she invokes her power through discourse, not commands, and believes that through discussion she can convince, not order, her husband to reconcile with Cassio, who, recognizing Desdemona's influence over Othello, vows to be her "true servant" (3.3.9).

As noted, critics have chastised Desdemona for embracing victimization at the end of the play as she submissively waits for her husband to kill her in her bedroom, and, with her last breath, insists that Othello is blameless for her demise. Deats (241; 243), Orlin (181–83), and Rose (221) all comment on the "contradictions" of Desdemona's character, whose final role as the martyr for her husband's jealousy and honor is incongruous to the young wife defying what we would identify as her subjective territory at the beginning of the play. Deats accounts for the disjuncture in her character by arguing that "Desdemona is never as unconventional as she first appears" (243), and that from her first appearance, she "accepts her subordinate role in society and defines herself in relation to men, as either a wife or daughter, but not as an independent individual" (244). Bevington's assessment of Desdemona is in accordance with Deats's: "The play as a whole confirms this wish in her to obey" (227), and, like Deats, Bevington refers to Desdemona's willingness to lie in wait for her wrathful husband as evidence for his argument (227).

Unlike Deats, Orlin, and Rose, however, Bevington does not refer to Desdemona's final line as a "lie" reflective of her willingness to accept a subordinate position as a woman in a patriarchal society,[26] but rather as one of the two instances of "disobedience, prevarication, or managerial assertion of control of which she is 'guilty' in the play" (227). Bevington's reading acknowledges Desdemona's agency, but he still identifies her last words as an act of self-betrayal and self-sacrifice, as her "attempt to take the blame for her own death" (227). We suggest instead that Desdemona's final words can be interpreted as more than a false statement issued from the mouth of a martyr in order to exonerate a guilty husband. As is evidenced by her behavior prior to her murder, Desdemona's last line implies that she recognizes herself as an active agent, not a passive victim, in what could be described as a masochistic scenario in which she has agreed to participate.

Prior to her death, Desdemona's behavior implies that she is keenly aware of the danger awaiting her, and her various attempts to orchestrate the scene of her own demise are suggestive of her masochistic role in her marriage. Although aware of her husband's murderous intentions, she obeys his order to dismiss Emilia, thereby allowing herself to be vulnerable to his attack. Having previously requested that her bed be adorned with her wedding sheets (4.2.107), she now asks Emilia to, following her death, "shroud me / In one of these same sheets" (4.3.22–23). Since she has created the opportunity for her husband to assault her unguarded, and purposefully decorated her deathbed in her nuptial garments, Desdemona's destruction at her husband's hands can be described as a suicidal scenario of which she is a willing participant. However, her actions during the attack suggest that Desdemona found the threat of death, or rather, the fantasy of punishment, more enticing than its actuality.

As noted, Deleuze identifies the contract as a specific feature of the masochist fantasy, as opposed to the anarchic institutional tendency of the sadistic scenario. According to Deleuze, "The masochistic contract implies not only the necessity of the victim's consent, but his ability to persuade, and his pedagogical and judicial efforts to train his torturer" (75). In *Othello*, the victim's failure to draw a contract with her torturer signals her demise, as well as the deconstruction of the notion of "divine justice." During her death scene, after she unsuccessfully tries to convince Othello of her innocence, she pleads for her life by offering him four alternatives: "O, banish me, my lord, but kill me not!" (5.2.77); "Kill me tomorrow, let me live tonight!" (5.2.79); "But half and hour!" (5.2.81); "But while I say one prayer!" (5.2.82).

Othello rejects her attempts at negotiation, and her last appeal, "O Lord! Lord! Lord!" (5.2.83), is not answered by God, but by the pillow that her husband uses to smother her. Desdemona's failure to adequately "train" her torturer prior to the moment of her death leads ultimately to her destruction, and demonstrates, within the context of the play, the absurdity of the concept of a "higher" court of law that protects the innocent.

Desdemona's actions prior to Othello's final entrance—her adornment of her bedchamber, her specificity regarding her death shroud, her dismissal of her waiting-maid—are evidence that Desdemona was aware of the possibility that Othello intended to harm her, and that she took active and purposeful steps to facilitate his desire. While this can be read as an affirmation of Bevington's reading of the "wish in [Desdemona] to obey" (227), it also suggests that she had more than a "wish," but rather a desire "to obey." If we accept the possibility that the manifestation of Desdemona's desires were outside official culture's codes of "proper" sexual practice, we can reassess her supposedly "contradictory" character without reducing her to a submissive woman who martyrs herself in her unyielding devotion to a man. Rather, we can read her final words as an acknowledgment of her responsibility in a masochistic scenario gone awry. This reading does not imply that Desdemona asked for what she got; on the contrary, our reading of Desdemona's final line and actions reclaim Desdemona's agency, empowering her with the possibility that her passivity in the final act was only seemingly so, and that her death was not merely the result of her retreat into a submissive, "female" role. As evidenced by her attempts to invoke the power of discourse within an official culture that privileges the male voice to negotiate a masochistic contract with her torturer in the final moments of her life, Desdemona did not go quietly or submissively to her grave.

Although Desdemona, as well as Othello and Iago, can be cast according to Deleuze's distinction between the pedagogical style and subversive intent of the sadist and the masochist, the play as a whole calls into question the stability of his definitions. According to Deleuze:

> In the contractual relation, the woman typically figures as an object in the patriarchal system. The contract in masochism reverses this state of affairs by making the woman into the party with whom the contract is entered into. Its paradoxical intention extends even further in that it involves a master-slave relationship, and one furthermore in which the woman is the master and torturer. The contractual basis is thereby implicitly challenged, by excess of zeal, a humorous acceleration of the clauses and a complete reversal of the respective contractual status of man and woman. (92)

Deleuze's account of the masochistic contract's capacity to overturn patriarchal law implies that the masochist moves transversally by disrupting official culture; however, this can only occur within the distinctly heterosexual and patriarchal framework within which Deleuze has constructed the masochist: the woman is always the torturer and the man the victim who, despite surface appearances, remains in control. In *Othello*, the Shakespearean heroine reverses the assumed gender of Deleuze's masochist by becoming a female "masochistic 'educator,'" who attempts to "train" her male abuser. In her masochistic role, Desdemona appears to reverse the patriarchal order of things: contrary to conventional gender constructs, she is the "captain's captain" (2.1.74). While her failure to generate a contract with her male torturer, along with Othello's rejection of her denials in favor of Iago's hearsay, reveal the cruel injustice of an official culture that privileges the male voice, they do not cause the overturning of the dominant, patriarchal judicial structures within the world of the play. On the contrary, the conclusion reaffirms those very structures, embodied in the forms of the Venetian noblemen, Lodovico and Gratiano, and the newly appointed governor of Cyprus, Cassio, all of whom are brought in at the end of the play to restore order and ensure that Iago receives his just desserts.

Othello further calls into question Deleuze's deconstruction of the "sadomasochistic unity" by suggesting that the role of the torturer and the tortured is subjective, and dependent upon the individual perspectives of those engaged in the fantasy. While Desdemona is the martyred victim in her own scenario, Othello identifies her as the torturer, "the cause" (5.2.1), in his. What is revealed by the relationship between Shakespeare's newlyweds are not affirmations of the "sadomasochistic unity," but an awareness of the subjective perspectives of the participants in either a masochistic or a sadistic scenario, and the subjective framework within which Deleuze has constructed his distinctions between sadism and masochism.

By comparing Iago's sadistic tendencies to Othello and Desdemona's masochist style, we have further elucidated differences among forms of subversion and transversal movement in *Othello*. Iago's successful destruction of Othello highlights the difference: more than just questioning the subjectivity of Venetian state ideology, Iago undermines it, asserting himself transversally. While we question Deleuze's construction of the masochist within a framework of gender roles and "norms" perpetuated by the sociopolitical conductors of various official cultures, we recognize the usefulness of his analysis to transversal theory, and apply it in this section not only to expand Fitzpatrick and Reynolds's exploration, but to exemplify the benefits of the investigative-expansive mode of analysis, which has allowed Reynolds and Segal to reparameterize the proceeding inquiry in response to new ideas that challenge common interpretations of Iago's intentions, Othello's odious cruelty toward his wife, and Desdemona's docile devotion to her husband.

Toward investigative-expansiveness

Having moved from critical history to ideological indoctrination, from Schlegel and Hunter to Althusser and Iago to Guattari and Deleuze to Reynolds and Segal, Fitzpatrick and Reynolds's transversal analysis defies the sort of holistic conclusion that is characteristic of the dissective-cohesive mode of analysis. More importantly, though, our view of the role that transversal power plays in Iago's machinations leads us to believe that a dissective-cohesive conclusion, one that seeks to explain Iago or to explain *Othello* itself, would necessarily misrepresent the play. *Othello*, as we have noted, breaks the world into binary oppositions, and encourages us to use these oppositions to categorize its characters, its plots, and its themes: thus, as Hunter tells us, the black foreigner turns out to be a good Christian, while the white Venetian is quickly seen to be an evil man who masks his true self with a veneer of honesty. Nevertheless, when we use these oppositions to assemble a holistic view of the play, we arrive at contradictions. Where Hunter sees the black man with a white soul being corrupted by the white man with a black soul, Schlegel sees a black foreigner who masquerades unsuccessfully as a Venetian before reverting to his original barbaric state. These readings use the binary oppositions found in *Othello* to describe each character at each point in the play (and beyond the play, during events we may assume took place before the beginning of the play); and then they use these definitions to determine how the characters move from point to point. By contrast, our analysis is concerned primarily with this movement, the process of change, rather than the initial and final states. Iago is postulated as an Althusserian state supporter only because the action of the play suggests ideological indoctrination; we accept Althusser's designation of objectivity as the opposite of ideology only because Iago's transversal movement out of his own subjective territory, symptomatically revealed when he exceeds the state-prescribed position on cuckoldry, requires some concept of exteriority. While *Othello* lays out a system of interrelated binary oppositions, it exists neither in nor between them, but in its own movement through them. To use an example from Hunter's essay, the play is not built on categorizations such as that of Othello as the black man with the white soul and Iago as the white man with the black soul; nor is *Othello* built on black and white distinctions between sadism and masochism. Rather, it is built on such actions as the audience's realization that their expectations have been contradicted. It is a play neither of seeming nor of being, but of becoming. And the character that best represents *Othello* is Iago, who shows us "motive hunting" rather than a motive, and whose transversal movements constitute the action of the play.

Notes

1. For more on "transversal theory," beyond the introduction to this book, see Bryan Reynolds, "The Devil's House, 'or worse': Transversal Power and Antitheatrical Discourse in Early Modern England," *Theatre Journal* 49.2 (1997): 143–67, and *Becoming Criminal: Transversal Performance and Cultural Dissidence in Early Modern England* (Baltimore: Johns Hopkins University Press, 2002); and Bryan Reynolds and Joseph Fitzpatrick, "The Transversality of Michel de Certeau: Foucault's Panoptic Discourse and the Cartographic Impulse," *Diacritics* 29.3 (1999): 63–80.
2. See also Karen Newman, "'And wash the Ethiop White': Femininity and the Monstrous in *Othello*," *Fashioning Femininity and English Renaissance Drama* (Chicago: Chicago University Press, 1991).
3. Of course, Othello's relationships with these two institutions are themselves both interdependent and intertwined, leading, for example, Allan Bloom and Harry Jaffa to comment: "How is he able to make himself so Venetian? It is surely, in part, his Christianity that is responsible" (48).
4. Though the first conclusive textual evidence for his conversion comes at 2.3.161–63 ("Are we turn'd Turks, and to ourselves do that / Which heaven has forbid the Ottomites? / For Christian shame, put by this barbarous brawl"), Othello's Christianity is implied both by his references to heaven during the trial scene (1.3.163; 266), and in his initial dialogue with Iago, in which Iago implies that the supposed theft of Desdemona was a legal marriage (1.2.11).
5. Here we side with A.C. Bradley in calling Iago a Venetian. Bradley cites 3.3.201 and 5.1.89 f. as evidence, and reads Cassio's comment, "I never knew a Florentine more kind and honest" (3.1.42–43) to mean, "not that Iago is a Florentine, but that he could not be kinder and honester if he were one" (200). M.R. Ridley appears to agree with Bradley in his 1958 edition of the text, in which his annotation of this line clarifies it as "i.e. *even* a Florentine (Cassio being from Florence)" (Ridley, ed., *Othello* 92).
6. Hunter may be paraphrasing A.W. Schlegel here, who writes in an essay later alluded to by Hunter, "While the Moor bears only the nightly colour of suspicion and deceit on his visage, Iago is black within" (Schlegel 196).
7. Bryan Reynolds and James Intriligator introduced this methodology in a paper entitled "Transversal Power," given at the Manifesto Conference at

Harvard University on May 9, 1998. For a detailed analysis of transversal methodology, see zooz's "Transversal Poetics: I.E. Mode" in the appendix of this book.

8. In Adelman, see especially pages 129–31 and 137–44.
9. Schlegel writes: "What a fortunate mistake that the Moor, under which name a baptized Saracen of the Northern coast of Africa was unquestionably meant in the novel, has been made by Shakespeare in every respect a negro! We recognize in Othello the wild nature of that glowing zone which generates the most raging beasts of prey and the most deadly poisons, tamed only in appearance by the desire of fame, by foreign laws of honour, and by nobler and milder manners" (195).
10. An even more recent example would be Walter S.H. Lim's comment: "That Iago finds it easy to undermine the Moor's emotional stability reveals that Othello can never fully obliterate his African self when he fabricates his Venetian/white identity" (62).
11. Julie Hankey has mentioned Marowitz's 1972 adaptation *An Othello* as an example of how a radically conceived reworking of the play, which Hankey herself labels "propaganda," demonstrates the deep-rooted investment of Shakespeare's original in problems of ideology (Hankey 113–14).
12. For an in-depth analysis of the power of storytelling in the play, see Stephen Greenblatt, "The Improvisation of Power," *Renaissance Self-Fashioning: From More to Shakespeare* (Chicago: University of Chicago Press, 1980), 222–54; especially pages 237–47.
13. Here we can explain in Althusserian terms Bloom and Jaffa's assertion quoted in note 4: it is because Othello has been inculcated in the state ideology through the "Ideological State Apparatuses" of religion (Christianity) that he is able to assimilate, to whatever extent, into Venetian society.
14. The explicit connection of the term heaven to Christian thought is even more clear in the folio and second quarto texts, where "God" is replaced by "Heaven."
15. Oliver Parker's film version presents an interesting interpretation of this scene, breaking Iago and Othello's discussion into segments that occur in different settings at different times of the day, suggesting that the conversation is either repeated almost verbatim a number of times, or dragged out for much longer than the audience is shown.
16. While Sinfield does not apply his "entrapment model" specifically to the type of interpretation given in this essay, he does note that, "entrapment... arises generally in functionalism, structuralism, and Althusserian Marxism" (39). For discussion of this subversion/containment paradigm, see, among other sources, Bryan Reynolds, "The Devil's House, 'or worse': Transversal Power and Antitheatrical Discourse in Early Modern England," *Theatre Journal* 49.2 (1997): 143–67, 148–49; Jonathan Dollimore, *Sexual Dissidence: Augustine to Wilde, Freud to Foucault* (Oxford: Clarendon Press, 1991), 81–91; Michael Bristol, *Shakespeare's America, America's Shakespeare* (London: Routledge, 1990), 189–211; and Louis Montrose, "Professing the Renaissance: The Poetics and Politics of Culture," *The New Historicism*, ed. Aram H. Veeser (New York: Routledge, 1989), 15–36, 20–24.

17. In fact, Althusser's system does not allow for such a creature as Iago, an ideologue who *opposes* the state, whose indoctrination of state citizens constitutes his own private *war* against the state.
18. Bryan Reynolds and Janna Segal have purposefully appropriated for their title of this coauthored section the phrase commonly used in *Monty Python's Flying Circus* to introduce a new skit that is supposedly a non sequitur, yet fits within the theme and contributes to the whole of a single episode. Borrowing Monty Python's framing technique, Reynolds and Segal offer this section as "something different," a new variable found through the process of the proceeding investigation that is not "completely" unrelated to the previous argument.
19. For an examination of desire and the role of marriage in *Othello*, see, among others: Catherine Belsey's "Desire's excess and the English Renaissance theatre: *Edward II, Troilus and Cressida, Othello*," *Erotic Politics: Desire on the Renaissance Stage*, ed. Susan Zimmerman (New York: Routledge, 1992), 84–102; David Bevington's "*Othello*: Portrait of a Marriage," *Othello: New Critical Essays*, ed. Philip C. Kolin (New York: Routledge, 2002), 221–31; Lawrence Danson's " 'The Catastrophe is a Nuptial': The Space of Masculine Desire in *Othello, Cymbeline*, and *The Winter's Tale*," *Shakespeare Survey* 46, ed. Stanley Wells (Cambridge: Cambridge University Press, 1994), 69–79; Sara Munson Deats's " 'Truly, an obedient lady': Desdemona, Emilia, and the Doctrine of Obedience in *Othello*," *Othello: New Critical Essays*, ed. Philip C. Kolin (New York: Routledge, 2002), 233–54; Géza Kállay's " 'To Die Upon a Kiss': Love and Death as Metaphors in *Othello*," *HJEAS* 3.2 (1997): 25–58; Lean Cowen Orlin's "Desdemona's Disposition," *Shakespearean Tragedy and Gender*, ed. Shirley Nelson Garner and Madelon Sprengnether (Bloomington: Indiana University Press, 1996), 171–92; Edward Pechter's " 'Have You Not Read of Some Such Thing?': Sex and Sexual Stories in *Othello*," *Shakespeare Survey* 49, ed. Stanley Wells (Cambridge: Cambridge University Press, 1996), 201–16; Mary Beth Rose, "The Heroics of Marriage in *Othello* and *The Duchess of Malfi*," *Shakespearean Tragedy and Gender*, ed. Shirley Nelson Garner and Madelon Sprengnether (Bloomington: Indiana University Press, 1996), 210–38; and Yoko Takakuwa's, "Diagnosing Male Jealousy: Woman as Man's Symptom in *The Merry Wives of Windsor, Othello, The Winter's Tale* and *Cymbeline*," *Hot Questrists After the English Renaissance: Essays on Shakespeare and His Contemporaries*, ed. Yasunari Takahashi (New York: AMS Press, 2000), 19–36. For an examination of masculine desire in *Othello*, see in particular Bevington, Danson, and Takakuwa. For a discussion of Desdemona's obedient devotion to Othello, refer especially to Bevington, Deats, Orlin, and Rose.
20. In this section, all citations from the play will be taken from *Othello*, ed. E.A.J. Honigmann (London: Arden Shakespeare, 2001).
21. Bevington (227), Deats (246–48), Orlin (181–83), and Rose (221–23) refer to Desdemona's passivity in the final moments of her life as a reflection of a patriarchal order. Deats (247), Orlin (181), and Rose (221–22) specifically refer to the line we have cited as a "lie" told by Desdemona in

order to exonerate Othello. Only Orlin identifies a possible subversive effect arising from Desdemona's death: "Her innocence exposes the predictive incapacity of patriarchalism and is in this respect subversive of its conceptual power" (187).

22. Reynolds and Segal disagree with the labeling of either masochism or sadism as a "perversion." It is their contention that "perversion" carries a negative connotation, and implies that there is a sex standard or norm against which a particular practice can be deemed "perverse." (Fitzpatrick was unavailable for comment on this issue.)

23. In "Morality, Ethics and the Failure of Love in Shakespeare's *Othello*," John Gronbeck-Tedesco recognizes Iago's instructive style, and attributes it to his military background: "No one understands the military ways better than Iago. He relies on the direct relationship between words uttered in the chain of command and the actions of authority" (*Othello: New Critical Essays*, ed. Philip C. Kolin [New York: Routledge, 2002], 255–70, 264). Gronbeck-Tedesco does not account for why Othello, who is as equally aware of "the military ways" as his ensign is, does not relate to others with the same commanding force.

24. In "*Othello*: Portrait of a Marriage," Bevington notes that Desdemona "gives us no reinforcement of what Othello has presumed to say about her motives, about her pitying him and loving him for the dangers he had passed" (225). However, asking the Duke for the right to accompany Othello to Cyprus, Desdemona states, "I saw Othello's visage in his mind, / And to his honours and his valiant parts / Did I my soul and fortunes consecrate" (1.3.253–55). Her insistence that she saw Othello as he perceived himself implies that she knows Othello from his vantage point, from how he presented himself to her during their courtship. Furthermore, the "honours and valiant parts" of Othello to which Desdemona has "consecrate[d]" herself can be read as a reference to the stories of "his honours and valiant parts" that Othello has admitted to wooing Desdemona with.

25. In "*Othello*: Portrait of a Marriage," Bevington agrees with our observation that Desdemona's admiration validates Othello, making him feel better about his own person (224–25). Unlike us, Bevington insists that Desdemona's love is able to affirm Othello's ego precisely because it comes from one who, as a woman, is in a lowly social position that is beneath Othello's: Desdemona's "admiration [is] made all the more keen by the great distance between her uneventful and constricted domestic life and the unlimited opportunity for adventure that he has enjoyed as a man. He finds her pity so gratifying because it dramatizes him as the protective male" (224–25). Of course, Othello, as a former slave, has not enjoyed an "unlimited opportunity for adventure," even though he is a man. By situating Desdemona as disadvantaged by her gender, Bevington disempowers her, ignoring her rebellion against the racist ideology of the play's official culture. Although his analysis of Desdemona and Othello's relationship neglects the role of race in their marriage or in their society, the play, with its constant allusions to Othello's "blackness" issuing from, among others, Othello himself, refuses to let the theater audience or reader disregard

the social codes that have been breached by Othello and Desdemona's interracial marriage.
26. Refer to note 21.

Works cited

Adelman, Janet. "Iago's Alter Ego: Race as Projection in *Othello.*" *Shakespeare Quarterly* 48.2 (1997): 125–44.
Althusser, Louis. "Ideology and Ideological State Apparatuses (Notes towards an Investigation)." *Lenin and Philosophy.* Trans. Ben Brewster. New York: Monthly Review Press, 1972. 127–86.
Bartels, Emily C. "Making More of the Moor: Aaron, Othello, and Renaissance Refashionings of Race." *Shakespeare Quarterly* 41.4 (1990): 433–54.
Belsey, Catherine. "Desire's Excess and the English Renaissance Theatre: *Edward II, Troilus and Cressida, Othello.*" *Erotic Politics: Desire on the Renaissance Stage.* Ed. Susan Zimmerman. New York: Routledge, 1992. 84–102.
Bevington, David. "*Othello*: Portrait of a Marriage." *Othello: New Critical Essays.* Ed. Philip C. Kolin. New York: Routledge, 2002. 221–31.
Bloom, Allan, with Harry V. Jaffa. *Shakespeare's Politics.* Chicago: University of Chicago Press, 1964.
Bradley, A.C. *Shakespearean Tragedy: Lectures on Hamlet, Othello, King Lear and Macbeth.* New York: Penguin, 1991.
Cavell, Stanley. *Disowning Knowledge in Six Plays of Shakespeare.* Cambridge: Cambridge University Press, 1987.
Coleridge, Samuel Taylor. *Coleridge's Shakespearean Criticism.* Vol. 1. Ed. Thomas Middleton Raysor. Cambridge: Harvard University Press, 1930.
Cummings, Peter. "The Making of Meaning: Sex Words and Sex Acts in Shakespeare's *Othello.*" *Gettysburg Review* 3.1 (1990): 75–80.
Danson, Lawrence. " 'The Catastrophe is a Nuptial': The Space of Masculine Desire in *Othello, Cymbeline,* and *The Winter's Tale.*" *Shakespeare Survey* 46. Ed. Stanley Wells. Cambridge: Cambridge University Press, 1994. 69–79.
Deats, Sara Munson. " 'Truly, an obedient lady': Desdemona, Emilia, and the Doctrine of Obedience of *Othello.*" *Othello: New Critical Essays.* Ed. Philip C. Kolin. New York: Routledge, 2002. 233–54.
Deleuze, Gilles. "Coldness and Cruelty." *Masochism.* New York: Zone, 1991. 9–138.
Deleuze, Gilles and Félix Guattari. *A Thousand Plateaus: Capitalism and Schizophrenia.* Trans. Brian Massumi. Minneapolis: University of Minnesota Press, 1987.
Dollimore, Jonathan. *Sexual Dissidence: Augustine to Wilde, Freud to Foucault.* Oxford: Clarendon Press, 1991.
Greenblatt, Stephen. "The Improvisation of Power." *Renaissance Self-Fashioning: From More to Shakespeare.* Chicago: The University of Chicago Press, 1980. 222–54.
Gronbeck-Tedesco, John. "Morality, Ethics and the Failure of Love in Shakespeare's *Othello.*" *Othello: New Critical Essays.* Ed. Philip C. Kolin. New York: Routledge, 2002. 255–70.

Hankey, Julie. "Introduction." *Othello*. By William Shakespeare. Ed. Julie Hankey. Plays in Performance series. Bristol: Bristol Classics Press, 1987. 1–147.
Hedrick, Donald and Bryan Reynolds. "Shakespace and Transversal Power." *Shakespeare Without Class: Misappropriations of Cultural Capital*. Ed. Donald Hedrick and Bryan Reynolds. New York: St. Martin's Press, 2000. 3–47.
Hunter, G.K. "Othello and Colour Prejudice." *Dramatic Identities and Cultural Tradition: Studies in Shakespeare and his Contemporaries*. New York: Barnes and Noble, 1978. 139–63.
Kállay, Géza. " 'To Die Upon a Kiss': Love and Death as Metaphors in *Othello*." *HJEAS* 3.2 (1997): 25–58.
Knight, G. Wilson. "The *Othello* Music." *The Wheel of Fire: Interpretations of Shakespearian Tragedy with Three New Essays*. London: Methuen, 1949. 97–119.
Leavis, F.R. "Diabolic Intellect and the Noble Hero: or The Sentimentalist's *Othello*." 1937; rpt. in *The Common Pursuit*. Harmondsworth: Peregrine, 1952.
Lim, Walter S.H. "Representing the Other: *Othello*, Colonialism, Discourse." *The Upstart Crow* 8 (1993): 57–78.
Little Jr., Arthur L. " 'An essence that's not seen': The Primal Scene of Racism in *Othello*." *Shakespeare Quarterly* 44.3 (1993): 304–24.
Montrose, Louis. "Professing the Renaissance: The Poetics and Politics of Culture." *The New Historicism*. Ed. Aram H. Veeser. New York: Routledge, 1989. 15–36.
Newman, Karen. " 'And wash the Ethiop White': Femininity and the Monstrous in *Othello*." *Fashioning Femininity and English Renaissance Drama*. Chicago: Chicago University Press, 1991.
Orlin, Lena Cowen. "Desdemona's Disposition." *Shakespearean Tragedy and Gender*. Ed. Shirley Nelson Garner and Madelon Sprengnether. Bloomington: Indiana University Press, 1996. 171–92.
Pechter, Edward. " 'Have You Not Read of Some Such Thing?' Sex and Sexual Stories in *Othello*." *Shakespeare Survey* 49. Ed. Stanley Wells. Cambridge: Cambridge University Press, 1996. 201–16.
Reynolds, Bryan. *Becoming Criminal: Transversal Performance and Cultural Dissidence in Early Modern England*. Baltimore: Johns Hopkins University Press, 2002.
——. "The Devil's House, 'or worse': Transversal Power and Antitheatrical Discourse in Early Modern England." *Theatre Journal* 49.2 (1997): 143–67.
Reynolds, Bryan, and Joseph Fitzpatrick. "The Transversality of Michel de Certeau: Foucault's Panoptic Discourse and the Cartographic Impulse." *Diacritics* 29.3 (2000): 63–80.
Ridley, M.R. Introduction. *Othello*. By William Shakespeare. Ed. M.R. Ridley. London: Methuen, 1958. xv–lxx.
Rose, Mary Beth. "The Heroics of Marriage in *Othello* and *The Duchess of Malfi*." *Shakespearean Tragedy and Gender*. Ed. Shirley Nelson Garner and Madelon Sprengnether. Bloomington: Indiana University Press, 1996. 208–38.

Schlegel, Augustus William. *A Course of Lectures on Dramatic Art and Literature.* Vol. 2. Trans. John Black. London: J. Templeman and J.R. Smith, 1840.
Shakespeare, William. *Othello.* Ed. E.A.J. Honigmann. London: Arden Shakespeare, 2001.
———. *Othello.* Ed. M.R. Ridley. London: Routledge, 1958.
Sinfield, Alan. *Faultlines: Cultural Materialism and the Politics of Dissident Reading.* Berkeley: University of California Press, 1992.
Stallybrass, Peter. "Patriarchal Territories: The Body Enclosed." *Rewriting the Renaissance: The Discourse of Sexual Difference in Early Modern Europe.* Ed. Margaret W. Ferguson, Margaret Quilligan, and Nancy J. Vickers. Chicago: Chicago University Press, 1986. 123–42.
Takakuwa, Yoko. "Diagnosing Male Jealousy: Woman as Man's Symptom in *The Merry Wives of Windsor, Othello, The Winter's Tale* and *Cymbeline.*" *Hot Questrists After the English Renaissance: Essays on Shakespeare and His Contemporaries.* Ed. Yasunari Takahashi. New York: AMS Press, 2000. 19–36.

CHAPTER 4

"WHAT IS THE CITY BUT THE PEOPLE?"
TRANSVERSAL PERFORMANCE AND
RADICAL POLITICS IN SHAKESPEARE'S
CORIOLANUS AND BRECHT'S *CORIOLAN*

Bryan Reynolds

BIG TIME, LOST
I knew, that cities were being built
I didn't go there.
That belongs to statistics, I thought—
Not to history.

But what are cities, built
Without the wisdom of the people?
     —Bertolt Brecht[1]

Shakespeare's *Coriolanus* and Brecht's adaptation of it were written during historical periods of high cultural anxiety and frustration.[2] Both works reflect and comment on social, economic, and political problems contemporaneous to their conception. Whereas Brecht makes his purpose for creating *Coriolan* explicit in his nonfictional writings, he does not reveal the direct correlations between his play and, as he calls them, the "dark times" in which he wrote. To do this would have been dangerous for Brecht while living in a newly formed East Germany that was indirectly governed and closely monitored by Soviet forces. Similarly, despite the obvious topicality of its subject matter, Shakespeare gives no clear indication of his political investment in *Coriolanus*, and censorship was an issue that Jacobean dramatists were compelled to consider.[3] In this chapter, I compare *Coriolanus* and *Coriolan* in light of their historical contexts in hopes of bettering our understanding of both the actions of the plebeians and Coriolanus's negotiation of his own subject position in response to the changing sociopolitical environment in Shakespeare's play. I aim to propose a new reading of *Coriolanus* that emphasizes performance rather than literal possibilities for the play-text.

Historical contexts, critical approaches

The Soviet Union made the years immediately following the establishment in 1949 of the German Democratic Republic (GDR) dire for East Germany's working class. It extracted resources and goods from the GDR and continued to do so until war reparations were formally settled in August 1953.[4] The extractions caused depression and the rationing of staple products like fat, sugar, and meat; consumer durables remained scarce and expensive. Although wages increased for people employed by the state, living standards for the majority were terribly low.

The predicament culminated on June 16, 1953. After several years of inefficient and disruptive negotiations between the workers and the government, the people's unrest escalated into an angry protest composed of approximately 10,000 demonstrators.[5] By noon of June 17, the protest approached intractability. Violence between the police and the malcontents became more frequent and unmanageable. Finally, the Soviet military commander proclaimed a state of siege, and heavy artillery was mobilized into the cities' centers. Confronted with overwhelming military power, the insurrection was soon suppressed.

During these difficult years, Brecht wrote *Coriolan*, completing it sometime between 1952 and 1955.[6] Given Brecht's dedication to the needs of the proletariat and to the promotion of Marxist ideology and aesthetics, it is not surprising that with his adaptation he planned to represent and champion Germany's working class by celebrating the communal empowerment of the plebeians depicted in the play.[7] Brecht says of *Coriolan*, "we must at least be able to 'experience' not only the tragedy of Coriolanus himself but also of Rome, and specifically of the plebs" (*Coriolan* 374). Brecht's partiality for the proletariat is confirmed by another work that he wrote shortly after June 17. As historian Henry Ashby Turner, Jr. states in his account of the June 17 workers' uprising, "the foremost Communist literary figure of East Germany, Bertolt Brecht, gave expression to sentiments of many in a poem he secretly circulated" (123). The sarcastic, propagandistic poem, ironically entitled "The Solution," illustrates Brecht's strong support of the proletariat at the time:

> After the uprising of the 17th of June
> The Secretary of the Writer's Union

Had leaflets distributed in the Stalinallee
On which one could read, that the people
Had forfeited the trust of the government
And could win it back only
By redoubled efforts. And so, would
It not be easier if the government
Dissolved the people
And selected another? (1009–10)[8]

Brecht's devout Marxism can be seen in all of his work,[9] and, as I hope to demonstrate, Shakespeare's *Coriolanus* was a ready vehicle for the promotion of his politics. By examining Brecht's *Coriolan* as an interpretation of *Coriolanus*, I want to argue that *Coriolanus* was predisposed to Brecht's purpose, both textually and historically, and that it could easily be performed, and possibly was originally performed in early modern England, in line with Brecht's politics, that is, *without* textual adaptation.

When writing his version of the Coriolanus legend, Shakespeare too considered the plight of the working class, or rather, the "common people," as they were referred to in early modern England and in his play. The potential for Shakespeare's *Coriolanus* to serve Brecht's enterprise without alteration is informed by the fact that it, like *Coriolan*, was written in an era of fervent sociopolitical conflict resulting in a revolt led by plebeians. In England, beginning in 1607 and culminating in 1608 (probably the year Shakespeare's *Coriolanus* was originally performed),[10] the increasing cost of corn and the accompanying dread of famine were serious, manifested concerns. A consequence of this distressing situation was the Midlands insurrection of May, 1607. "A great number of common persons"—up to five thousand, says John Stow in his *Annals*, gathered in various Midlands counties, including the county of Warwickshire where Shakespeare usually resided, to dig up and level enclosures and to protest the dearth of food (890).[11] Like the East German protesters, the early modern English protesters damaged properties and threatened wealthy landowners (whose position resembled that of political conformists in the GDR, where there were no landowners).[12] And, as in the workers' revolt in East Germany, the Midlands crisis was eventually quelled by military forces and trailed by executions.

The Midlands insurrection differed from the riots that took place in the preceding century, under Queen Elizabeth's rule, in that it "was purely and nakedly a demand for economic redress," while the earlier riots were provoked by "economic and social grievances" mixed with "religious and political issues" (Pettet 34). As Karl Marx and Friedrich Engels describe what they call "the proletarian movement"—a concept that anticipated the June 17 uprising—the Midlands revolt was wholly a "self-conscious" action taken *by* "common people" for the benefit of the "common people" (*Communist Manifesto* 92). Thus, the Midlands insurrection may have been a significant event foreshadowing the revolution of the 1640s, a war that was largely motivated by the people's desire for a more equitable distribution of wealth and a form of government allowing them greater representation.

Yet, unlike Brecht, who openly, though only to a select few, discloses his partiality for the workers in his post–June-17 poem, there is no record of Shakespeare having so frankly expressed his opinions on any sociopolitical issue. This is why it is difficult to discern from Shakespeare's writings his relationships to various matters of his day. In the case of *Coriolanus*, however, the evidence showing a correlation between the play and the people's uprising of 1607 is substantial. This can be found in Shakespeare's departures from Livy's and, more importantly, Plutarch's version of the Coriolanus legend, the latter being the main source from which Shakespeare contrived his play.[13]

Whereas there is often ambiguity regarding the reasons for Shakespeare's divergences from his sources (such as in the abundance of departures in the three parts of *Henry VI*), the evidence here seems straightforward. As noted by Pettet, "This particular divergence is a very marked and peculiar one: Shakespeare has reduced the grievances of the plebeians to the one matter of scarcity of corn (which he entwines with the fear of Coriolanus's absolutist temper), and he has set this grievance prominently before us in the first scene of the play, declining to follow Plutarch's account of a rising against the oppression of usury" (37). Shakespeare's reduction of the people's complaints to a single, highly charged point of contention establishes an association with the recent socioeconomic problems and subsequent revolt. By giving *Coriolanus* this conspicuous topicality, Shakespeare constructed a relationship between the play and issues important to its early modern audience, making the play more compelling, and the audience ready for a relevant political statement on the Midlands crisis. Regardless of Shakespeare's actual intentions, doing this transposed the theater into a political arena, and the stage into a platform from which the ideology delineated in the play was expressed without pretext, subterfuge, or mystification. This means that any framing of the performance, as a play, say, mainly about the historical figure Coriolanus and ancient Rome, was breached. In effect, the audience was keyed into another (additional) framing of the play as germane political commentary.

Examining criticism published on *Coriolanus* since Charles Gildon (1710) and Samuel Taylor Coleridge (1815), and especially over the last fifty years, quickly leads to the discovery of what appears to be an interpretive consensus. Of the claims shared by this consensus, the following assertions are most pertinent to the issues I am discussing here: (1) Shakespeare's *Coriolanus* vents hatred for the plebeians, or it is at least unsympathetic to their needs (Gildon 1710; Coleridge 1817; Brockbank 1976; Farnham 1963; Grass 1964; Huffman 1971; Ide 1980; MacCallum 1967; Pettet 1950; Rabkin 1967; Rossiter 1961; Simmons 1973; Zeeveld 1962); and (2) the play condemns the notion of democracy, or it is at least in favor of an absolute monarchy (Gildon 1710; Coleridge 1817; Farnham 1963; Grass 1964; Honig 1951; Huffman 1971; Kishlansky 1986; Phillips 1940; Rossiter 1961; Simmons 1973; Zeeveld 1962.)[14] Recently published alternative readings to the ones just listed are by Adelman (1980), Bristol (1987), Dollimore (1984), Jagendorf (1990), Patterson (1989), Scofield (1990), and Sorge (1987); except for those by Jagendorf and Scofield, which do not address in any relevant way the concerns of this essay, I will refer to these readings as my analysis progresses.[15] For now, however, it is sufficient to note that these readings, insofar as they deal with issues relevant to the assertions listed here, are alternative mainly in that they either sympathize with the plebeians, suggest that the play sympathizes with the plebeians (if only in passing), do both of these things, or maintain that the play is ambiguous, politically, in its handling of the plebeians' situation.

In my view, *Coriolanus* welcomes a performance-oriented interpretation that differs greatly from the apparent consensus and affirms some of the suggestions made by the alternative readings. As many critics have pointed out, Shakespeare's *Coriolanus*, like Brecht's *Coriolan*, is primarily about two powerful conflicts: (1) the tragedy of a great pathological individualist (Coriolanus) existing in contradiction to his social world, and (2) the struggle for solidarity between the "common people" and the ruling oligarchy. In analyzing what seems to be *Coriolanus*'s endeavor to resolve these two conflicts, I will argue that Shakespeare's play parallels Brecht's *Coriolan* in its criticism of the notion of an irreplaceable leader or absolute monarch. This parallelism is especially clear when the play is analyzed as a performed text situated within the historical context of its initial performance for the early modern English audience. It is my hypothesis that the original production accomplished this critical feat by prompting the audience to identify and thereby empathize, rather than sympathize, with the plebeians. As a result, Shakespeare's *Coriolanus* encouraged the audience's support of the plebeians' wish for a more democratic form of government, both within and beyond the Roman world represented in the play. Moreover, I want to make the corollary argument that through its treatment of the concept of indispensability in its depiction of Coriolanus, the play most powerfully posits as viable sociopolitical enterprises not only replaceability, but also the changing and expanding of identity through what I call "transversal movements."

Communist commoners?

The field of sociopolitical forces operating in both *Coriolanus* and *Coriolan* is explicitly exemplified in the first scene, which sets the plot and tone for the plays. In a lively parley among the First Citizen, the Second Citizen, and a chorus of others, we are invited to share the plebeians' craving for justice. From their conversation we learn that the plebeians are willing to sacrifice their lives while fighting to free themselves from the deprivation imposed on them by the upper class; they are "all resolved rather to die than to / famish" (1.1.3–4; cf. *Coriolan* 59).[16] They all agree that "Caius Martius [Coriolanus] is chief enemy / to the people" (1.1.6–7; cf. *Coriolan* 59), but settle upon this conclusion only after they consider "what services he has done for / his country" (1.1.29–30). We are told that the plebeians are motivated by their "hunger for bread, not in / thirst for revenge" (1.1.24–25) and by their desire to be self-governing: "we'll have corn at our own / price" (1.1.9–10; cf. *Coriolan* 59). Despite the fact that their logic during this initial scene is a little fuzzy, "the people's belief that the death of Coriolanus would allow them to have corn at their own price is," as Janet Adelman observes, "eventually sustained by the plot, insofar as Coriolanus opposes the giving of corn gratis" (3.1.113–17; cf. *Coriolan* 100–01; Adelman 136).

The plebeians' declarations, which also reflect Brecht's point of view, are echoed by Marx and Engels in a decisive statement about the Communists: "They openly declare that their ends can be attained only by the forcible overthrow of all existing conditions" (120). Like the Communists, the plebeians of *Coriolanus* and *Coriolan* want "To unbuild the city and to lay all flat" and therefore equal (3.1.196; cf. *Coriolan* 104). Furthermore, when Menenius inquires of the angry citizens: "Will you undo yourselves?" (1.1.63)—asking them if they will risk everything in an attempt to take on the "Roman state" (1.1.68)— the First Citizen wittily yet honestly responds, "We cannot, sir, we are undone already" (1.1.64; cf. *Coriolan* 61). Marx and Engels in

The Communist Manifesto similarly argue the First Citizen's contention: "The proletariat have nothing to lose but their chains. They have a world to win" (121). Like the proletariat that Marx and Engels are concerned with, the proletariat of Shakespeare and Brecht, heading for starvation, has nothing to lose by revolting except the chains of their oppression (1.1.83; *Coriolan* 61). Although these similarities between Shakespeare's *Coriolanus* (and Brecht's *Coriolan*) and *The Communist Manifesto* are not elaborate, I note them because they testify to an ideological compatibility between *Coriolanus* and *The Communist Manifesto* with regard to governmental matters and the concept of an indispensable leader, at least insofar as the plebeians represent the particular ideology promoted by Shakespeare's play.

Initially, Menenius tries to shift the responsibility for the "soaring prices" of food and the people's accompanying "misery" (61) away from the government and onto the gods: "Your suffering," he tells them, "in this dearth, you may as well / Strike at the heaven with your staves . . . For the dearth, / The gods, not the patricians, make it" (1.1.66–72; cf. *Coriolan* 61). However, Menenius is not taken seriously. The First Citizen ignores his mention of the gods and angrily responds by stating exactly how the patricians oppress the "common people." He says that they "Suffer us to famish" while "their store- / houses crammed with grain"; they "support usurers" and "repeal daily any wholesome act / established against the rich, and provide more / piercing statutes daily, to chain up and restrain the poor" (1.1.79–84; cf. *Coriolan* 61). These allegations demonstrate sharply the gravity of the situation, especially as they are contrasted against Menenius's feeble attempt to hold the gods responsible for the starving people, while the rich have plenty of food.

Without addressing the complaints of the First Citizen, complaints that are incidentally never addressed by the patricians throughout the plays, Menenius goes on to deploy a new strategy; he endeavors to assuage their animosity with a parable. But, as Marx explains in his own critical commentary on Shakespeare's *Coriolanus*, "when the Roman plebeians struck against the Roman patricians, the patrician Agrippa [Menenius] told them that the patrician belly fed the plebeian members of the body politic. Agrippa failed to show that you feed the members of one man by filling the belly of another" (106). Furthermore, as Thomas Sorge writes in his astute account of the failure of Menenius's "pretty tale" (1.1.89; cf. *Coriolan* 61):

> a second look at the way the First Citizen parades the conventional rhetoric of the fable's underlying analogy shows how he turns the tale in

the plebeians' favor by stressing that the different members are after all oppressed "by the cormorant belly" [1.1.120]—a move which forces Menenius both to interrupt him and to beg for more time and "Patience awhile" [1.1.125] before he can tell them the belly's already well-known answer. And it is no surprise that the answer eventually comes as an unsatisfactory anticlimax. (234-35)

The First Citizen responds sarcastically to Menenius's answer, "It was an answer. How apply you this?" (1.1.146; cf. *Coriolan* 63). The tale does not influence the First Citizen, and thus this is the plebeians' first victory; they successfully resist this first effort to suppress their insurrection. But the entrance of Coriolanus quickly diminishes the power they gain in this rhetorical battle. Apart from one satirical comment made by the First Citizen in response to Coriolanus's denigrating the plebeians (1.1.165), Shakespeare's plebeians, unlike Brecht's, are silent for the duration of the scene. But this silence, I think, does not signify defeat or concession.

In Brecht's *Coriolan*, Coriolanus enters "unnoticed except by Menenius" and "escorted by two armed men" (63), and he does this some twenty lines earlier than indicated by the stage direction in Shakespeare's text, which only calls for the entry of "Caius Martius [Coriolanus]." Brecht provides an explanation for this alteration in his brilliant analysis of the first scene:

> I'd suggest having Marcius and his armed men enter rather earlier than is indicated by Agrippa's [Menenius's] 'Hail, noble Marcius!' [1.1.163]....
> The plebeians would then see the armed men looming up behind the speaker, and it would be perfectly reasonable for them to show signs of indecision. (*Brecht on Theatre* 258)

Staging the action as Brecht proposes would also explain Menenius's "sudden aggressiveness" toward the First Citizen (*Brecht on Theatre* 258). Having Menenius willing to risk acting in this manner because he feels empowered as well as safeguarded by the presence of Coriolanus and his soldiers would stress that his ideology, like that of Coriolanus, welcomes the threat of violence, and must be reinforced by this threat in order for the plebeians to submit to it.

Although Brecht's work on the opening scene reveals how he developed successfully for his purposes, with much textual alteration, what is already in Shakespeare's text, I aim to show that it was unnecessary for him to change any text, including the stage direction, to achieve the effects he describes. In fact, this scene could be staged in a variety of ways without any alteration to the text. It could be staged just as Brecht proposes. Menenius could be made to see Coriolanus in the distance (with or without the armed men), yet only acknowledge Coriolanus when his presence is verging on the obvious, as Brecht would have it. Or, to give another possibility, Menenius's sudden hostility toward the First Citizen (and his fellow mutineers) could be staged as a defensive measure caused by the embarrassment and anxiety he experiences as a result of his failed parable. In this case, Coriolanus's entrance comes as a relief to him, and may prevent him from being injured by the angry rabble of citizens.

Moreover, Shakespeare's plebeians, notwithstanding their silence following the First Citizen's sarcastic remark, can be seen as continuing actively to protest Coriolanus even after his arrival. They mock him with their silence—by ignoring his authority. This act of dissidence could be made blatant with gesture, such as with smirks, nudges, backslapping, and cupped whispers. Furthermore, by remaining silent like the audience, the plebeians' dilemma is more identifiable; the audience becomes aligned with the plebeians, both being spectators to Coriolanus's ruthlessness. Such a staging is actually reinforced by Shakespeare's and Brecht's texts: when Coriolanus commands the plebeians to leave ("Go get you home, you fragments!" [1.1.221; cf. *Coriolan* 65]), they do not respond to him verbally, and there is no stage direction calling for their departure. Finally, after Coriolanus ridicules the plebeians with his request that they be permitted to go to the "Capital" to prepare for war with the nobles, we are told by the stage direction that the "Citizens steal away" (Brockbank 114). This stage direction is frequently cited to illustrate the play's belittlement of the "common people." For example, John Dover Wilson says that Shakespeare has the plebeians steal away in order to demonstrate their pusillanimity (Wilson cited in Brockbank 114). Nevertheless, as in the instance of Coriolanus's entrance, the text invites more than one interpretation.

Perhaps the plebeians exit stealthily not out of fear but as protest; they actively oppose, with much irreverence, Coriolanus's enticement ("The Volsces have corn.... Worshipful mutiners...follow" [1.1.248–50]) by departing quietly, indifferently. This is plausible considering, as noted earlier, that they ignore his command for them to leave. To broaden our perspective on the implications of this stage direction, we must consider the positioning of the plebeians on the stage throughout the whole first scene. Shakespeare's Coriolanus delivers the news of the newly appointed Tribunes for the people solely to Menenius; consequently, he and Menenius are the only ones to discuss the ramifications of these appointments. So, to insure that the plebeians are denied access to this private conversation they must be positioned (within the stage illusion) out of hearing distance from Coriolanus and Menenius. After all, why would they stand close enough to Coriolanus—an enemy, reputed warrior, and physical threat to them—to be able to hear the talk regarding the Tribunes? It makes sense to have them spatially marginalized (as they are socially, economically, and politically) by the presence of the formidable figure now dominating the stage.

Like Brecht's, Shakespeare's play directs our attention toward the mistreatment of the "common people" under the elitism and selfishness of the ruling body. Even if the stage direction ("Citizens steal away") is only meant to acknowledge that the situation has changed since the beginning of the scene—when the stage direction reads: "Enter a company of mutinous Citizens, with staves, clubs, and other weapons"—the entire scene suggests that it is dangerous for a country to have leaders who do not consider the welfare of its commoners. As Brecht eloquently puts it, "The wind has changed, it's no longer a favorable wind for mutinies; a powerful threat [the approaching war with the Volscians that was just made known] affects all alike, and as far as the people goes this threat is simply noted in a purely negative way" (*Brecht on Theatre* 261). But, unlike the commoners, Coriolanus does not have a negative perception of war: "I am glad on't," he declares, "then, we shall ha' means to vent / Our musty superfluity" (1.1.224–25; cf. *Coriolan* 66). He looks forward to the war as an opportunity to dispose of the unhappy "common people" whom he sees as the country's dispensable surplus.

Coriolanus is often described in *Coriolanus* and *Coriolan* as a great warrior, even godlike, especially after he defeats the attacking Volsces almost single-handedly; thus many characters see him as a tremendous asset to his country. Cominius holds that, "The man I speak of [Coriolanus] cannot in the world / Be singly counter-pois'd" (2.2.86–87, see also 2.2.168; cf. *Coriolan* 92). Yet, to Coriolanus's dismay, being "prov'd best man i'th'field" (2.2.97; cf. *Coriolan* 92) and being able to "Turn terror into sport" with "his sword, death's stamp" (2.2.105–07; cf. *Coriolan* 92) are not enough; they are not the qualities required to be an adequate ruler of Rome and to gain the majority's approval. As the Third Citizen says, "if he would incline to the people, there was never a worthier man" (2.3.38–40); but for him to do this, because it would mean going against his own beliefs, his own code, he would have to "perform a part" (3.2.109) in a "wolvish toge" (2.3.114)—a false "napless vesture of humility" (2.1.232)—as the fabled wolf wears the sheep's clothing (Brockbank 1976, 185). Rather, Coriolanus prides himself as someone "never / [to] be such a gosling to obey instinct, but stand / As if a man were author of himself / And knew no other kin (5.3.34–37; cf. *Coriolan* 140). He insists uncompromisingly on his own self-fashioning and self-reliance, and his irreplaceability.

Thus, unlike in Plutarch, Shakespeare's and Brecht's Coriolanus refuses to display his wounds for the people. It would be traumatizing for him to "stand naked" and entreat people whom he sees as inferior to him, whom he calls "curs," "rats," "dissentious rogues," "quarter'd slaves" (2.2.137; cf. *Coriolan* 93). Doing so would require admitting to himself as well as revealing to the masses that he too has to work according to society's conventions to achieve his goals; that he too is a human being susceptible to injury; that "he too has a mouth, that he is a dependent creature" (Adelman 137). And, therefore, in his view, that he too is replaceable like the "common people." Symbolically, by submitting to and thereby aligning himself with the "common people," which would require him to act just as they desire, he would also be revealing to the audience that he too is an actor that, like a professional actor, pretends to be someone else for profit. As Donald Hedrick puts it, " 'performance for pay' would for Coriolanus constitute the most abject realm of loathing, like mercenary soldiership."[17] Moreover, having been identified as an actor by an audience living in Jacobean England, a social world that discriminated against people for being actors even while it simultaneously supported them at the theater, Coriolanus would be revealing that he too (like the actor playing him) is capable of being wounded by society, that he is subject to its laws and prejudices. Hence, for the early modern audience, the performance frame would be disrupted, the fourth wall broken, and the actor playing Coriolanus would emerge prominently as an actor, a real person. However, I believe this situation occurs nevertheless, and more strikingly, through Coriolanus's emphatic refusal to act. His controversial, adamant refusal to act as the "common people" want makes the concepts of acting and identity performance the focus of the scene. His refusal breaches the performance frame and compels the audience's awareness and contemplation of performance as an operative sociopolitical mechanism in the real world. In effect, the audience is also encouraged to consider, inversely, its role in the world of the theater.

Chief enemy to the people

Adelman's observation about today's American audience's participation in this drama is especially apt, the second clause of which is applicable to the early modern audience as well: "Nor is it only our democratic sympathies that put us uncomfortably in the position of the common people throughout much of the play: Coriolanus seems to find our love as irrelevant, as positively demeaning, as theirs; in refusing to show the people his wounds, he is at the same time refusing to show them to us" (144). In his refusal to perform, Coriolanus aligns the audience with the "common people," thereby disparaging the audience as well. As a result, he sets himself apart physically, morally, and spiritually from all people (the characters, the actors, the audience), as so many monarchs and dictators have done throughout history, including those who ruled over Shakespeare and Brecht (James I and Stalin, respectively).

Shakespeare's Coriolanus seems to correspond to Elizabeth I, who distinguished herself by her claim to virginity and her excellence in leadership. He might even be compared to Sir Walter Raleigh, who was said to have been an exceptional soldier, but may have had treasonous transactions with Spain, as Coriolanus has with the Volscians. Yet the figure that most likely occurred to the early modern audience is James I, who was in power when the play was written, and stands out because of his renowned erudition, his steadfast belief in the doctrine of Divine Right, and his resistance to public appearances. Because James was educated in Roman rather than in English constitutional law, and because he wrote several theoretical works intended to legitimize and foster the absolute power of Christian kings, such as *The Trew Law of Free Monarchies* (1598) and *The Basilicon Doron* (priv. pr., 1599; public ed., 1603), as Albert H. Tricomi explains, "James awakened suspicions early on that he had little regard, foreigner that he was, for English rights and liberties": instead, he "inspired fears that he might become tyrannical, or had tyrannical designs" (54, 56).

In *The Trew Law of Free Monarchies*, one of many places in which James's views are documented, James maintains:

> [The] allegeance of the people to their lawful King, their obedience, I say, ought to be to him, as to Gods lieutenant in earth, obeying his commands in all thinges, except directly against God, as the commands of Gods Minister, acknowledging him as a Iudge set by God ouer them, hauing power to judge them, but to be judged onely by God. (*Minor Prose* 69)

Later he adds that the "King make daily / statutes & ordinances, inoyning such paines thereto as he thinks meet, without any advise of parliament or estates" (*Minor Prose* 71). James's perception of the "Nature of a King" and "Christian Monarche" is similar to Coriolanus's own self-concept. And the fear of James shared by members of the parliament and English populace is comparable to the fear of Coriolanus experienced by many of the play's characters. Although James's style of monarchy fell far short of the autocratic ideal articulated in his writings, his writings caused much anxiety in the realm. On March 21, 1610, for example, James was compelled to respond to the people's fear of his alleged tyrannical aspirations with a speech to parliament delineating his ideas on Divine Right. Instead of placating the unrest among its members, however, the address provoked parliament to confront James with the Petition of Right (May 1610): a declaration of their rights as English citizens, which James ignored. From a letter written by Lord Chamberlain dated May 24, 1610, we learn some of the effects of James's speech:

> I heare yt bred generally much discomfort; to see our monarchicall powre and regall prerogative strained so high and made so transcendant every way, that yf the practise shold follow the positions, we are not like to leave to our successors that freedome we receved from our forefathers. (301)

Inasmuch as the conflict between James and his subjects was serious when Coriolanus was written, and it was very serious, it is easy to compare James to Coriolanus and see *Coriolanus* as an anticourt drama.[18]

In regards to *Coriolan*'s topicality, it is similarly easy to imagine Joseph Stalin and Adolf Hitler as models for Brecht's Coriolanus, the former serving for Brecht's audience as a dissident perspective on Coriolanus, whereas the latter serves as a conservative one. Like Coriolanus, Stalin and Hitler were both despotic pathological egotists who tried to give the impression that their greatness surpassed all. According to historian Isaac Deutscher, "Each established himself as an unchallengeable master ruling his country in accordance with a rigid *Führerprinzip*" (566). Considering the historical context in which Brecht wrote *Coriolan*, as Darko Suvin points out, "now, in that era of High Stalinism, the great leader [Coriolanus] running berserk and believing himself to be indispensable could not fail to be associated with Generalissimo Joseph Stalin (and possibly the lesser Stalins that Stalinism bred)" (200). Unfortunately, because *Coriolan* did not reach the stage for the Berliner Ensemble until 1964, eight years after Brecht's death, and in a very different Germany, we can only speculate on the kinds of response it would have gotten a decade earlier. In the communist Germany of 1964, the play was a wonderful success.

Like Coriolanus, Stalin was a great military leader, famous for his accomplishments in both the Russian Civil War and World War II, and he was eponymous to the city of Stalingrad, as Coriolanus is to Corioli. Most significant, however, is that Stalin not only believed himself irreplaceable, but also like Coriolanus—who wishes for the opportunity to kill "thousands" of rebellious plebeians (1.1.97–99; cf. *Coriolan* 64)—he sought the destruction of all people who represented the potentiality of an alternative government; and, like Hitler, he murdered millions of his country's own citizens.

Hitler's perception of war and his role in the army are also much like those of Coriolanus (recall, 1.1.224); John Keegan notes from Hitler's diary, "Hitler found the war [WWI—the only war in which he fought] 'the greatest of all experiences,'" and he thought himself "'the first soldier of the Reich'" (236, 235). In relation to Hitler, as Margot Heinemann writes:

> [Brecht's potential] spectators many of whom were still under the influence of Nazi myth and glamour, brought up in SA or Hitler Youth, could all too easily see the story [of *Coriolan*] in terms of the true patriot and military hero [Coriolanus or Hitler], stabbed in the back by the cowardly masses under Red labor leaders. The production must show that no leader, however talented, is indispensable. (221)

Brecht's textual changes ensure that his play cannot be misread as profascist or right-wing, that it cannot be readily appropriated as Shakespeare's *Coriolanus* has been, such as by T.S. Eliot in his overtly fascist poem *Coriolan*. This safeguarding is most pronounced in the short but nonetheless powerful final scene that Brecht adds to Act 5. The scene makes it clear that "the world goes on without the hero" (125), who "aimed to make himself dictator" (126), by having the "Consul, senators, [and] tribunes" deliberating together over "current business" (126). He even has them forbid the public mourning of Coriolanus (146). This change in endings stresses the perspective, as Brecht explains, that "society can defend itself" from the individual who has risen up against it—"who has wrongly thought of himself as irreplaceable" (Brecht qtd. in Suvin 192).[19]

Similar to Stalin and Hitler, as well as James I, Coriolanus desperately endeavors to keep his fantasy alive: he works to maintain his dream of being self-reliant, greater than all other men, and an irreplaceable necessity to his country. Trying to maintain a self-image on such grandiose terms, nevertheless, places Coriolanus in an awkward contradiction to himself (as a member of a society) and to his overall social world. Coriolanus directs himself against the people, he establishes a binary opposition between himself and those people from whom he differentiates himself, and he does this to secure what he sees as his own extraordinary identity. For him to be the "heroic" politician the Romans want he would need to negotiate and make compromises; he would have to be capable of playing the necessary roles. But since such performing would cause him to lose touch with his own resolute sense of self, a self that can no longer exist if he must flatter and "be rul'd" (3.2.90) by commoners—or at least treat them congenially—he can never be their consul. Instead, as an antiperformer of sorts, a kind of antitheatricalist, he holds that he would rather "play / the man I am" than feign to be the man the people desire (3.2.15–16; cf. *Coriolan* 105). In fact, it is impossible for him to do otherwise, as when he fails to perform the part of, and thereby pass as, a commoner in Antium (4.5). His predicament reflects critically on the artificial nature of theater and the theatricality of politics and social life in general. It puts into question the relationship between self and social performance, asking if there is an inherent or constant true self. It suggests implicitly that one's social identity might contradict this true self; it suggests that identity and self, however interconnected or similar, are always constrained and must be performed to be socially affirmed. Hence Coriolanus's revelation near the play's conclusion: "Like a dull actor now / I have forgot my part and I am out, / Even to a full disgrace" (5.3.40–42).

Transversal territory

In the plays, Coriolanus is criticized as "too unbending" (*Coriolan* 106) and "too absolute" (3.2.39) in his devotion to his ideals and idealized self-image; and it is this persevering characteristic that most informs his inability to empathize with and relate to the "common people." This characteristic also explains why Coriolanus could not act successfully as a commoner in Antium; he cannot think and feel beyond the rigid boundaries of what I call his own "subjective territory." Subjective territory is my term for the scope of the conceptual and emotional experience of those subjectified by what I call "state machinery," which includes all inculcating forces in society, such as the educational, juridical, religious, and familial structures that support the beliefs of the dominant classes and the government. In agreement with its investment in cultural, social, political, and economic fluctuations and determinations, this organizing machinery functions over time and space, sometimes consciously and sometimes unintentionally, to consolidate social and state powers in order to construct a society of subjects and thus "the state": the totalized state machinery. Subjective territory is therefore delineated by conceptual and emotional boundaries that are normally defined by the prevailing science, morality, and ideology. These boundaries bestow a spatiotemporal dimension, or common ground, to an aggregate of individuals, and work to ensure and monitor the coexistence of this social body. In short, subjective territory is the existential and experiential realm in and from which a given subject of a given hierarchical society perceives and relates to the universe and his or her place in it.[20]

Coriolanus's subjective territory is constructed less in the interest of a greater social body of "common people" than in a ruling oligarchy, and mostly in the interest of consolidating his own personal power in the world. As one of his own men, the First Officer, says of him: "That's a brave fellow; but he's vengeance / proud, and loves not the common people" (2.2.5–6; cf. *Coriolan* 90); "he seeks their hate with / greater devotion than they can render it him, and leaves nothing undone that may fully discover him their opposite" (2.2.18–21). Coriolanus's compulsion to maintain the integrity of his subjective territory, which is of course an unconscious undertaking, extends well beyond his endeavors to define himself against the plebeians to a more wholesale self-defining against the patricians, whom he includes with the plebeians as people of lesser worth than himself:

> You [patricians/senators] are plebeians
> If they [plebeians] be senators; and they are no less
> When, both your voices blended, the great'st taste
> Most palates theirs.
>   By Jove himself,
> It makes the consuls base; and my soul aches
> To know, when two authorities are up,
> Neither supreme, how soon confusion
> May enter 'twixt the gap of both. (3.1.100–10)

Launched moments before this passage, Brutus's accusation against Coriolanus, "You speak o'th'people / As if you were a god to punish, not / A man of their infirmity" (3.1.79–81), reveals that others are aware that Coriolanus thinks himself of divine stature as well as one of Rome's "Real necessities" (3.1.146). Coriolanus informs Sicinius that he himself, "Shall remain! / Hear you this Triton of the minnows? Mark you / His [meaning his own] absolute 'shall?'" (3.1.87–88). Yet to claim absolute power like James I was unacceptable in ancient Rome. This probably explains Cominius's abrupt response to Coriolanus's comment: "'Twas from the canon" (3.1.89). As Samuel Johnson says, and Brockbank agrees with his view, Coriolanus's remark "was contrary to the established rule; it was a form of speech to which he has no right" (Brockbank 200).

Coriolanus's determination always to present himself superior to others through the arbitrary negation and denigration of others' works to assert a self-constancy that he imagines presumptuously for himself. It is this steadfast determination, however, that makes it both impossible and possible for Coriolanus to engage in what I call "transversal movement." People move transversally when they cross the boundaries of their subjectification through the experience and deliberation of alternative perspectives and emotions, that is, when they breach and move beyond the parameters imposed onto them socially, conceptually, and/or physically by any organizational social structure. Such movements require engagement with and at the very least a passing through "transversal territory," which is "the non-subjectified region of one's conceptual territory [of one's imaginative range]. It is entered through the transgression of the conceptual boundaries and, usually by extension, the emotional boundaries of subjective territory" (Reynolds 149). Coriolanus's transversal movement occurs in the contradictory space of incompatibility between his investment in the Roman state (his duty) and his own self-interestedness. The more constraints are imposed onto him the more he actively slips unpredictably into this space in-between, this liminal space that is transversal to official culture, and thus challenges the subjective territories of both the plebeians and the Tribunes. Yet, he is consistent in his articulation of his self-concept, a self-concept he could only achieve by moving transversally beyond the parameters of his state-prescribed identity as devoted warrior/politician and consequently in opposition to Rome's organizational structure. Despite the tragic consequences in the play, it is precisely this consistency that some critics celebrate, such as Bradley (1912), Ide (1980), and Rossiter (1961). For these critics, the appeal of Coriolanus's resolute individualism supersedes its ultimate destructiveness to both Coriolanus and the state (I will return to this point). Nevertheless, with the character of Coriolanus, regardless of however praiseworthy his consistency might be, Shakespeare's play criticizes the idea of monarchical supremacy and the doctrine of Divine Right. Coriolanus is representative of a monarch or dictator whose egotism makes him dangerous to his nation. His obvious military usefulness is undermined by his arrogance.

The "common people" of *Coriolanus* and *Coriolan* initially give Coriolanus their vote for consul, and they do this because of his impressive military achievements. Then, after considering what they have done, Shakespeare's plebeians realize that they permitted, as Sicinius declares, an "ignorant election" (2.3.217), and Brecht's citizens never allow the election to become final (101). When Coriolanus opposes the offering of corn gratis and verbally assaults the "common people" for the last time without repercussions (3.1.120–70; cf. *Coriolan* 101–02), his egotism and sheer hatred for them are made explicit; it is confirmed that he is a "Fast foe to th'plebii" (2.3.182), their "fixed enemy" (2.3.248). "The people are incens'd against him" (3.1.30) and they determine that he deserves "present death" (3.1.209; cf. *Coriolan* 104). The contradictions in Coriolanus's character are thus reflected in the people's conflicting needs. They want protection against alien armies, and this is why they want Coriolanus. However, freedom from war means little to people who are starving and oppressed, circumstances that Coriolanus actively promotes.

By causing the audience to identify with the "common people," as discussed earlier, the audience is made to move transversally into the contradictory space as well. The subjective territories of its members, as the plebeians would have it, are made to "lay all flat" (3.1.196; cf. *Coriolan* 104). They are equalized and blurred into the subjective territories of the plebeians to the extent that the audience's members expand their own conceptual and emotional range. This allows them to relate to the direness of the situation. Thus, the audience also censures Coriolanus for not acting as the people want, for not being transversal enough. Acting is transversal. It requires movement outside of one's own subjective territory and entrance into the subjective territory of someone else, even if that someone is only a fictional character. Such engagement exposes the actor to thoughts and emotions that might differ hugely from those he or she experiences in everyday life. With this exposure, comes challenge, expansion, and possibly significant transformation of the actor's own subjective territory. To uphold his radically individuated subjective territory and possess state-sanctioned sociopolitical power, Coriolanus must simultaneously maintain his own sense of self and yet transverse it to meet, much less comprehend, the demands of the plebeians and the Tribunes.

It is this paradox that prevents Coriolanus from achieving the disciplinary control of the Roman populace that he so aggressively pursues. The people realize that he "affects / Tyrannical power" (3.3.1–2) and charge him appropriately:

> you have contriv'd to take
> From Rome all season'd office, and to wind

Yourself into a power tyrannical;
For which you are a traitor to the people. (3.3.63–66; cf. *Coriolan* 111)

The people decide that Coriolanus is a traitor and determine his punishment by the democratic method: their conclusions are "Set down by th'poll" (3.3.9) and presented through "chairmen of the electoral districts" who carry "a list of all the voters they represent" (*Coriolan* 109; cf. *Coriolanus* 3.3.7–11). Again, as in the first scene, the audience is invited to experience the uniting of the plebeians in their communal action against their oppression and in their progress toward establishing a democratic republic. All the plebeians exclaim, "Our enemy is banish'd! He is gone! Hoo! Hoo!" and the stage direction reads, "They all shout, and throw up their caps" (3.3.137; cf. *Coriolan* 113). This marks the third victory for the "common people." Cominius, Menenius, and the other patricians—who make many remarks suggesting that Coriolanus is out of line—now accommodate (though not reluctantly in Brecht) the people's wish to banish Coriolanus, just as they do earlier by appointing the people Tribunes: they "all forsook me" says Coriolanus to Aufidius (4.5.77; cf. *Coriolan* 123).

Only Coriolanus does not acknowledge his defeat. Instead, pathetically and desperately, he retorts: "I banish you" (3.3.123, 113). He predicts: "I shall be lov'd when I am lack'd" (4.1.15), which can be interpreted as Brecht's Coriolanus says, "They'll love me when they need me" (114). To prove this, Coriolanus joins the Volscians and marches with them against Rome, and in Shakespeare, as well as in Brecht, this further illuminates the plebeians' conflicting attitudes toward Coriolanus. They are panic-stricken and some regret Coriolanus's banishment. Yet before Volumnia convinces Coriolanus not to fight, the majority of Brecht's plebeians report for military duty and are given arms by Cominius (137). In Shakespeare, Coriolanus is quickly persuaded by his mother and we are never permitted to see the outcome of the people's confusion. Still, Coriolanus's behavior shows that his pathological individuality, exacerbated by his quest to prove his masculinity—that he does not have "a pipe / as Small as an eunuch" and "schoolboy's tears" (3.2.113–16; cf. *Coriolan* 108)—has far exceeded society's expectations. To demonstrate his indispensability and magnanimous manliness, Coriolanus betrays his family, others who care about his well-being, and the nation's "common people"; his wilful jeopardizing of their lives—given that such jeopardizing contradicts his sociopolitical ideals—testifies to the great extent to which he has moved transversally beyond the parameters according to which he is expected to live and operate.

In the end, however, we learn once and for all that Coriolanus is hardly the individual he imagines himself to be. As Brecht's Volumnia tells him, "You are no longer indispensable" (142), and Shakespeare's Aufidius keenly observes that he is not the quintessential embodiment of masculine prowess but rather a mere "boy of tears" (5.6.100). The case of Coriolanus suggests that martial expertise, sociopolitical acumen, and emotional intelligence are both related and evaluated relatively within the world of the play. In addition to *Coriolanus*'s sympathizing with the "common people" and its critique of sociopolitical indispensability, as Dollimore argues, "a radical *political* relativism is advanced" (229). In his soliloquy concerning the fortunes of Coriolanus, Aufidius concludes,

> our virtues
> Lie in th'interpretation of our time,
> And power, unto itself most commendable,
> Hath not a tomb so evident as a chair
> T' extol what it hath done. (4.7.49–53; cf. *Coriolan* 132)

Here Aufidius speaks of people's "virtues" (rather than, say, reputation) as socially constructed (Dollimore 229). This opposes the view held by the early modern Church of England that people's virtues are essential attributes bestowed by God. Again, Shakespeare seems to have anticipated Karl Marx. In his preface to *A Contribution to a Critique of Political Economy*, Marx maintains, "It is not the consciousness of men that determines their existence, but on the contrary it is their social existence that determines their consciousness" (11–12). Marx reiterates this in *The German Ideology*: "Consciousness is from the very beginning a social product, and remains so as long as men exist at all" (*Marx-Engels Reader* 158). For Marx, consciousness is produced socially and exists in our relationships to our environment; it is through consciousness that we interpret these relationships, relatively, as individuals who live in society. Both Marx and Aufidius claim that our virtues are sociohistorically determined rather than inherently possessed or furnished by God.

However controversial Aufidius's view would seem to have been, because there is no decisive proof for a production of *Coriolanus* during Shakespeare's own lifetime, there is no way for us to know how Coriolanus was originally received by its Jacobean audience. Albeit the stage directions show the play ready for the theater, the play may not have been staged then. It may have been censured for its politics. The radical potential we have seen in the play, a potential that exists in

Coriolanus prior to any alterations made by Brecht—consider Brecht's own query regarding *Coriolanus*, the last entry in his research notes, "Couldn't one do it just as it is, only with skilful direction?" (qtd. in Heinemann 219)—powerfully emphasizes the value of reconsidering established readings of Shakespeare's plays. As we have seen in the case of *Coriolanus*, it is especially important to (re)examine those plays that have been used to support political ideologies. To further stress this point, I want to conclude this analysis by returning briefly (as I promised several pages ago) to the celebration of the character Coriolanus by some critics. Rather than discuss the varying opinions of critics, however, I want to conclude with the words of one influential critic, Nazi supporter H. Hüsges, from his introduction to a 1934 German edition of *Coriolanus*. Hüsges's reading epitomizes the kinds of political employment of Shakespeare about which I am most concerned; it justifies my pleas that political criticism and use of Shakespeare needs to be taken seriously. According to Hüsges:

> The meaning of this last and most mature of Shakespeare's works for the new Germany lies in the heroic features inherent in it. The poet treats the problem of *Volk* [people] and *Führer* [leader]; he portrays the true nature of a Führer in contrast to the indiscriminating masses: he portrays a misled *Volk*, a false democracy whose representatives give in to the wishes of the *Volk* for the sake of egotistical goals. The figure of the true hero and Führer Coriolanus towers high above these weaklings; he wants to lead the misdirected *Volk* to recovery, as Adolf Hitler wants to lead our beloved German Fatherland today. (157)[21]

Notes

1. Bertolt Brecht, "Big Time, Lost" in *Die Gedichte von Bertolt Brecht in einem* (Austria: Suhrkamp, 1984), 1010. Translated by Nina Venus and Bryan Reynolds.
2. To avoid confusion, I will refer to Shakespeare's play and its protagonist as Coriolanus (like the traditional Latin spelling in Shakespeare), while Brecht's version will be called *Coriolan* (as in German), yet its protagonist will still be referred to as Coriolanus.
3. On censorship and the politics of early modern English drama, see: Martin Butler, *Theatre and Crisis 1632–1642* (Cambridge: Cambridge University Press, 1984); Janet Clare, "'Greater Themes for Insurrection's Arguing': Political Censorship of the Elizabethan and Jacobean Stage," *RES* 38 (1987): 169–83; Philip J. Finkelpearl, "'The Comedians Liberty': Censorship of the Jacobean Stage Reconsidered," *ELR* 16 (1986): 123–38, and "The Role of the Court in the Development of Jacobean Drama," *Criticism* 24 (1982): 138–58; and Alvin Kernan, *Shakespeare, the King's Playwright: Theater in the Stuart Court 1603–1613* (New Haven: Yale University Press, 1995).

4. See V.R. Berghahn, *Modern Germany: Society, Economy and Politics in the Twentieth Century* (Cambridge: Cambridge University Press, 1989), 217–19.
5. For a thorough description of the uprising of June 16, 1953 see Henry Ashby Turner, Jr., *The Two Germanies Since 1945* (New Haven: Yale University Press, 1987), 116–24.
6. Though the exact date that Brecht stopped revising *Coriolan* is unknown, most scholars agree that it was between 1952 and 1955. See Darko Suvin, *To Brecht and Beyond: Soundings in Modern Dramaturgy* (New Jersey: Barnes and Noble, 1984), 200; Margot Heinemann, "How Brecht read Shakespeare" in *Political Shakespeare: New Essays in Cultural Materialism*, ed. Jonathan Dollimore and Alan Sinfield (Ithaca: Cornell University Press, 1985), 221; and John Willett, *The Theatre of Bertolt Brecht* (London: Metheun, 1977), 63, 121.
7. In his 1979 biography of Brecht, *Brecht* (London: Marion Boyars), Klaus Volker provides an insightful investigation of Brecht's affinity for Marxism and his interest in establishing a Communist regime in Germany. See also Willett, *The Theatre of Bertolt Brecht*.
8. "Die Lösung," *Die Gedichte von Bertolt Brecht in einem Band* (Frankfurt: Suhrkamp, 1981), 1009–10. Translated by Nina Venus and Bryan Reynolds.
9. See note 6, and Bertolt Brecht, *Brecht on Theatre: The Development of an Aesthetic*, ed. and trans. John Willett (New York: Hill and Wang, 1964).
10. For a discussion of the date of composition and original production of Shakespeare's play, see E.C. Pettet, "Coriolanus and the Midlands Insurrection of 1607," *Shakespeare Survey* 3 (1950): 34–42; Philip Brockbank in Shakespeare, *Coriolanus*, ed. Philip Brockbank (London: Methuen, 1976), 24–29; and Annabel Patterson, *Shakespeare and the Popular Voice* (Cambridge: Basil Blackwell, 1989), 135–46.
11. As noted by E.C. Pettet in "*Coriolanus* and the Midlands Insurrection of 1607," the insurgents that participated in the Midlands revolt were "commonly described as 'Diggers' and 'Levellers,'" which are the same names that "not more than a generation later" were perpetuated by those radicals whose "ideas so shocked the Commonwealth leaders" (35). For an account of the roles that people called Diggers and Levellers played in the English revolution of the 1640s, see Christopher Hill, *The World Turned Upside Down: Radical Ideas During the English Revolution* (Harmondsworth: Penguin, 1975), 126–28.
12. For more on the damage caused by the early modern protesters, see Pettet, "*Coriolanus* and the Midlands Insurrection of 1607," 34.
13. For a detailed investigation of Shakespeare's sources, see Philip Brockbank, ed., William Shakespeare, *Coriolanus*, 29–35.
14. Philip Brockbank, ed., William Shakespeare, *Coriolanus* (London: Methuen, 1976); Samuel Taylor Coleridge, *Biographia Literaria* (New York: J.M. Dent, 1947); Willard Farnham, *Shakespeare's Tragic Frontier* (Berkeley: University of California Press, 1963); Günter Grass, *The Plebeians Rehearse the Uprising*, trans. Ralph Manheim (New York: Harcourt, Brace and World, 1966); Edwin Honig, "*Sejanus and Coriolanus:* A Study in Alienation," *Modern Language Quarterly* 12 (1951): 407–21; C.C. Huffman, *Coriolanus* in

Context (Lewisburg: Bucknell University Press, 1971); Richard Ide, *Possessed with Greatness: The Heroic Tragedies of Chapman and Shakespeare* (London: Scholar Press, 1980); Samuel Johnson, ed., *The Plays and Poems of Shakespeare* (Philadelphia: Bioren and Madan, 1795–1796); M.W. MacCallum, *Shakespeare's Roman Plays, and Their Background* (London: Macmillan, 1967); E.C. Pettet, "*Coriolanus* and the Midlands Insurrection of 1607," *Shakespeare Survey* 3 (1950): 34–42; James Emerson Phillips, *The State of Shakespeare's Greek and Roman Plays* (New York: Columbia University Press, 1940); Norman Rabkin, *Shakespeare and the Common Understanding* (New York: Free Press, 1967); A.P. Rossiter, *Angel With Horns*, ed. Graham Storey (London: Longmans, 1961); J.L. Simmons, *Shakespeare's Pagan World: The Roman Tragedies* (Charlottesville: University Press of Virginia, 1973); W. Gordon Zeeveld, "*Coriolanus* and Jacobean Politics," *Modern Language Review* 57 (1962): 321–24.
15. Zvi Jagendorf's "*Coriolanus*: Body Politic and Private Parts," *Shakespeare Quarterly* 41 (1990): 455–69, discusses the play's preoccupation with body metaphors without making any explicit claims about the play's politics. Martin Scofield's "Drama, Politics, and the Hero: *Coriolanus*, Brecht, and Grass," *Comparative Drama* 24 (1990–1991): 322–41, discusses the aesthetic merits of Brecht's adaptation without addressing the politics of Shakespeare's play.
16. All quotations from Shakespeare's *Coriolanus* follow the new Arden edition, ed. Philip Brockbank (London: Methuen, 1976), and all quotations from Brecht's *Coriolan* are taken from *Collected Plays*, ed. Ralph Manheim and John Willett (New York: Random House, 1973). In addition, from here on, when citing Shakespeare's text I will give the act, the scene, and the line numbers, and I will refer to Brecht's play by page numbers only. Also, because Brecht modernizes Shakespeare's language for his play, to avoid needless repetition of lines I will cite Shakespeare and then give the page number(s) of the corresponding text in Brecht; for example (2.2.137; cf. *Coriolan* 93).
17. Quoted from the following unpublished conference paper: Donald K. Hedrick, "Male Surplus Value." Presentation at Folger Library Shakespeare Colloquium, Washington, D.C., April, 1999.
18. For discussion on similarities between Coriolanus and various people of Shakespeare's day, see P.A. Jorgensen, "Shakespeare's Coriolanus: Elizabethan Soldier," *PMLA* 64 (1949): 221–35.
19. Suvin translated these lines from Henning Rischbieter, *Brecht II* (Velber, 1966): 75.
20. See Bryan Reynolds, "The Devil's House, 'or worse': Transversal Power and Antitheatrical Discourse in Early Modern England," *Theatre Journal* 49.2 (1997): 143–67, and my introduction to this collection.
21. Translated by John Rouse and Bryan Reynolds.

Works cited

Adelman, Janet. "'Anger's My Meat': Feeding, Dependency, and Aggression in *Coriolanus.*" *Representing Shakespeare*. Ed. Murray M. Schwartz and Coppelia Kahn. Baltimore: Johns Hopkins Press, 1980. 129–49.

Berghahn, V.R. *Modern Germany: Society, Economy and Politics in the Twentieth Century*. Cambridge: Cambridge University Press, 1989.

Bourdieu, Pierre. *The Field of Cultural Production: Essays on Art and Literature*. Ed. and trans. Randal Jonson. New York: Columbia University Press, 1993.

Bond, Ronald B., ed. *Certain Sermons or Homilies (1547) and A Homily against Disobedience and Wilful Rebellion (1570): A Critical Edition*. Toronto: University of Toronto Press, 1987.

Bradley, A.C. "*Coriolanus*: British Academy Lecture 1912." *A Miscellany*. London: Macmillan, 1929.

Brecht, Bertolt. "Big Time, Lost." *Die Gedichte von Bertolt Brecht in einem*. Austria: Suhrkamp, 1984. 1010.

———. "Die Lösung." *Die Gedichte von Bertolt Brecht in einem Band*. Frankfurt: Suhrkamp, 1981. 1009–10.

———. *Coriolanus. Collected Plays*. Vol. 9. Ed. Ralph Manheim and John Willett. Trans. Ralph Manheim. New York: Random House, 1973. 57–146.

———. *Brecht on Theatre: The Development of an Aesthetic*. Ed. and trans. John Willett. New York: Hill and Wang, 1964.

Bristol, Michael D. "Lenten Butchery: Legitimation Crisis in *Coriolanus*." *Shakespeare Reproduced: The Text in History and Ideology*. Ed. Jean E. Howard and Marion F. O'Connor. London: Methuen, 1987. 207–22.

Brockbank, Philip, ed. *Coriolanus*. By William Shakespeare. London: Methuen, 1976.

Butler, Martin. *Theatre and Crisis 1632–1642*. Cambridge: Cambridge University Press, 1984.

Chamberlain, John. *The Letters of John Chamberlain*. 2 Vols. Ed. N.E. MacClure. Philadelphia: American Philosophical Society, 1939.

Clare, Janet. " 'Greater Themes for Insurrection's Arguing': Political Censorship of the Elizabethan and Jacobean Stage." *RES* 38 (1987): 169–83.

Coleridge, Samuel Taylor. *Biographia Literaria*. New York: J.M. Dent, 1947.

Deutscher, Isaac. *Stalin: A Political Biography*. New York: Oxford University Press, 1949.

Dollimore, Jonathan. *Radical Tragedy: Religion, Ideology and Power in the Drama of Shakespeare and His Contemporaries*. Chicago: University of Chicago Press, 1984.

Farnham, Willard. *Shakespeare's Tragic Frontier*. Berkeley: University of California Press, 1963.

Finkelpearl, Philip J. " 'The Comedians Liberty': Censorship of the Jacobean Stage Reconsidered." *ELR* 16 (1986): 123–38.

———. "The Role of the Court in the Development of Jacobean Drama." *Criticism* 24 (1982): 138–58.

Gay, E.F. "The Midlands Revolt and the Inquisitions of Depopulation of 1607." *Transactions of the Royal Historical Society*. N.S.18 (1904): 195–244.

Gildon, Charles. "The Argument of *Coriolanus*." *Coriolanus: Critical Essays*. Ed. David Wheeler. New York: Garland Press, 1995. 7–10.

Grass, Gunter. *The Plebeians Rehearse the Uprising*. Trans. Ralph Manheim. New York: Harcourt, Brace and World, 1966.

Hedrick, Donald K. "Male Surplus Value." Presentation at Folger Library Shakespeare Colloquium, Washington, D.C., April, 1999.
Heinemann, Margot. "How Brecht read Shakespeare." *Political Shakespeare: New Essays in Cultural Materialism.* Ed. Jonathan Dollimore and Alan Sinfield. Ithaca: Cornell University Press, 1985. 202–30.
Hill, Christopher. *The World Turned Upside Down: Radical Ideas During the English Revolution.* Harmondsworth: Penguin, 1975.
Honig, Edwin. "*Sejanus* and *Coriolanus*: A Study in Alienation." *Modern Language Quarterly* 12 (1951): 407–21.
Howard, Jean E. and Marion F. O'Connor, eds. *Shakespeare Reproduced: The Text in History and Ideology.* New York: Methuen, 1987.
Huffman, Clifford Chalmers. *Coriolanus in Context.* Lewisburg: Bucknell University Press, 1971.
Hüsges, H., ed. *Coriolanus.* By William Shakespeare. Braunschweig/Berlin/Hamburg, 1934.
Ide, Richard. *Possessed with Greatness: The Heroic Tragedies of Chapman and Shakespeare.* London: Scholar Press, 1980.
Jagendorf, Zvi. "*Coriolanus*: Body Politic and Private Parts." *Coriolanus: Critical Essays.* Ed. David Wheeler. New York: Garland Press, 1995. 229–50.
———. "*Coriolanus*: Body Politic and Private Parts." *Shakespeare Quarterly* 41 (1990): 455–69.
James I. *Minor Prose Works of King James VI and I.* Ed. James Craigie. Edinburgh: The Scottish Text Society, 1982.
Johnson, Samuel, ed. *The Plays and Poems of William Shakespeare.* Philadelphia: Bioren and Madan, 1795–1796.
Jorgenson, P.A. "Shakespeare's Coriolanus: Elizabethan Soldier." *PMLA* 64 (1949): 221–35.
Keegan, John. *The Mask of Command.* New York: Viking Press, 1987.
Kernan, Alvin. *Shakespeare, the King's Playwright: Theater in the Stuart Court 1603–1613.* New Haven: Yale University Press, 1995.
Kishlansky, Mark. *Parliamentary Selection: Social and Political Choice in Early Modern England.* Cambridge: Cambridge University Press, 1986.
MacCallum, M.W. *Shakespeare's Roman Plays, and Their Background.* London: Macmillan, 1967.
Marx, Karl. *A Contribution to the Critique of Political Economy.* London: International Library Publishing Co., 1904.
———. *The Marx-Engels Reader.* Second edition. Ed. Robert C. Tucker. New York: W.W. Norton and Company, 1978.
Marx, Karl, and Friedrich Engels. *The Communist Manifesto.* Intro. A.J.P. Taylor. Penguin: Harmondsworth, 1982.
———. *Wage-Labour and Capital and Value, Price, and Profit.* New York: International Publishers, 1976.
Patterson, Annabel. *Shakespeare and the Popular Voice.* Cambridge: Basil Blackwell, 1989.
Pettet, E.C. "*Coriolanus* and the Midlands Insurrection of 1607." *Shakespeare Survey* 3 (1950): 34–42.

Phillips, James Emerson. *The State of Shakespeare's Greek and Roman Plays.* New York: Columbia University Press, 1940.

Rabkin, Norman. *Shakespeare and the Common Understanding.* New York: Free Press, 1967.

Reynolds, Bryan. "The Devil's House, 'or worse': Transversal Power and Antitheatrical Discourse in Early Modern England." *Theatre Journal* 49.2 (1997): 143–67.

Rossiter, A.P. *Angel With Horns.* Ed. Graham Storey. London: Longmans, 1961.

Scofield, Martin. "Drama, Politics, and the Hero: Coriolanus, Brecht, and Grass." *Coriolanus: Critical Essays.* Ed. David Wheeler. New York: Garland Press, 1995. 251–72.

Simmons, J.L. *Shakespeare's Pagan World: The Roman Tragedies.* Charlottesville: University Press of Virginia, 1973.

Sorge, Thomas. "The Failure of Orthodoxy in *Coriolanus.*" *Shakespeare Reproduced: The Text in History and Ideology.* Ed. Jean E. Howard and Marion F. O'Connor. New York: Routledge, 1987. 225–39.

Stow, John. *Annales.* London: 1631.

Suvin, Darko. *To Brecht and Beyond: Soundings in Modern Dramaturgy.* New Jersey: Barnes and Noble, 1984.

Tillyard, E.M.W. *The Elizabethan World Picture.* London: Chatto and Windus, 1943.

Tricomi, Albert H. *Anticourt Drama in England 1603–1642.* Charlottesville: University Press of Virginia, 1989.

Turner, Henry Ashby Jr. *The Two Germanies Since 1945.* New Haven: Yale University Press, 1987.

Volker, Klaus. *Brecht.* London: Marion Boyars, 1979.

Wilson, John Dover, ed. *Coriolanus.* By William Shakespeare. London: New Shakespeare, 1960.

Zeeveld, Gordon W. "*Coriolanus* and Jacobean Politics." *Modern Language Review* 57 (1962): 321–34.

CHAPTER 5

UNTIMELY RIPPED: MEDIATING
WITCHCRAFT IN POLANSKI AND
SHAKESPEARE[1]

Bryan Reynolds

Shortly after midnight on August 9, 1969, as instructed by cult leader Charles Manson, Susan Atkins, Linda Kasabian, Katie Krenwinkel, and Charles Watson unlawfully entered the Hollywood Hills estate of director Roman Polanski and his movie star wife, Sharon Tate. To promote "Helter Skelter," a horrific dream scheme designed by Manson to effect ultimately a worldwide racial war ending in a white elite ruling over a black population, the four Manson family members set out to rob and brutally murder the inhabitants of the Polanski residence. Steven Parent, a friend of the estate's caretaker, was the first to die. As Parent attempted to leave the property in his car, he was suddenly, and in short order, shot, stabbed, and killed by Watson. Then, while Kasabian kept a lookout for people coming at the end of the driveway, Atkins, Krenwinkel, and Watson invaded the house, forced its occupants at knife-point into the living room, tied them together with nylon rope by their necks, and began stabbing and shooting them to death. One hundred and two stab wounds scattered the mangled corpses of Abigail Folger, Voityck Frykowski, Jay Sebring, and Sharon Tate, who was eight and a half months pregnant with her and Roman's first child. When the assassins completed their mission, they returned to Manson who was waiting for them at a bar.

In the spring of 1970, Polanski decided to make a cinematic adaptation of *Macbeth*, his first film since the 1969 tragedy. Already internationally acclaimed for an assortment of macabre films that culminated in the critical and financial success of *Rosemary's Baby* (1967), Polanski, and friend Andrew Braunsberg, had little difficulty producing *Macbeth*.

Well-known critic and playwright Kenneth Tynan collaborated with Polanski in writing the screenplay, and *Playboy* owner Hugh Hefner agreed to fund the project. The provocative production relationship of Hefner, Tynan, and Polanski, the nudity and gruesome violence in the film, and especially the film's association with recent events, along with a variety of less conspicuous factors, all caused *Macbeth* to be extremely controversial and surprisingly popular for a Shakespeare film. *Macbeth* was rated "R": for the first time Shakespeare, the traditional symbol of human greatness and poetic genius, was not only restricted culturally to those allegedly possessing the special ability to understand and appreciate Shakespeare, but also, via Polanski, restricted by law to those of an age presumed less susceptible to Shakespeare's adulterating or traumatizing potential. Shakespeare came to stand for something modern and ominous. Hailed by the press prior to its release as an ingenious and obscene bloodbath, "Macbeth" was viewed by an audience with a peculiar horizon of expectations. By writing to magazines across the country, for instance, some "Shakespeare lovers" even attempted to inspire a national boycott of the film.[2]

Through an examination of its cultural situation, narrative style, and cinematic structure, I hope to explain the controversy that surrounded Polanski's film. With both complex cinematic semiology and poignant sociohistorical mediation, *Macbeth* brings to a critical juncture the philosophies of the 1960s peace-love-revolution hippies, Vietnam war protestors and civil rights activists, the reified mainstream American populace, and the ruling conservatives, as well as society's preoccupation with aestheticization. Specifically, I want to argue that Polanski's *Macbeth* manifests what Gilles Deleuze calls the "crystal-image," and, by extension, realizes Antonin Artaud's "Theater of Cruelty," constituting, in effect, a terrorist intervention into a discursive cultural and ideological struggle. Ultimately, I shall argue that *Macbeth* itself, in totality, is a "crystalline narration" that is shot through with various allusions to actual and virtual circumstances particular to the cultural environments from which it initially emerged in Renaissance England and then reemerged in the United States of 1971. In the first of what is a three-part analysis, I compare these cultural environments. The second section explicates Deleuze's theories on cinema and discusses them in conjunction with an analysis of the film's Theater of Cruelty. Finally, I contemplate the film's sociopolitical implications.

Comparable contexts

In 1563, when Queen Elizabeth I passed her statute against "Conjurations, Enchantments, and Witchcrafts," there was more concern over witchcraft than ever before in England's history.³ Unlike Elizabeth, who was troubled chiefly with the treasonous implications of witchcraft and political prophecy (as the witches vaticinate in Shakespeare and Polanski), religious figures of both Left and Right, Calvinists and Romanists, perceived witchcraft and pagans, just as the 1960s conservatives perceived recreational drug use and the diversified 1960s counterculture, as encroaching threats to the ethical and sanctified fabric of human existence. King James I shared Elizabeth's dread of witchcraft as heretical and treasonous. In 1590, while James was king of Scotland, one Dr. John Fian was convicted for being the ringleader of a coven of witch-terrorists (as Charles Manson was frequently described in the media after his arrest). Fian and several other people, mostly women, were executed for conspiring with Satan to sink the ship carrying James to Norway to fetch his future queen.⁴ In 1597, James published his *Daemonologie*, which refutes Reginald Scot's skeptical account of witchcraft and witch trials, entitled *The Discoverie of Witchcraft* (1584).⁵ Scot believed in the existence of witches, but not in the corporeal relationship between witches and the Devil that was a reality for James. James thought *Daemonologie* so important that after he became king of the united realms of Scotland and England, he republished it in 1603, 1604, 1607, and 1616. And, in 1604, he imposed in a new witchcraft statute, more austere punishments for the same offenses delineated in the Elizabethan act.⁶ James's belief in witches and the necessity of their extermination was reflected in the hundreds of witch hunts and witch trials performed throughout his reign. In a letter to his friend Robert Cecil (1605), James alludes to the predicament: "My little beagle... I have been out of privy intelligence with you since my last parting for having been ever kept so busy with hunting witches."⁷

In the midst of England's witch craze, Shakespeare's *Macbeth* was written (between 1603 and 1606) and performed, among other places, in the presence of King James.⁸ To perform a play during the English Renaissance and especially before royalty was an enterprise not to be taken lightly, for politically offensive material could yield harsh punishments, including death.⁹ Shakespeare, however, seems to have considered more than censorship; *Macbeth* is a straightforward appeal to James's sensibilities. Apart from the connection between the witch craze and *Macbeth*, as Kenneth Muir observes, the play's topicality and allegiance to James are exemplified by its allusions to the Gunpowder Plot, which was a conspiracy to blow up the English Parliament and James on November 5, 1605.¹⁰ A notable deviation from Shakespeare's main source, Holinshed's *Chronicles of Scotland*, reveals another obvious effort to please James: James's ancestor, Banquo, is not depicted as Macbeth's conspirator as recorded by Holinshed.[11] Yet Shakespeare does include the implication that Banquo's descendants "shall gouern the Scottish kingdome by long order of continuall descent."[12] Superseding these somewhat superficial alliances or concessions are the far more significant sociopolitical connotations of Shakespeare's witches, particularly as their relevance is refigured in Polanski's *Macbeth*.

The public's heterogeneous responses to certain aspects of *Macbeth* are eclipsed by the Manson murders as a focus for interpretation. Every review of *Macbeth* of which I am aware makes reference to the butchery of Polanski's wife, unborn child, and friends. To quote a few: Roger Ebert says, "It is impossible to watch certain scenes without thinking of the Charles Manson case. It is impossible to watch a film directed by Roman Polanski and not react on more than one level to such images as a baby 'untimely ripped from his mother's womb'" (32); Paul Zimmerman points out that "parallels between the Manson murders and the mad, bloody acts of these lost Macbeths keep pressing themselves on the viewer" (59); and Kenneth Rothwell notices that "hovering in the background, obscuring the Bard slightly, were the gruesomeness of the Manson cult murders . . . and the new interest in witchcraft (shades of James I!), helped along by Polanski's own 'Rosemary's Baby'"—a chilling movie about a woman whose husband involves her in a witchcraft cult and uses her body against her will to give birth to Satan's child (71). "Rosemary's Baby," and many other 1960s horror films, mirror the revival of interest in witchcraft during the decade (as depicted in Oliver Stone's film *The Doors* [1991]). Leslie Fiedler writes at the end of the 1960s:

> But not until recently has there reappeared a growing sect of witches ["whether in the interests of black mysticism, practical magic, women's liberation, or some combination of the three"], propagandizing, proselytizing, committing what seem to alien and unsympathetic eyes atrocities, mass orgies, and ritual murders, in order to establish their identity and express their contempt for the nonwitch world. Once more witchcraft has moved from the occasional Sunday supplement to daily headlines, becoming, as it was in Shakespeare's day, news and the stuff of living literature. (64)

"I show people something so obviously impossible as witchcraft," states Polanski in a 1971 interview, "and I say to them, 'Are you certain it is not true?' Too many people accept things for certainty. . . . I want people to question certainties" (*New York Times Magazine* 79). But why does Polanski want people to "question certainties"? Why choose "something so obviously impossible as witchcraft" as the questionable subject matter? And by what means does he expect his film to provoke questions? Michel Foucault's conceptualization of the elusive network of power–knowledge relations integral to the maintenance of most social structures reverberates in Stephen Greenblatt's provocative assertion that "one of the highest achievements of power is to impose fictions upon the world and one of its supreme pleasures is to enforce the acceptance of fictions that are known to be fictions" (141). The thematic parallelism between Greenblatt's statement and Polanski's stated intentions encourages inquiry into the actual role played by Polanski's *Macbeth* in the distribution of power and cultural anxiety. Regardless of whether Shakespeare shared Polanski's motivation, for a Jacobean audience living in a world in which witchcraft was a terrible reality, the conceivability of *Macbeth* must have been frightening. Did Polanski want his audience to experience a comparable reality? If so, for what purpose?

Renaissance England's alleged witches were usually women with unconventional characteristics. Like the eclectic 1960s counterculture, they may have been weird or rebellious in their demeanor or ideas, introverted, conventionally unattractive, or socially stigmatized for any number of reasons; they may have even in their innermost self-consciousness considered themselves witches. The following lyrics of Manson's speak for many (like Manson) who have had eccentricity biologically and/or socially thrust upon them, but (unlike Manson) are unjustly persecuted for their idiosyncrasies:

> People say I'm no good
> But never never do they say
> Why their world is so mixed up
> Or how it got that way
> They all look at me and they frown
> Do I really look so strange
> If they really dug down in themselves
> I know they'd want to change
> Everybody says you're no good
> 'Cause you don't do like they think you should. (qtd. in Wizinski 199)

Both Polanski's film and Shakespeare's play invite their audiences to question the dangers of the unconventionalities represented by their witches. Actually, they both appear to be trying to affirm the existence of witchcraft and augment the terror associated with this possibility. The supernatural powers and magic potions utilized by the witches of Polanski and Shakespeare reinforce conservative conceptions of the dangers of alternative lifestyles and marginalized people, as well as the fears of adulteration and identification that inform them (Manson said to the public in his trial testimony: "I am just a reflection of every one of you" [qtd. in Wizinski 142–43]).

In Polanski, the witches might also stand for 1960s arts, drugs, and countercultural politics gone too far. They suggest that counterculture will corrupt or destroy the established social structure if not controlled or crushed. If so, *Macbeth* was not alone in its apparent sentiments toward counterculture, specifically occultism, hippiedom, and feminism. There were numerous films (such as John Wayne's *The Green Berets* [1968] and Clint Eastwood's *Dirty Harry* [1971]) and television programs (such as the space-hippie "The Way to Eden" episode of Star Trek [1969]) that were hostile toward and mocked counterculture. A mainstream literary example is a 1969 *Time Magazine* article, "New Feminists: Revolt against Sexism," that describes patronizingly the consciousness-raising groups of the "new feminists" as meeting in "covens" (54). Although the article makes no mention of them, there was in fact a national militant guerrilla-theater squad called WITCH: Women's International Terrorist Conspiracy from Hell. Many countercultural groups (like the Manson family) promulgated an affinity for witchcraft and occultism, but hardly any of them protested or committed crimes in their names.

Manson and his followers have been recurrently portrayed by journalists, talk-show hosts, politicians, and various spokespeople, as an inevitable consequence of the radical 1960s (Macbeth is described similarly by Macduff and Malcolm as influenced or possessed by a presumably already existing "Devil" and "Evil" [cf. 4.3]). Consider journalist Joan Didion's recount of the historical moment:

> On August 9, 1969, I was sitting in the shallow end of my sister-in-law's swimming pool in Beverly Hills when she received a telephone call from a friend who had just heard about the murders at Sharon Tate Polanski's house on Cielo Drive. . . . Black masses were imagined, and bad trips blamed. I remember all of the day's misinformation very clearly, and I also remember this, and I wish I did not: *I remember that no one was surprised.* [. . .]. Many people I know in Los Angeles believe that the Sixties ended abruptly on August 9, 1969, ended at the exact moment when word of the murders on Cielo Drive traveled like brushfire through the community, and in a sense this is true. The tension broke that day. The paranoia was fulfilled. (42; 47)

Conservatives said "I told you so," and blamed a supposedly ignorant population for making the heinous crimes possible. They claimed that preventive measures should have been taken; stricter drug laws should have been enforced; someone should have cracked down on the "hippie-freaks": as King Duncan permits the infiltration (through Macbeth) of the witches, the Devil, and evil into his Scottish realm, the American populace allegedly brought Manson upon themselves. And if the public was not responsible, who was? Some people even had the audacity to suggest that Polanski and Tate somehow inspired the murders.

The murders were reviewed originally in light of Polanski's films. Movie critic Pauline Kael recalls:

> Even though we knew that Roman Polanski had nothing whatever to do with causing the murder of his wife and unborn child and friends, the massacre seemed a vision realized from his nightmare movies. (76)

Prosecutor of the Tate-LaBianca trials and coauthor of *Helter Skelter: The True Story of the Manson Murders*, Vincent Bugliosi, writes:

> In the news stories following the Tate murders, reporters were quick to note that in *Repulsion* [an earlier Polanski film] Miss Deneuve went mad and murdered two men, while in *Cul de Sac* [Polanski's first international hit] the inhabitants of an isolated castle each meet a bizarre fate until only one man is left alive [as the caretaker remained alive at the Polanski residence, which was coincidently located in a cul de sac]. (27)

Outraged, Polanski says in an interview:

> There's no way out, is there? What happened was reviewed in terms of my films. Now it's vice versa. Now my films are reviewed in terms of what happened. If I did a comedy, aha! (*New York Times Magazine* 64)

When details were made known about the extravagance of Polanski's personal life and the drugs discovered on the Polanski premises after the murders, along with the information that two of the victims, Folger and Frykowski, had a substantial amount of the hallucinatory drug MDA in their systems when they were murdered, the press shifted responsibility for the murders onto the victims. Rumors of Polanski's own Satanism and associated perverted practices performed at his home also contributed to the blame-the-victims conception. After all, for most people, this information aligned Polanski and his friends with the feared counterculture. Polanski remembers:

> It was a modern version of a superstition story. Suddenly Sharon became responsible for her own death. The press was despicable. They talked

about orgies at the house, sadism. They sensationalized something that was already sensational. . . . It is like the story of the beautiful girl in the village. It's alright with villagers as long as there's no hoof-and-mouth disease. But once an epidemic breaks out, she's blamed. (*New York Times Magazine* 64)

As far as the majority of America was concerned, it was not "alright" to be different; it was not "alright" to be a member of any countercultural group in the 1960s. As English Renaissance women were accused of witchcraft and held accountable for almost anything that had gone wrong in the community, Polanski and Tate were held responsible for an atrocity that was not even intended for them; Manson was not aware that Polanski was renting the house at the time.

There was a blurring of borders between theatricalized terrorism and Manson's real-life terrorism. An imaginary correlation was fabricated between art and reality. The horror of the reality of the Manson murders became something apprehensible and familiar, however elided, displaced, or reconstructed. The commodification of this horror was symptomatic as a culmination of its particular styles of aestheticization and fetishization. It was transformed into journalistic capital, which desensitizes through sensationalism and tautological reproduction. It quickly acquired gossip status, and it was reproduced and distorted in a variety of artistic and animated forms (such as in paintings, sketches, comic-strips, posters, songs, and on t-shirts).[13] This aestheticization converted the once empathy-arousing nature of that horror into a fetishized product for popular consumption. Yet, in a strange way, Polanski's *Macbeth* seems to frustrate and reverse this process. By re-aestheticizing the horror in an extraordinary manner, *Macbeth* aspires to resensitize its audience by de-reifying the horror as commodity-fetish. It does this through its use of the crystal-image and correlative creation of a cinematic Theater of Cruelty.

The crystalline effect

In his cinematic theory, Deleuze analyzes "images" (camera shots and shot sequences), and differentiates between the "movement-image," the "action-image," and the "time-image." Movement-image refers to the interaction among variable elements (characters and objects) that are not the focus in a scene, of the camera's gaze: we watch background characters snicker at a romantic moment between two lovers. Action-image refers to the interaction between elements that are the camera's focus and other elements, those that are periphery or simply not the focus: we watch a character (an action-hero) single-handedly overcome an onslaught of "bad guys." Movement-images and action-images, as the mobile sections of a cinematographic duration, typically indicate sequential or causal connections between the objects or events depicted: a character shoots a gun and another falls, objects or characters depart from point A and arrive at B. Both movement-images and action-images represent time as actual, as contiguous, passing, and measurable. To do this, they must achieve a plausible linkage and seemingly chronological coherence of sensory-motor relations within the image (among objects and characters) and between the image and the viewer (the viewer must not be confused or doubtful of the order of events represented).

In contrast, time-image refers to a virtual presentation of time, when the image's elements and events appear simultaneous, random, or simply without chronology. This is achieved only through a frustration or breakup of the image's sensory-motor schemata (the expected and tractable interactions among characters and objects) in the form of a purely optical and sound situation. Composed of apparently chance relations, the time-image does not extend logically or comfortably into action or reaction, any more than it is stimulated by these processes: we see incoherent montage cuts and sequences, dissolves, deframings, superimpositions, intricate camera movements, and so on. Whereas there can be, and usually is, interaction among focused and periphery elements within a time-image, the image is predominately nonchronological and amorphous or hallucinatory, such that the internal movement cannot be centered or contextualized with certainty in relation to other elements (cuts or objects) within the time-image. Because of its inherent ambiguity, the time-image must be "read" as much as it is seen (and heard); it produces what Deleuze calls "any-space-whatever": disconnected space that demands contemplation and allows for greater freedom of interpretation, and thus challenges the viewer's analytical capacity.

The effect accounted for by Deleuze's "any-space-whatever" has the potential to achieve what I call "transversal territory," which is the nonsubjectified region of one's conceptual and emotional range.[14] It is entered through the transgression of the conceptual boundaries, sometimes imagined as tangible ones, and, by extension, the emotional boundaries of what I refer to as "subjective territory": the scope of the conceptual and emotional experience of an individual within any hegemonic society or subsociety. Transversal territory is where people go conceptually and emotionally when they venture, through what I call "transversal movements," beyond the boundaries of their own subjective territory and experience alternative thoughts and feelings. A common means by which transversal movements occur is through "subjunctive space," the hypothetical space of both "as if" and "what if," which is an in-between space operative between subjective territory and transversal territory.[15] In subjunctive space, unlike in transversal territory, people necessarily maintain agency and can self-consciously imagine scenarios and experiences, thereby self-activating their own transversal movements. Subjunctivity enables processes of empathizing (or attempting to emphasize) with others, which is to say, thinking and feeling atypically and "like" someone else. Similar to empathizing, people can engage in transversal movements by performing the unfamiliar social identity of another, by acting "as if" they were someone else. While moving transversally, people pass into, and usually through, transversal territory, in their journeys into and out of the subjective territories of others. The time-image, insofar as it creates the conditions for an any-space-whatever, engages with alternative subjectivities and encourages transversal movements.

As a potent mode of expression, the time-image frequently emerges at pivotal moments in the filmic narrative. This is often done, as in Polanski's *Macbeth*, in the form of a "recollection-image" or a "dream-image," such as in a flashback, premonitory, fantasy, or hallucinatory sequence. Recall Macbeth's hallucinatory observance of the "air-drawn dagger" (3.4.61) and his nightmare in which he is murdered by Banquo and his son; both of these images create any-space-whatever, the former by means of a feeling of timelessness conveyed through suspense and eerie music—as though the scene is a disjunctive dramatic aside, and the latter through nonchronological montage. Recollection- and dream-images are virtual images of events (often related to a character's past, future, or unconscious experiences) that are made accessible by the actual sensory–motor schemata of the corresponding movement- and action-images. For instance, the dagger reappears as the actual murder weapon in the action-image in which Macbeth kills Duncan. However, recollection- and dream-images are not by definition time-images. For them to be time-images, as in Macbeth's dagger hallucination and murder dream, the virtual must interact with the actual such that the image is double-sided and mutually engaged. And, also as in the cases of Macbeth's hallucination and dream, if this engagement becomes convoluted or united to the point that the actual and the virtual are blurred, and the meaning of the relationship between the two is unclear, the time-image is in fact a crystal-image.

It is with the crystal-image that the present analysis is most concerned. The any-space-whatever crystallized by the coalescence of the actual and the virtual is a gateway for what Deleuze calls "the powers of the false" with which all images are potentially imbued. Put differently, the crystal-image influences the viewer with the idea that things are not as they appear, that the narration may be false: truth is questionable. According to Deleuze,

> By raising themselves to the indiscernibility of the real and the imaginary, the signs of the crystal go beyond all psychology of the recollection or dream, and all physics of action. . . . What is in play is no longer the real and the imaginary, but the true and the false. And just as the real and the imaginary become indiscernible in certain very specific conditions of the image, the true and the false now become undecidable and inextricable: the impossible proceeds from the possible, and the past is not necessarily true. (274–75)

It is in this inherently undecided and unstable any-space-whatever in which the past and present are possibly just as uncertain as the future that the powers of the false allow for the articulation of Artaud's Theater of Cruelty through Polanski's film. It is through the crystalline effect that Polanski forces his viewers to question certainties.

Antonin Artaud claims that the Theater of Cruelty, like Deleuze's crystal-image, breaks through the barriers typically constructed by theatricalization; it no longer permits its audience the opportunity to suspend disbelief willingly or to become mere "Peeping Toms" passively observing an imitation of events that are usually distressing (*Theater and Its Double* 84). But, unlike the crystal-image per se, Theater of Cruelty makes "theater a believable reality which gives the heart and the senses that kind of concrete bite which all true sensation requires" (*Theater and Its Double* 85). To put it another way, as Polanski expressively professes: "If you don't show violence the way it is. If you don't show it realistically, then that's immoral and harmful. If you don't upset people, then that's obscenity" (*New York Times Magazine* 64). As in Polanski's *Macbeth*, Theater of Cruelty should concentrate "around famous personages," such as Shakespeare, Polanski, Tate, and Manson, and "atrocious crimes," such as the Manson murders and the Vietnam war, and "superhuman devotions," such as witchcraft and the Manson cult's commitment to Helter Skelter (*Theater and Its Double* 85). Because "images of a crime presented in the requisite theatrical conditions" of Theater of Cruelty can realistically involve the audience, says Artaud, the images are "infinitely more terrible for the spirit than that same crime when actually committed" (*Theater and Its Double* 85). However, as Artaud himself points out, theater and cinema are very different mediums, and, as in the Theatre of Cruelty, he maintains that cinema must achieve "purely visual situations whose drama would come from a shock designed for the eyes, a shock drawn, so to speak, from the very substance of our vision and not from psychological circumlocutions of a discursive nature which are merely the visual equivalent of a text" (*Selected Writings* 151). Clearly, as Deleuze explains, Artaud privileges the image over other cinematic narrative devices and anticipates the crystal-image: "He says that cinema is a matter of neuro-physiological vibrations, and that the image must produce a shock, a nerve-wave which gives rise to thought" (167). In the words of Polanski, "art starts being art when it moves you, when it involves you. Art is involving an audience through aesthetic technique" (*New York Times Magazine* 68). Polanski's chosen technique, as we shall see, was the crystal-image, and the use of this technique combined with the appropriateness of the historical and cultural situation for a Theater of Cruelty in the form of cinematic *Macbeth* directed by Polanski were perfect for the terrorist action that the film inevitably executed.

Shakespeare's *Macbeth* neither calls for nor seems to solicit theatrical representations of the dreadful murders of Duncan and his attendants, nor of the murders of anyone in Macduff's castle other than his boy, but Polanski's version includes all of these heinous murders and many more. From the very beginning, the film's imaging is terrorizing, and this terror is progressively intensified until manifested in full-blown terrorism. The opening sequence of movement-images and close-ups showing three haggard witches burying a severed hand, a dagger, a noose, and herbs, and then pouring blood over the covered hole is mysterious and frightening; it conveys an obscure sense of irrationality and sets a precedent for the confounded relationship between the virtual and the actual that haunts the film. Moreover, the sight of three witches probably caused an audience already looking for recollection-images of Manson to recall the three female cult members (Susan Atkins, Leslie Van Houten, and Patricia Krenwinkel) convicted with Manson, all of whom wore jean jackets embroidered with the insignia "Devil's Witches, Devil's Hole, Death Valley" and were commonly referred to as witches and Devil worshippers in the media.[16] Two sanguinary scenes portentous in nature quickly complement the formidable ambiguity of this introductory spell-casting scene. First, we witness a mute soldier pitilessly bludgeon (loudly!) a wounded enemy soldier who was feigning death on the clammy beach-battlefield; second, a man is shown (the Captain in Shakespeare), with blood dripping down his ragged face, enthusiastically telling Duncan that Macbeth "unseam'd" the enemy leader, Macdonwald, "from the nave to th'chops, / And fix'd his head upon our battlements" (1.2.22–23).

These initial images are our first tastes of the terror and terrorism to come. Terrorism, meaning systematic covert warfare to produce terror for political coercion, such as the calculated violence perpetrated by an airplane hijacker and the bomber of innocent citizens and (supposedly) the Manson murders, is frequently performed on the stage and on the screen.[17] Terrorism is thus aestheticized, objectified, commodified, and occasionally criticized. Consequently, these theatrical implementations or representations (depending on their efficacy) of terrorism are usually at the expense of the terror immediately related to the act. As Anthony Kubiak eloquently puts it:

> Here we are faced with a paradox, however, because in performance what cannot be articulated must be shown, and when it is shown, it ceases to be what it was. Thus when terror enters the information systems of performance, it ceases, in a sense, to be terror—which is unspeakable, and unrepresentable—and becomes a mask of itself. Terror is transformed into the imaging system of terrorism. (11)[18]

Theatrical terrorism is seemingly always already theoretical; it is to add an *ism* to terror, to transform terror into an ideological event. It becomes ideological inasmuch as it is no longer terror (that which cannot enter freely into language), but is now a politicized, coercive action. At the very least, this terrorism endeavors to force the audience into accepting its authenticity as it displaces the camouflaged terror that may have stimulated its existence.

In other words, as Elaine Scarry says of pain, terror, as we human beings experience it, is ordinarily unthinkable and incommunicable as a result of its relative inexpressibility and our individual subjectivity. Terror, like pain, is a phenomenon that resists verbal objectification: it is typically unsignifiable. Therefore, as pain is commonly misrepresented and supplanted hermeneutically by what appears to be its physical signification (as depicted by a bodily wound or symptomatic behavior), so is terror affected by the tangibility of terrorism.[19] For terror to maintain its terrific and tyrannical power, terrorism is regularly and often effectively portrayed theatrically and figuratively by allusion or as a conspicuous absence. We learn secondhand of its source, or its conclusion, as with the murder of Duncan in Shakespeare's version of *Macbeth*, yet our imagination is compelled, whether consciously or not, to improvise, to account for what we ourselves neither saw nor experienced. And it is in this void, what Deleuze calls an any-space-whatever, because of our inevitable inability to realistically supplement what we cannot know, that a vigil of fear, a conundrum of terror thrives within us. Whereas most straightforward theatrical terrorism in effect fails to tap this fear because its displacement of terror allows for observer-rejection, the Theater of Cruelty does not. It is never theoretical. It instead pretends to confer the option of rejection to the audience, and it is this believed illusion of an imaginary option that permits the Theater of Cruelty to surreptitiously strike its audience in a way that is deeply disturbing.

While Polanski's *Macbeth* incorporates the requisite historical and cultural ingredients for a realization of Artaud's Theater of Cruelty, it is difficult to determine whether a Jacobean *Macbeth* could have been one. This possibility is worth pursuing, if only briefly, because of the residual impact it may have had on Polanski's version. The Jacobean productions of *Macbeth* seem to have also operated in the proper setting for a Theater of Cruelty and Shakespeare's text undoubtedly contains the necessary substance and framework. Apart from the noted allusions to famous personages (James and his ancestors), atrocious crimes (witchcraft, heresy, and treason) and superhuman devotions (occultism and regal aspirations, such as the Gunpowder Plot), Shakespeare includes regicide (which may have been staged, but probably not in front of James), he emphasizes the innocence of those who are murdered (see, for instance, 1.7.21–25), he has a young boy ruthlessly slain before his own mother (4.3.86), he depicts the murder of an ancestor of royalty explicitly identified with James (Banquo), he reinforces beliefs in not only the existence and malevolence of witches but also in ghosts, the Devil, and evil, and he does all of these things in a play literally saturated in blood. In fact, there are more references in *Macbeth* to blood and bleeding than in any other play by Shakespeare.

Like witchcraft, bloody murder was a topic of much popular literature in Elizabethan and Jacobean England. And one of the most grotesque of the surviving murder pamphlets, *The Most Cruell and Bloody Murther Committed by an Inkeepers Wife*, describes explicitly how a murderess, like the killers of Sharon Tate, murdered a pregnant woman: "shee ript her up the belly, making herselfe a tragicall midwife, or truly a muderesse, that brought an abortive babe into the world" (A3). The circulation of this pamphlet at a time when Shakespeare may have been writing or revising *Macbeth* causes me to wonder whether Shakespeare had this probably well-known murder case in mind when contriving certain lines, particularly: "Macduff was from his mother's womb / Untimely ripp'd"

(5.8.15–16). More important is whether members of Shakespeare's audience would have recollected the murder case because of the potential allusions to it in the play, and what effects the association between the case and the play might have had on them. Did it make the likelihood and reality of barbarous murders, like the one committed by the innkeeper's wife, more terrorizing?

Tracts about witches were, perhaps, more fashionable than even the most scandalous and horrifying murder pamphlets. Considering that *Macbeth* contains some of the most atrocious forms of murder imaginable and clearly implies, as in Hecate's speech (3.4), that the play's witches are ultimately to blame for the murders (and not the ambitiousness of Macbeth or Lady Macbeth), it seems probable that the play influenced its audience's sentiments toward eccentric women living in their neighborhoods, particularly those women unlucky enough to have had circulating beforehand rumors of their witch status. Insofar as *Macbeth* was legitimized by its performance at court, it seems possible that the play possessed the power to perpetuate the transference and displacement of responsibility for crimes and other unfortunate circumstances onto these unusual members of society. In his essay "The Power of Evil," Charles Manson explains:

> An illusion to some may be a death reality to others. A play on stage may invoke madness somewhere else as it may circle the stage and be in the streets behind the stage plays—There are looks that kill and motions of a finger that can destroy much. (qtd. in Schreck 135)

So, for some people *Macbeth*'s witches remain invariably within the safely contained, thespian world of illusion, yet for others the witches represent a horrible out-of-theater reality that threatens incessantly the ethical, physical, and spiritual sanctity of human life. Indeed, the witches challenge that life itself. The power of theater is therefore not to be perceived frivolously, for this power may be a product of madness and other things, just as it may produce or reproduce madness and other things. Manson's understanding of the power of theater sheds light on the causality and potential ramifications of Polanski's wish to use the power of cinema to make people "question certainties," such as something "so obviously impossible as witchcraft."

Interestingly, no recourse is taken against the witches either in Shakespeare's *Macbeth* or in Polanski's adaptation. It seems, then, that the dangerous and crucial, and for some, gratifying task of plucking the dreadful witches from society and punishing them is left to the impressionable and socially responsible members of the audience. To be sure, it is conceivable that a cinematic Theater of Cruelty, articulated through a crystalline narration, would influence its audience as Manson describes. In other words, as Deleuze explains, the crystal-image causes the prevalent idea (or "thing") behind the image to emerge multi-paradoxically, both virtually and actually and truly and falsely: "a pure optical-sound image, the whole image without metaphor, brings out the thing in itself, literally, in its excess of horror or beauty, in its radical or unjustifiable character, because it no longer has to be 'justified,' for better or for worse" (20). The image is so powerful and contradictory that it temporarily transcends judgment, and it is through this any-space-whatever that the "thing," whether it be a good or bad idea or emotion, penetrates and affects the viewer profoundly.

In a recollection-image horrifically allusive to the Manson murders, Polanski's Macbeth mercilessly butchers King Duncan with a dagger: he repeatedly thrusts the blade in and out of Duncan's naked chest as Duncan pitifully looks with a desperate, aghast gaze. The montage of close-ups shifting periodically from the faces of Macbeth and Duncan (from both of their viewpoints), to Duncan's bloody and lacerated body, and then to Duncan's crown as it slips from his grasp and awkwardly rolls to inertia, are very powerful. They trenchantly emphasize Duncan's vulnerability and innocence and Macbeth's guilt. But a close-up of Macbeth's confounded and compunctious countenance immediately before he begins, however reluctantly, to hack Duncan illustrates what appears to be a profound lack of agency on his part in the crime: the powers of the false are at work. Akin to Shakespeare's *Macbeth*, despite the absence of Hecate, Polanski's film leads its audience to believe that the crimes committed by Macbeth and Lady Macbeth are brought about by the spells cast by the witches on the beach-battlefield and during their cabalistic sabbath.

In pursuit of truth, Polanski's Macbeth nervously descends into the dark and mysterious witches' lair, as if from the security and rectitude of his privileged masculine reality, and ingenuously participates in their arcane ceremony (as in Shakespeare 4.1). This scene is the film's crystalline climax. Encompassed by dozens of naked hippie-looking witches of assorted ages and appearances (I say hippie-looking because of their long disheveled hair, the style of their jewelry, and the nature of their gestures and dance), Macbeth willingly gulps down their magic drug-beverage (not in Shakespeare). He becomes part of an infamous and clandestine ritual that has been perceived as fascinating, terrifying, and titillating. Polanski proffers a film manifestation of not only the eroticized male fantasy of centrality in an otherwise exclusively female society, but also the human desire to access supernatural knowledge. Yet, as the mythologized male fear of annihilation by woman and the taboo against divine aspiration are often artistically represented, the witches bewitch and ruin Macbeth, and, via association, his wife too. They turn Macbeth on to an LSD-like trip: as Deleuze notes, "Some characters, caught in certain pure optical and sound situations, find themselves condemned to wander about or go off on a trip" (41). (See photo 5.1.)

In this "trip-image" (my term), Polanski masterfully constructs with an astonishing assortment of cinematographic techniques (montage-cuts, close-ups, distance shots, mirrorings, superimpositions, dissolves, and disembodiments), Macbeth and the viewer experience, in rapid succession, a phantasmagoria of enlightening, ambiguous, and ghastly hallucinations. Crystalline effectuation occurs with tremendous magnitude: almost instantly the virtual and the actual and the real and the imaginary become indiscernible (as in the earlier crystalline scene in which Macbeth sees Banquo's ghost), an any-space-whatever is established, and the powers of the false reign on both Macbeth's and the viewer's quest for the true. To be sure, the information Macbeth acquires of two seemingly true but inevitably false impossibilities is indicative of this crystalline *tour de force*: he is told that "none of woman born / Shall harm Macbeth" (4.1.80–81) and "Macbeth shall never vanquish'd be, / until Great Birnam would to high Dunsinane hill / Shall come against him" (4.1.92–94). When we return with Macbeth from the trip-image to an action-image (we watch Macbeth depart) and the "reality" that the action-image's sensory–motor schemata implies, uncertainty and anxiousness still prevail. For Macbeth the outcome is unpredictable, but he has a strategy for improving the odds: he has Macduff's entire family and servants "Savagely slaughter'd" (4.3.205). On drugs of ambition, Macbeth, like Charles Manson, is either demonically possessed or severely intoxicated, at the least.

Throughout the massacre at Macduff's castle, quick montage-cuts and abrupt close-ups are again employed to stress the innocence of the victims and the terrorism of the episode. Unlike Shakespeare, who includes only one onstage murder, that of Macduff's boy, Polanski highlights five victims, just as there were five people murdered by the Manson devotees at his residence. The band of assassins first kill Macduff's (approximately seven-year-old) boy in front of his mother, and, like Duncan, the boy is naked (with the exception of a blanket wrapped around him) when he dies. On the "Macbeth" set, Tynan recalls that when he "queried Roman's estimate of the amount of blood that would be shed by a small boy stabbed in the back," Roman replied,

Photo 5.1 Witches' lair.

"'You didn't see my house in California last summer . . . I know about bleeding'" (*Esquire* 189). A relationship between the Manson murders and the murders at the Macduff castle is obvious, and this relationship becomes more impressive as the nightmarish scene continues. When Lady Macduff tries to flee from her assailants, the camera shows a few of the other murderers gang raping a screaming woman in a dingy corridor of the castle. Next, Lady Macduff, still scuffling to escape the killers, looks in on her two smaller children who have already been slain, their bodies mangled and drenched in blood. Finally, like Shakespeare, Polanski leaves the termination of Lady Macduff to the wild imagination of his audience; the audience is transformed into both the victim and the psychic instrument for terrorism.

Timely mediation

Whereas it is widely documented that members of the 1971 audience interpreted this harrowing film in relation to the Manson murders, which would suggest the bewitched Macbeth or the hippie-looking witches as symbolic of Charles Manson, it is unclear to what extent the film could have been regarded as an effort to exonerate Manson or his followers, to shift the responsibility for their crimes away from them and onto the drugs, the witchcraft, the counterculture with which they were associated. It is also unclear whether the film was ever seen to blame the state apparatus of which Manson is largely a product: when Manson was arrested for the Tate-LaBianca murders he had already spent seventeen of his thirty-four years in state penitentiaries. If King Duncan is to blame for the infiltration of the witches, and not some kind of absent god, then Macbeth is primarily a vehicle in the business of an already existent contention between the witches and the state. Although Macbeth seems totally out-of-control while the witches seem to have total control, there are moments in the film when Macbeth's struggling will emerges and is emphasized. The most significant of these calculated moments is when Macbeth self-consciously responds to the criminal portentousness of the illusory dagger floating before him, when he pauses ruminatively and in quandary just prior to killing Duncan, and when he reports regretfully the murder of Duncan to his wife. It is insinuated that if Macbeth were a stronger and better person he would have resisted the witches' charms and the persuasion of Lady Macbeth—who starts off headstrong and enthralled but becomes weaker and weaker until "by self and violent hands / Took off her life" (5.9.36, 37).

Photo 5.2 Macbeth's severed head.

Yet, since this is not the case, Macbeth, like Manson in the media, becomes the personification of evil. And Macbeth, like Manson in the courtroom, is sentenced to death; Macduff decapitates Macbeth, and his "cursed head" (5.9.21) is fixed upon a pole, as the Captain says Macbeth displayed the head of Macdonwald. (See photo 5.2.)

But if Macbeth is indeed an instrument and victim of complex circumstances beyond his control, then does it really follow that Manson or his devotees are too? Manson's actual role in the Tate-LaBianca murders is similar to that of Polanski's witches in the murders of Duncan, Banquo, and Macduff's household. Manson was the influential force behind the actions of his assassins; he did not participate with his own hands in the slaying. So it seems, then, that Susan Atkins, Katie Krenwinkel, and Charles Watson were in a comparable situation to that of Polanski's Macbeth and Lady Macbeth. They were bewitched stooges assisting Manson's master plan, as many people, including the jurors, concluded at the time (Manson was convicted in the Tate-LaBianca trial on seven counts of conspiracy to commit murder).[20] In an ironic statement, Manson acknowledges this situation:

> We as Americans have been taught to believe in *Life, Liberty, and the Pursuit of Happiness*. While I was free and out there *Pursuing Happiness* they took my *Liberty* and gave me *Life*. I was convicted of witchcraft in the Twentieth Century. (qtd. in Schreck 26)

The irony, however, lies neither in Manson's literal reading of the *Declaration of Independence*, nor in a meditation on free will and coercion, which this discussion certainly invokes but chooses not to pursue, but rather in the public's acceptance of the Tate-LaBianca murders as the culmination of 1960s radicalism. Manson was not affiliated with any readily identifiable countercultural group of the 1960s. In fact, he was openly antagonistic toward beatniks, folkniks, hippies, radicals, and civil rights activists. Manson could not relate to these people. They had not spent half of their lives in prison. Inasmuch as Manson is an identifiable product of the U.S. penal system, as he himself recognizes that he was socialized while incarcerated within this system ("My father is the jail. My father is the system"), Manson serves only to consolidate the power structure behind this system (qtd. in Gilmore and Kenner 11). The Manson family and the Tate-LaBianca murders were thus appropriated and constructed to represent the allegedly anticipated and unavoidable result of the diversified 1960s counterculture. To recall once more the telling words of Joan Didion, "the Sixties ended abruptly on August 9, 1969, ended at the exact moment when word of the murders on Cielo Drive traveled like brushfire through the community, and in a sense this is true. The tension broke that day. The paranoia was fulfilled."

We have seen that Polanski's *Macbeth* emanates instability and uncertainty from beginning to end. Its crystalline narration imbues the viewer with the any-space-whatever between the virtual and the actual, the real and the imaginary, and the past and the present that allows for its Theater of Cruelty to constitute a terrorist action. Indeed, it is this remarkable situation that led me to choose the line "Untimely ripp'd" for the title of this essay. *Macbeth* is untimely, and is viewed as such, not only because it does not occur at any particular time, and is not tied to one historical moment, but also because its use of crystal-images combined with its numerous, multilayered historical and cultural allusions to both the English Renaissance and 1960s America, reveal the film to be what we might call a "crystalline artifact." This is a cultural–filmic artifact that was ripped from time, just as it rips its viewer from rationality. My use of "ripped" here is twofold. First, the film seems to have been ripped from the fabric of history, such that its threads dangle in multiple directions. Second, I am thinking of the slang use of ripped, as in "get ripped," meaning intoxicated. The film's crystalline effect intoxicates its viewer, transports the viewer with its crystal-images, such that the viewer experiences a timelessness or nonchronological time that is characteristic of some drugs, especially hallucinogens. Like Macbeth's hallucinogenic experience in the witches' lair, this film is an awesome trip: as Deleuze puts it, "It is a matter of something too powerful, or too unjust, but sometimes also too beautiful, and which henceforth outstrips our sensory-motor capacities.... something has become too strong in the image" (18).

Notes

1. I am especially grateful to James Andreas, editor of *The Upstart Crow*, for his comments on my essay, "The Terrorism of Macbeth and Charles Manson: Reading Cultural Construction in Polanski and Shakespeare" (*The Upstart Crow* 13 [1993]: 109–29), on which this essay is based. "The Terrorism of Macbeth and Charles Manson" focuses on the comparable cultural contexts of early modern productions of *Macbeth* and Polanski's adaptation, while the present analysis is mainly concerned with the effects of Polanski's cinematic technique. An earlier version of this essay, "Untimely Ripped," was published in *Social Semiotics* 7.2 (1997): 201–18.
2. *Time Magazine*, January 25, 1971.
3. A reprint of Elizabeth's statute and background information can be found in Rossell Hope Robbins, *The Encyclopedia of Witchcraft and Demonology* (New York: Crown Publishers, 1959), 156–59. For information about the witch craze in Renaissance England, see Alan Macfarlane, *Witchcraft in Tudor and Stuart England: A Regional and Comparative Study* (New York: Harper and Row, 1970), and *The Damned Art*, ed. Sidney Anglo (London: Routledge and Kegan Paul, 1977).
4. See *Newes From Scotland* (1591) (London: Curwen Press, 1924).
5. Reginald Scot, *The Discoverie of Witchcraft* (London: 1584); Scot explains his position on page 472.
6. Unlike the Elizabethan Act of 1563, the Jacobean Act of 1604 mandated capital punishment for a first offence of conjuring evil spirits (see Robbins 156–59). As noted by social historian Stuart Clark in his essay "King James's *Daemonologie*: Witchcraft and Kingship" in *The Damned Art: Essays in the Literature of Witchcraft*, ed. Sidney Anglo (London: Routledge and Kegan Paul, 1977), 156–81, the subsequent rate of prosecution dropped below that of Elizabeth's reign (161). James's harsher statute may have been a deterrent for those people interested in conjuring evil spirits, or it may have encouraged those already attempting to conjure evil spirits to be more covert in their operations.
7. James I, *Letters of King James VI & I*. G. P.V., ed. G.P.V. Akrigg (Berkeley: University of California Press, 1984), 250.
8. See xv–xxv of Kenneth Muir's introduction to William Shakespeare's *Macbeth* for a thorough discussion of the date of *Macbeth* (London: Methuen, 1984), xiii–lxv. All references to Shakespeare's *Macbeth* are to this edition, and will be cited in the text.
9. See Annabel Patterson, *Censorship and Interpretation: The Conditions of Writing and Reading in Early Modern England* (Madison: University of Wisconsin Press, 1984).
10. See Muir's introduction, xx–xxiii, and *Macbeth*, 2.3.7–12 and 4.2.46.

11. Holinshed in Muir 172.
12. Holinshed in Muir 171; cf. Shakespeare, 4.1.111–24.
13. For examples of Manson paraphernalia, see *The Manson File*, ed. Nikolas Schreck (New York: Amok Press, 1988).
14. For more on "transversal theory," beyond the introduction to this book, see Bryan Reynolds, "The Devil's House, 'or worse': Transversal Power and Antitheatrical Discourse in Early Modern England," *Theatre Journal* 49.2 (1997): 143–67, and *Becoming Criminal: Transversal Performance and Cultural Dissidence in Early Modern England* (Baltimore: Johns Hopkins University Press, 2002); and Bryan Reynolds and Joseph Fitzpatrick, "The Transversality of Michel de Certeau: Foucault's Panoptic Discourse and the Cartographic Impulse" *Diacritics* 29.3 (1999): 63–80.
15. For further discussion of "subjunctive space," beyond the introduction to this book, and in relation the "Mary/Mollspace" of Thomas Middleton and Thomas Dekker's *The Roaring Girl*, see Bryan Reynolds and Segal, "The Reckoning of Moll Cutpurse: A Transversal Enterprise," *Rogues in Early Modern English Literary Culture*, ed. Craig Dionne and Steve Mentz (University of Michigan Press, forthcoming 2004).
16. See photograph in Schreck 151.
17. Manson and his followers were a source of inspiration for several other radical factions who lauded them for their terrorism. To give two well-known examples: regarding the Manson murders, Bernadine Dohrn, leader of the Weather Underground, exclaimed, "First they killed those pigs, then they ate dinner in the same room with them, and they even shoved a fork into the victim's stomach! Wild!" (Dohrn qtd. in Jane and Michael Stern, *Sixties People* [London: Macmillan, 1990], 197); and Yippie Jerry Rubin, after visiting Manson in prison, wrote, "I fell in love with Charlie Manson. . . . His words and courage inspired [me]" (Rubin qtd. in Todd Gitlin, *The Sixties: Years of Hope, Days of Rage* [New York: Bantam, 1987], 404).
18. For a relevant discussion of the relationship between tragedy, terror, *Katharsis* and theatrical terrorism, see Anthony Kubiak's *Stages of Terror* (Indianapolis: Indiana University Press, 1991), 18–22.
19. See Part One of Elaine Scarry's *The Body and Pain: The Making and Unmaking of the World* (New York: Oxford University Press, 1985).
20. He was convicted of another murder and given the death penalty, but was not executed because capital punishment was declared unconstitutional.

Works cited

Anglo, Sidney, ed. *The Damned Art: Essays in the Literature of Witchcraft*. London: Routledge and Kegan Paul, 1977.

Artaud, Antonin. *Selected Writings*. Ed. Susan Sontag. Trans. Helen Weaver. Berkeley: University of California Press, 1988.

———. *The Theater and Its Double*. Ed. and trans. Mary Caroline Richards. New York: Grove Press, 1958.

Bugliosi, Vincent and Curt Gentry. *Helter Skelter: The True Story of the Manson Murders*. New York: Norton, 1974.
Clark, Stuart. "King James's *Daemonologie*: Witchcraft and Kingship." *The Damned Art: Essays in the Literature of Witchcraft*. Ed. Sidney Anglo. London: Routledge and Kegan Paul, 1977. 156–81.
Deleuze, Gilles. *Cinema 2: The Time-Image*. Minneapolis: University of Minnesota Press, 1995.
Didion, Joan. *The White Album*. New York: The Noonday Press, 1991.
Ebert, Roger. Rev. of *Macbeth*, dir. Roman Polanski. Chicago Sun-Times April 3, 1972: 32.
Fiedler, Leslie. *The Stranger in Shakespeare*. New York: Stein and Day, 1972.
Gilmore, John and Ron Kenner. *The Garbage People*. New York: Omega Press, 1971.
Gitlin, Todd. *The Sixties: Years of Hope, Days of Rage*. New York: Bantam, 1987.
Greenblatt, Stephen. *Renaissance Self-Fashioning: From More to Shakespeare*. Berkeley: University of California Press, 1980.
James I. *Letters of King James VI & I, G. P. V*. Ed. G.P.V. Akrigg. Berkeley: University of California Press, 1984.
Kael, Pauline. *New Yorker* February 5, 1972: 76.
Kubiak, Anthony. *Stages of Terror*. Indianapolis: Indiana University Press, 1991.
Macfarlane, Alan. *Witchcraft in Tudor and Stuart England: A Regional and Comparative Study*. New York: Harper and Row, 1970.
Manson, Charles. *The Manson File*. Ed. Nikolas Schreck. New York: Amok Press, 1988.
The Most Cruell and Bloody Murther Committed by an Inkeepers Wife. London, 1606.
Muir, Kenneth. Introduction. *Macbeth*. By William Shakespeare. Ed. Kenneth Muir. London: Methuen, 1984. xiii–lxv.
"New Feminists: Revolt Against Sexism." *Time*. November 21, 1969: 54.
Patterson, Annabel. *Censorship and Interpretation: The Conditions of Writing and Reading in Early Modern England*. Madison: University of Wisconsin Press, 1984.
Polanski, Roman, dir. *Macbeth*. Perf. Jon Finch and Francesca Annis. Columbia/Tristar Studios, 1971.
Reynolds, Bryan. *Becoming Criminal: Transversal Performance and Cultural Dissidence in Early Modern England*. Baltimore: Johns Hopkins University Press, 2002.
———. "The Devil's House, 'or worse': Transversal Power and Antitheatrical Discourse in Early Modern England." *Theatre Journal* 49.2 (1997): 143–67
———. "Untimely Ripped." *Social Semiotics* 7.2 (1997): 201–18.
———. "The Terrorism of Macbeth and Charles Manson: Reading Cultural Construction in Polanski and Shakespeare." *The Upstart Crow* 13 (1993): 109–29.
Reynolds, Bryan and Joseph Fitzpatrick. "The Transversality of Michel de Certeau: Foucault's Panoptic Discourse and the Cartographic Impulse." *Diacritics* 29.3 (1999): 63–80.

Reynolds, Bryan and D.J. Hopkins. "The Making of Authorships: Transversal Navigation in the Wake of Hamlet, Robert Wilson, Wolfgang Wiens, and Shakespace." *Shakespeare After Mass Media*. Ed. Richard Burt. New York: St. Martin's Press, 2002. 265–86.
Robbins, Rossell Hope. *The Encyclopedia of Witchcraft and Demonology*. New York: Crown Publishers, 1959.
Rothwell, Kenneth. "Roman Polanski's Macbeth: Golgotha Triumphant." *Literature Film Quarterly* 1.1 (1973): 71–75.
Scarry, Elaine. *The Body and Pain: The Making and Unmaking of the World*. New York: Oxford University Press, 1985.
Scot, Reginald. *The Discoverie of Witchcraft*. London: 1584.
Shakespeare, William. *Macbeth*. Ed. Kenneth Muir. London: Methuen, 1951.
Stern, Jane and Michael Stern. *Sixties People*. London: Macmillan, 1990.
Tynan, Kenneth. "The Polish Imposition." *Esquire* 76.3 (1971): 122–25, 180–89.
Weinraub, Bernard. Interview with Roman Polanski. *New York Times Magazine*. December 12, 1971: 36–37, 64–83.
Wizinski, Sy. *Charles Manson: Love Letters to a Secret Disciple*. Indiana: Moonmad Press, 1976.
Zimmerman, Paul. Rev. of *Macbeth*, dir. Roman Polanski. Newsweek January 10, 1972: 59.

Chapter 6
Nudge, Nudge, Wink, Wink, Know What I Mean, Know What I Mean?[1]
A Theoretical Approach to Performance for a Post-Cinema Shakespeare

D.J. Hopkins, Catherine Ingman, & Bryan Reynolds

The wave of success that has established William Shakespeare as the most popular and prolific screenwriter currently working in Hollywood shows no sign of abating.[2] But, as evidenced by the disparity in style and quality of recent productions, there is more than one way to make a "Shakespeare movie." In an article titled "Shooting Shakespeare," written for the popular magazine *Madison*, movie columnist Graham Fuller sums up the last ten years of film versions of Shakespeare: "It has been a noble collective effort, a testament to the most inventive literature in existence—and, by extension, to the absence of invention in modern cinema" (83). Fuller notes a lack of inventiveness on the part of Hollywood writers who continue to raid Shakespeare's collected works for their screenplays. We agree that there is an "absence of invention" in contemporary approaches to "shooting Shakespeare," but we experience the problem less in the screenwriting than in the direction and acting.

Defining a cultural moment by redefining Shakespeare is a practice that began with the publication of the first Folio, has continued ever since on both stage and screen, and will certainly mark cultural production for generations to come. Yet it is not Hollywood's ongoing urge to *do* Shakespeare that we question. We wonder why so few filmmakers seem to know what to do *with* Shakespeare. There is nothing either good or bad in the measure of a film's proximity, filiation, or "faithfulness" to its Shakespearean source. Our interest is not in making an obvious qualitative statement like, "Some Shakespeare movies just aren't very good." Rather, we want to interrogate the relationship between the written versions of Shakespeare's plays—available to readers in numerous editions and collections (hereafter referred to as the "source text")—and the performances of the individual actors called upon to embody (or, in the case of filmed media, *dis*embody) Shakespeare's characters when these plays are presented on film.

Overall, at this millennial juncture, we want to propose a new theoretical approach to film performance, one that focuses on techniques of acting and framing, and that we have imagined for, but is not limited to, cinematic Shakespeare. By extension, we also want to propose, if not prophesize, a new direction for the future of film versions of Shakespeare. In a millennial preview of his own, entitled "Premature Synthesis," theater theorist Patrice Pavis makes some insightful observations about the future of French playwriting and acting that are especially relevant to our purpose here:

> From now on, instead of the actor embodying and characterizing his character, or the actor distancing himself according to Meyerholdian or Brechtian technique, one finds an actor-dramaturg participating in dramaturgic choices and changes.... [T]his actor-dramaturg links the lines, suggests the network of echoes ... thus contributing to the development of the score of the spectacle and its meaning. (79)[3]

To extend Pavis's argument for actors of filmed Shakespeare, we offer a theory of performance that positions the individual performer less as a mere mouthpiece for poetry and pathos, and more as a self-aware, self-referential participant in the representational process itself, a participant whose role would also involve both inspiring and welcoming the audience's cognizance of, and participation in, the constitution of the dynamic "score" of the filmed "spectacle." This approach to performance is, at this time, an intellectual endeavor; however, it is our contention that this theory has a practical application to Shakespeare and beyond that is waiting to be realized on screen.

Because of Shakespeare's long-standing stature as an icon for elite culture, today's moviegoing audience is especially self-conscious about their own sociocultural status when "seeing" Shakespeare on screen. Today's techno-media-inundated audience is also more alert to the mediated-constructedness of all filmic representation than ever before, and thus this audience is increasingly less likely to indulge in the willing suspension of disbelief, particularly when confronted with "realism." For these reasons, Shakespearean film acting should resist the traditional conventions of realism, and be demonstratively hyper-aware of itself.

It is time for the actor in performance to openly acknowledge the contrived nature of every filmic representation and incorporate this awareness into a performance practice that is therefore more honest, more "real," about its own theatricality and framing. In effect, the filmic representation should appeal more readily to the predispositionally skeptical audience, capitalizing on, as well as satisfying, this audience's desire to have "behind-the-scenes" information and thereby participate in the making—and by extension, the performing—of the filmic construct. At its most basic level—a level that most undergraduate students of theater studies will appreciate—our theoretical approach to film performance, that "transversally"[4] relates acting and framing, posits that film versions of Shakespeare should place a critically important emphasis on the performing actor's own self-aware textual analysis, the simultaneous interpretation and execution implied by Pavis's term "actor-dramaturg." A theory of performance practice with a practical application, we want to suggest that the theoretical dimensions of this kind of acting can be extended, and are greatly complemented by, narrative and cinematographic framing that correlate, and are thus conducive to, this approach to the enterprise of performing Shakespeare.

To illustrate our argument, we will first provide a brief account of today's mass-mediatized and commercialized Shakespeare by highlighting certain recent "Shakespeare-related" movies, particularly *Shakespeare in Love* (dir. John Madden 1998), in light of Richard Burt's concept of "the loser." Next, we will situate our analysis vis-à-vis some of the foundational texts of performance studies and game theory: the performance of lies, deceit, and "play" will be discussed in relation to examples drawn from *The Taming of the Shrew*. Then, we will return to recent films, and see how these same theories are or are not at work in *Never Been Kissed* (dir. Raja Gosnell 1999), *10 Things I Hate About You* (dir. Gil Junger 1999), and the much-discussed *Hamlet* (dir. Michael Almereyda 2000). Our focus in this section will be on *10 Things I Hate About You,* Junger's adaptation of *The Taming of the* Shrew. Finally, while discussing Almereyda's *Hamlet*, with an aside on Godard's *Contempt* (*Le Mépris* 1964), we will return to the particular mode of transversal performance we are proposing, which is to say the acting and framing style we call "Nudge, Nudge, Wink, Wink, Know What I Mean, Know What I Mean?".

"This time it's by [that loser] Shakespeare"

Pavis takes the turn of the millennium as an excuse for prematurely synthesizing contemporary movements in French theater in order to propose a trajectory for future theatrical evolutions. Although we too are writing at the same millennial juncture, our impetus is not so much a chronological turn as a cultural one. In our title we state that we are writing about a "post-cinema" Shakespeare. "Post-cinema" means "after film": we are writing about the many media (mechanically reproduced and otherwise) that are proliferating in the aftermath of the era that saw the overwhelming dominance of cinema and the primarily U.S.-centered industry (Hollywood) that made the medium of movies ubiquitous. The term "post-cinema" implies an explosion of forms that includes video, animation, CD-ROM, the Internet, and the resuscitation and revitalization of live performance/theater as a valid and vital art that in its "liveness" provides a radical response to a mediatized society.[5]

In this section, we sketch some of the parameters that govern post-cinematic treatments or responses to Shakespeare. To a greater or lesser degree, these responses participate in what W.B. Worthen calls "Shakespearean performativity": "the sense that the performance of a Shakespeare play can, or should, evoke the pastness of the text and what the text represents—early-modern values, behaviors, subjects" (118). Different "Shakespeare movies" negotiate their relationships with Shakespeare in different ways, and there is a range of "Shakespeare-relatedness" in popular cinema. Some films move on a course that takes them away from an actual confrontation with Shakespeare; other films proudly if awkwardly embrace Shakespeare, perhaps even going so far as to use the original titles of his plays and much of his dialogue.[6] Of course, there are also those films that are actually, to use Fuller's phrase, "shooting Shakespeare": films made directly and specifically out of one of Shakespeare's plays. Within this last category of films, there is variation too: some adhere more closely than others to Shakespeare's "script," most name Shakespeare as a source on screen and in promotional advertisements and posters (authorial credit that Shakespeare infrequently received in his own life), and some even incorporate Shakespeare's name and his play's title together in the title of the film.[7]

Five recent films map an interesting trajectory along this range of Shakespeare-relatedness. There is the Hong Kong style action-film, *Romeo Must Die* (dir. Andrzej Bartkowiak 2000), with Hong Kong star Jet Li as the Romeo character caught between Chinese, African American, and Anglo-American families and mob lords in a struggle over waterfront property sought for the development of a sports arena. *Romeo Must Die* is similar to *Never Been Kissed* (starring Drew Barrymore), which, as Linda Boose and Richard Burt explain, extends the already existing subgenre of Shakespeare production that positions

Shakespeare in "a wholly different representational economy and market strategy" than that which characterized the production of most twentieth-century Shakespeare films (Boose and Burt 1). This genre is articulated most notably by the Arnold Schwarzenegger film *The Last Action Hero* (dir. John McTiernan 1992), in which Shakespeare's *Hamlet* is a distant representational trace, much like the relationship that *Never Been Kissed* has with *As You Like It*. Nearer to the actual source material is *10 Things I Hate About You*, a prime example of a burgeoning subgenre that could be called "adaptations of Shakespeare's plays set in high school" (in this case, *The Taming of the Shrew*). And among those films that are unabashed screen adaptations of Shakespeare, Ethan Hawke takes the title role of the cinematic version of *Hamlet* directed by Almereyda, with Julia Stiles as Ophelia. Stiles's Ophelia followed closely her role as "Kat" in *10 Things* and preceded that of "Desi" in *O* (dir. Tim Blake Nelson 2001).[8]

There is also the film that threatened to overturn the conventions of Shakespearean performativity and cinematic production: *Shakespeare in Love* (dir. John Madden 1998). Rather than participating in the fugitive relationship with Shakespeare that characterizes the majority of popular cinematic appropriations, and not merely content to film one of the plays and mention the author in the title, *Shakespeare in Love* pursues Shakespeare himself, inventing a personal history that intersects with the composition of, and parallels closely, the story of *Romeo and Juliet*.[9] This is neither a film "inspired by" nor "based upon" one of Shakespeare's plays. It is a film inspired by the idea of "Shakespeare," a film based upon an author-function.

Shakespeare in Love appeals to exactly the kind of "loser" that Richard Burt discusses in his book *Unspeakable ShaXXXspeares: Queer Theory and American Kiddie Culture*. Burt proposes a new usage of the term "loser," not in a pejorative sense, but in a culturally equivocal one: "Being a loser is not just about cool indifference ... it is about setting, even raising standards as well as ignoring them" (Burt 9). This paradox governs a culture that, Burt argues, is increasingly determined by a popular, commercial, mediatized youth culture. In such a cultural climate, Shakespeare "is both cool and uncool, both a signifier of elite and of popular culture, in part because of the way he is now positioned inside and outside of academia" (6). In response to this contemporary cultural bifurcation of Shakespeare, a new kind of academic has emerged: the Shakespearean loser.

For Burt, a loser is "a highly self-conscious, paradoxical figure whose practices deconstruct oppositions between the creative and (self-)destructive, intelligent and stupid... between success and failure" (17). A Shakespearean loser, then, is someone who "tends to assume that he or she can include all reproductions in the database of Shakespeare studies and that he or she can adopt both low and high positions inside and outside of academia, can both get all the elite in-jokes and appreciate the general view," but in fact cannot do so (204). Such a litany of contradictory aspirations cannot be synthesized into a unified, stable form; and the Shakespearean loser knows this. As a film, *Shakespeare in Love* succeeds not only by appealing to "loser academic intellectual[s]" (239), those postmodern Shakespeareans who will get all its in-jokes and references, but also by positioning the character of "Will" as the ultimate Shakespearean loser.

We first hear Shakespeare's name mentioned when his producer, Philip Henslowe, desperate to end the torture (literally) inflicted on him by an impatient moneylender, defends the unoriginal scenario of a "nearly finished" play by proclaiming: "This time it's by Shakespeare!" Next we see Will Shakespeare (Joseph Fiennes) hunched over a writing desk, quill pen in hand. Ink-stained fingers crumple page after rejected page, but not of Henslowe's script. A close-up on the paper reveals that he's practicing his signature. Even before the first of many comparisons to "Kit" Marlowe's success as a playwright, the audience of *Shakespeare in Love* is introduced to Will Shakespeare: loser. Obsessed with his own as-yet-incomplete success, he works on his (now famous) signature rather than his play.

Shakespeare in Love functions on two different registers: first, it presents a historical Elizabethan London; second, it provides a transhistorical parallel to contemporary Hollywood. Will's struggles to succeed can be read on both registers. Will pursues a love affair doomed to failure by the strictures of Elizabethan society, and the premiere performance of *Romeo and Juliet* serves as the last remaining site for a union of these two "star-cross'd lovers." The film also presents Will's struggle to succeed in a commercial entertainment industry where actors, producers, and fans all constantly hector him with the established success of his nemesis/colleague/hero Marlowe—of whose success Will is already acutely aware. At his lowest point, early in the film, Will runs into Marlowe in a tavern. When Marlowe admits that his *Faustus* is getting restaged, Will confesses, "I love your early work," and then lovingly delivers the line, "Is this the face that launched a thousand ships." With this quote, Will reinforces his status as a Shakespearean loser: aware of divisions in Marlowe's output, selecting a line that only passing centuries of critical approval would single out for particular celebration, and painfully conscious that he speaks not as "Shakespeare," future icon of authorship itself, but merely as "Will," a relatively unknown and inconsistent writer. Marlowe proceeds to feed Will a plot for his new play, even offering a name for the main character: *Romeo*.

"The loser may not be able to complete a project," explains Burt, "for example, destroying successive drafts along the way, but he (or she) will produce a record of these failures. This record may in turn be regarded by others, if not by the loser, either as a work of art or as garbage" (17). The end of *Shakespeare in Love* marks the failure of his romance with Viola de Lesseps (Gwyneth Paltrow) but it simultaneously portrays the beginning of Will's success as a writer. The film proposes that the tragic end of *Romeo and Juliet* and the not-entirely hopeless beginning of *Twelfth Night* be seen as a veiled record of Will's past and Shakespeare's future. Whereas early in the film Will's work is portrayed as something like "garbage" (particularly the early draft of *Romeo and Ethel the Pirate's Daughter*, casually discarded on the street), the end of the film marks the beginning of success, and the reception of this particular loser's record as "art."[10]

Does the *Shrew* fit?

Many of the plays of Shakespeare describe the theater as having great powers to affect its audiences. Hamlet uses the players' performance of "The Mousetrap" to "catch the conscience of the King" (2.2.605). In the comedies, performance within the play is sometimes less, or differently, effective. The mechanicals' performance of "Pyramus and Thisbe" in *A Midsummer Night's Dream* provokes mockery from its onstage audience and laughter from its "real" audience. Yet in the Induction to *The Taming of the Shrew*, a messenger reports that a play can "bar a thousand harms and lengthen life" (Ind. 2.136). Such medicinal faith in the power of performance is in a sense embodied by the play. *Shrew*'s play-within-a-play encompasses most of the text itself—five full acts after the "frame" set by the Induction. However, unlike the play-within-a-play in *Midsummer*, *Shrew* does not reveal the effect that its play-within-a-play has on its onstage audience: it is unclear whether watching the play changes Christopher Sly.[11] One of our aims here is to show how *Shrew* substantiates the power and purpose of performance.

Twentieth-century criticism of *Shrew* has addressed (or redressed) the play for its "problem," the patriarchal ideals preached in Katherina's final speech. As Felicia Londré notes, *Shrew* is considered problematic because it "challenges or offends the sensibilities of large segments of the theatregoing public" with its "presentation of certain no-longer-acceptable assumptions about gender" (85). Even though the play "remains a popular favorite with audiences" (88), solving the patriarchal "problem" has become a focus of much of *Shrew*'s stagings, as well as of its literary criticism.[12] Various ways of softening Katherina's final speech have been attempted on stage and on screen. In the 1929 film starring Douglas Fairbanks and Mary Pickford, a wink in the final speech counters the subservient content of the speech by introducing a degree of dramatic irony (Haring-Smith 125–26).

From Lunt's 1935 production to Andrei Serban's 1988 production, directors of *Shrew* have placed the text in a world of high farce to prevent an audience from feeling that the story deserves serious thought (130). John Barton's production of the play (and several others since) added the epilogue from the "bad quarto," *The Taming of a Shrew*, in order to bring back the Induction's characters, and thus distance the plot of Katherina and Petruchio (149). On paper, Leah Marcus has defended this idea with her argument that it is patriarchal editors who ruled that *A Shrew* is a "bad" quarto in the first place. Other critics, such as Juliet Dusinberre and Michael Shapiro, have cited the play's metatheatrical references as evidence that the audience should be aware of the distance and performed nature of Katherina's speech. Marianne Novy tempers the misogyny in the play by pointing out the coexisting ideals of mutuality in Kate and Petruchio's relationship. All of these readings, be they critical interpretations or theatrical appropriations, offer responses, if not exactly solutions, to the *Shrew* problem.

Rather than expanding on, or adding to the body of work that analyses this "problem," we will instead focus our attention on the relationship between identity transformation and performance in the play. Katherina's "taming" is a *kind of* metamorphosis from shrew to obedient wife, and we intend to reexamine, in transversal terms, the way in which Shakespeare's play depicts this transformation, as well as the concept of identity in general. After taking stock of these practices of identity and transformation, we will turn to *10 Things I Hate About You* for a consideration of the moral ambiguity in this contemporary "taming," and its implications for a theoretical approach to cinematic performance.

The text of *The Taming of the Shrew* negotiates identity in a variety of ways. Most of the characters in the play assume another role at some point: some (such as Tranio, Lucentio, and the Induction's Page) *choose* to pretend to be someone else; others (like Katherina or Christopher Sly) are *induced* into their transformations. Either way, their performances require the crossing of the boundaries of their subjective territory, which is the culturally constructed, ideologically constrained existential and experiential realm in and from which a given subject of a society perceives and relates to the world; and, by moving outside of their established identity, they pass, though usually only temporarily, into the subjective territory of another or others, although the lasting effects of their movements must differ. In an especially insightful reading, Gwyn Williams identifies the induced metamorphoses as the major theme of *Shrew*. For instance, by the Induction's end, the drunkard Sly is convinced that he is actually a lord. In this way, as Williams puts it, the Induction "warns us to look out for changes of identity in the play... And changes of identity we will find, at more than one level; that is the theme of the play, not just the subjugation of a woman by a man" (Williams 25). As Williams suggests, the Induction signals for the audience the deconstruction of social identity through the act of transformation that continues throughout the play.

Barry Weller adds another point to the examination of identity presented by the play: "When the entire text of the play is performed, its subsidiary actions, with their abortive, incomplete, or explicitly theatrical attempts to reshape identity, put the finality and fixity... of the 'new' Katherina, in question" (297). The emphasis on the theatrical aspect of Katherina's transformation functions as a commentary on identity: "If the characters of the play seek to individuate themselves and the spectators judge that they have failed, either the playwright has failed in some rather fundamental way, or *he is commenting on the very enterprise of self-individuation*" (345, our emphasis). In this way, Weller suggests that the text itself is drawing attention to a concept of human identity that is inextricably linked to theater and performance.

"A Christmas gambol or a tumbling-trick?"

Christopher Sly does not know what a play is. He asks, "Ist not a comonty / A Christmas gambol or a tumbling-trick?" and the Page responds that a play "is a kind of history" (Ind. 2.137–40). This exchange draws attention to the difficulty of defining a "play," or, for that matter, "theater" or "performance." After all, the Page's words "kind of" betray the imprecision of his own definition, and the invocation of "history" suggests the mediated citational process through which all performances are infused with meaning. In a theater, because the actions onstage are marked by the context as "theater," it is usually obvious to both spectators and performers that a given action is a "performance." However, the wide-ranging scope of performance theory has demonstrated that performance occurs outside the theater as well, in the forms of ceremony, ritual, games, and daily human interactions. It is also common for scholars of sociology, anthropology, and linguistics to use theatrical metaphors and models within their respective fields of study.

When sociologist Victor Turner examines rites of passage in the pattern of separation, transition, and reincorporation, he recognizes this pattern as performance, calling his historical examples "social dramas." In "Semiotics of Theatrical Performance," Umberto Eco considers the "performance" in semiotician C.S. Pierce's example of the Salvation Army's placement of an unaware drunk person on a platform to promote temperance. These examples are outside of a "theater" proper, but are still considered performances. "Social dramas" (outside the theater) and "aesthetic dramas" (in a theater) affect one another in a model developed by Turner, and expanded by Richard Schechner.[13] The social drama is formed through "underlying aesthetic principles," and aesthetic drama draws from the "underlying processes of social interaction" (Schechner 190). Social and aesthetic dramas are shown to be intrinsically related, but the "kind of" behavior (to apply the Page's phrase) needed is not clarified. If, as Victor Turner argues, social rituals are performances, and if Umberto Eco's example of C.S. Peirce's drunkard can be considered a performance, how do we determine whether an action is a performance or not?

In "Universals of Performance," Herbert Blau explains: "What is universal in performance is the consciousness of performance" (148): if someone is aware of a certain behavior as performance, then their actions constitute a performance. To clarify Blau's idea, Marvin Carlson alludes to Hamlet's defense of mourning: "These indeed seem, / For they are actions that a man might play"; "But," Hamlet goes on to note, "I have that within which passes show" (2.2.83–85). Hamlet admits that these visible actions could "seem" to be a performance, a display of false grief, but Hamlet argues that in this case his "that within" is identical to his "seeming." The agency and awareness of displaying the signs of grief must also include an awareness of a self set apart from the grieving role: "that within." A player's double consciousness of the difference between role and self creates this "seeming," creates a performance. If the consciousness of this separation does not occur, or if it disappears, then the action is no longer performance, but "actual reality." In performance, consciousness of a preexisting self is inescapable, even definitional. Konstantin Stanislavski, whose acting techniques strive for concentration on the role rather than the self, also alludes to this requisite self-consciousness: "Yet all the time the actor can easily continue to play his part even while he is observing himself" (Stanislavski 267).

The self-consciousness that marks performance does not however imply a lack of authenticity or freedom. Even in the theater, a script does not fully determine the action on stage, but instead offers a set of rules that the performance is either confined by or exceeds. Schechner claims that such a boundary is necessary for play (or "a play") to exist, and thus play is not "free": "Playful activity constantly generates rules, and although these may change swiftly, there is not play without them. In other words, to use terms developed earlier, all play is 'scripted'" (101). The script is a performative boundary, in relation to which an improvisational authenticity exists.[14]

Such a scriptural boundary, which constitutes and contains its own set of rules, can be described as a "frame," a term that generally refers to a system organizing meaning, human interaction, and behavior. In *Frame Analysis*, Erving Goffman explains, "all social frameworks involve rules" (24). A "frame" governs the limits of any behavior, as well as the interpretation of any sign. This semiotic concept, as Gregory Bateson argues in "A Theory of Play and Fantasy," is intrinsically related to the consciousness of performance: the subject of this consciousness is the frame. Bateson identifies three types of communication:

> a) Messages of the sort which we here call mood-signs [denotative signs]; (b) messages which simulate mood-signs (in play, threat, histrionics, etc.): and (c) messages which enable the receiver to discriminate between mood-signs and those other signs which resemble them. The message "This is play" is of this third type. (189)

The third kind of message is the communication of the existence of a "frame." Within this "play frame," the second type of simulated message exists. For example, in the theater, we can observe that an audience will not interfere when one character "kills" another because the audience is aware that the murder is simulated, a species of play. This example holds because the actor and audience are conscious of the alternative frame, the "play frame," where the action does not represent murder as it would outside the theater. The frame governs, and is governed by, a specific system of interpretation, and thus those within it—both actors and audience members alike—respond in accordance to its rules. Performers perform with consciousness of the rules both inside and outside the frame, what Marvin Carlson calls a "double consciousness" (Carlson 4), while the audience members perform their role with a similar awareness of the "real" and the "theatrical" frames of the performance.

In our example of a staged murder, the frame is communicated to, and interpreted by, the audience, which plays a key role in constructing the frame by recognizing its existence and understanding its meaning. However, this is not always the case. Goffman focuses on the communication of these frames and introduces two terms, "keying" and "fabrication." He suggests that the introduction of a new frame is basically a form of "transformation." For instance, in a staged murder, a body lying motionless on the ground within the frame signifies a dead body. Communicating this change of meaning is what Goffman labels "keying": "a systematic transformation is involved across materials already meaningful" in which "participants in the activity are meant to know and to openly acknowledge that a systematic alteration is involved" (45). This keying is the process of establishing the frame of new rules or meanings for another person or group. The audience then knows that the motionless body is not actually dead, but *means* dead inside the frame. "Fabrication," on the other hand, is "the intentional effort of one or more individuals to manage activity so that a party of one or more others will be induced to have a false belief about what it is that is going on" (Goffman 83). In this process, the "deceiver" purposefully communicated wrongly (or more precisely, differently) a rule of the frame. Difference of perception is the result of fabrication: "For those in on a deception, what is going on is fabrication; for those contained, what is going on is *what* is being fabricated" (84).

In defining keying and fabrication, we have discussed consciousness with regards to both the audience and the performers. Yet, just whose consciousness is necessary for "performance" to occur, that of the performer or that of the audience? This point is much debated. In most performance situations, especially in the theater, the performers and the audience are both conscious of the frame. Yet, in fabrication, the "deceived" does not possess the double awareness. The object of the deceiver's performance, the "audience" of this deception, is not conscious of the frame. The audience's consciousness of the play frame is not required to call this fabrication a performance. Performance relies on the consciousness of the performer, and the presence, though not necessarily the consciousness, of an audience.

Our discussion of the terms "performer" and "audience" has remained general, but performance practice is invariably local, and the identities of performer and audience accordingly specific. Further discussion of performance requires us to define precisely our conception of "identity," or, more appropriately, "social identity." In our view, all identity is social insofar as humans are social beings. Although identity is formed by the acts of a subject, and preserves a semblance of autonomous creation, it is the social codes surrounding that subject that define the acts and verify their meaning. In the case of gender, for instance, there are social rules informing an identity referred to as "woman" which prescribe certain acts that compose and define "woman"—rules that have little to do with the biological sex of the subject—and thus work to consolidate a given woman's subjective territory.[15] Identity is a perceived notion of self that is shaped and constrained by the ideology of the society that develops and monitors the individual's subjective territory; identity is socially constructed and framed, and must be performed or displayed to be affirmed. The process of fulfilling or altering a social identity is rarely conscious, and is most often effected by the influence and recognition of others. Identity transformations occur through transversal movements, and these movements can be either conscious (self-enacted) or inadvertent undertakings.

The performative nature of social identity is a thematic concern of *Shrew*, where many of the play's characters find ways of harnessing the transversal power that makes identity mobility and transformation possible, which is to say they harness the power, from whichever source, that induces or fuels movement outside of a subjective territory, and they do this precisely for the purposes of performance and identity transformation. In fact, most of the characters in the play change their "role" and perform in some way. In the Induction, the Lord is the main performer. With his servants and players performing as part of his plot to transform Sly into a lord, the Lord functions as the "director" (to borrow a term from contemporary theater and cinema), the framer of the performance, and an actor within the performance event. The Lord's invention is "sport," which he instructs his servants to join in, most notably during the long monologue at the end of the Induction's first scene, in which the Lord provides the Page with a "how to" for performing a female role. These instructions not only demonstrate the Lord's consciousness and agency in the performance—"Sirs, I will practise on this drunken man" (Ind. 2.34)—but the flexibility of social identity. If the Page can be taught to play the part and "become" a lady, then the patriarchal female identity that Katherina will later be induced into "becoming" is no more stable than the Page's performance. In the Induction, gender is framed as a role that can be learnt and performed, and the play-within-the-play functions to frame the identities of the characters within a field of subjective territory that can be transversed through "play."

"He that knows better how to tame a shrew," or, "Don't say shit like that to me, people can hear you"

No one would criticize *Never Been Kissed* for misappropriating the Forest of Arden when Drew Barrymore, whose character distantly corresponds to Shakespeare's Rosalind, stands amid the grassy expanse of a baseball field, tears in her eyes, waiting for her "boyfriend" to show up. In *As You Like It*, Rosalind holds supreme command in Arden, and there the "green world" empowers her to escape, disguise, transform, take sexual license, and even summon a pastoral deity to assist Rosalind in the performance of her own wedding. But *Never Been Kissed* is so far removed from the codes and structures that govern *As You Like It* that neither departure from nor adherence to the source text affects the success or failure of the film. This is not the case with *10 Things I Hate About You*. When confronted with the question of proximity to its Shakespearean source text, *10 Things* is inclined both to *stay* and to *stray*: this is a film suspended between formulaic commercial comedy and self-conscious appropriation of a canonical text.

Since *Never Been Kissed* is only subtextually related to *As You Like It*, any relevance to the source text is only a temporary revelation of the film's latent content and is quickly submerged and re-disguised. The film draws on Hollywood formulae for slapstick romantic comedy, and not any performance model generated from *As You Like It*. But *10 Things I Hate About You* is almost uncloseted in its relationship to the source text. As a result, the debt owed to *The Taming of the Shrew* necessitates obvious structural parallelism: the degree to which *10 Things* tries to strike a balance between Hollywood formula and Shakespearean model reflects the degree to which the film makes sense or starts to break down. The decision to eliminate the Induction of *Shrew* threatens to reduce *10 Things* to a substanceless film with an inevitable conclusion, far removed from *Shrew*'s examination of the constructed nature of performed identity. Any notion of revising the misogynistic content of *Shrew* simply by placing the same plot in a contemporary setting fails: *10 Things* relies on a structural model that reinscribes rather than revises a patriarchal system. Without the Induction—which we argue is the key to any interpretation of the source text that mitigates the sense that Kate has been merely subjugated—*10 Things* reinforces the very system that it seemingly seeks to condemn.

Instead of employing an Induction, *10 Things* begins by introducing the "Shrew," and defining her through a contrasting image of "normal" teenagers. A Volkswagon convertible filled with young women is pulling up to a stoplight, the teens bouncing along unself-consciously to a cheery pop song. But their cheer is cut short by a single withering glare from the next car: Kat Stratford (Julia Stiles) has pulled alongside, all alone in a self-consciously ugly sedan. The cheery pop song is overwhelmed by Joan Jett's hard rock anthem, "I don't give a damn about my bad reputation," blaring from Kat's stereo. The juxtaposition of music, clothing, and cars clearly casts Kat as a "Shrew": the visual, behavioral, and ideological antithesis of what the film readily identifies as the social "norm."

The setting of *10 Things* has been transposed from Shakespeare's Padua to the castle-like "Padua High." A series of short scenes—in the school's quad, an English class, and the office of the oversexed guidance counselor, Miss Perky (Allison Janney)—quickly introduces the main characters. Cameron James (Joseph Gordon-Levitt) is a new student and the counterpart to *Shrew*'s Lucentio. While being shown around Padua High by Michael Eckman (David Krumholtz)—the Tranio figure and a loser—Cameron catches sight of the beautiful and popularity-obsessed Bianca Stratford (Larisa Oleynik). He instantly falls in love, whispering: "I burn, I pine, I perish."[16] Cameron immediately begins to plot a way to woo Bianca. Michael points out that it won't be easy: "It's a widely known fact that the Stratford sisters aren't allowed to date."

In *10 Things*, both Katherina and Petruchio are rendered as rebellious teens testing institutional authority. Patrick Verona (Heath Ledger), the Petruchio equivalent, is first seen in the guidance counselor's office, where he is being scolded for allegedly exposing himself in the cafeteria. The first encounters with Kat are similarly antiauthoritarian. Kat argues with her English teacher over the lack of women novelists on the syllabus, while rich pretty-boy Joey Donner (Andrew Keegan) casts gender-specific aspersions at her. Kat is more than able to hold her own with the slow-witted Joey and even with her teacher, but the latter sends Kat to the office.

In this introductory sequence, the film establishes a series of clearly defined social cliques into which every character in the high-school setting is pigeonholed, but subsequent performances and transformations undermine these seemingly rigid identities. After a comic interlude in the Stratford household, the sisters' father decides to relax his rule forbidding his daughters from dating: he introduces a new rule that Bianca will be allowed to date, but only when the staunchly anti-dating Kat decides to date. Bianca notes that this is practically the same as the previous system since her sister is "a particularly hideous breed of loser." Cameron joins forces with the otherwise execrable Joey in order to find an adequate partner for Kat. Joey, unaware that Cameron is his rival for Bianca, agrees to pay Patrick to do the job, despite the hyperbolic urban legends that surround the loner. Patrick promptly walks up to Kat, but his awkward advances are summarily rebuffed. Cameron, witnessing this first attempt, fears failure from the outset.

Following this initial contact with Patrick, the film begins to question the stability of Kat's identity as an aggressively misanthropic outsider. Bianca, who had previously revealed to Cameron that Kat "used to be really popular, and then it was like, she got sick of it," confronts Kat about her unstudied appearance and antisocial demeanor. Kat's defenses waiver, but she rebuts, "You don't always have to be who they want you to be, you know." Under the influence of Patrick and Cameron, both Bianca and Kat experience transformations that moderate their (polar and opposite) approaches to social interaction.

Though her goal is to go out with Joey, Bianca spends time with Cameron, helping him "research" her sister's tastes so that Patrick can perform the role of her ideal date better ("So you're telling me I'm a non-smoker?" Patrick asks. "Just for now," Michael replies). Patrick arranges to be seen by Kat at a concert by her favorite band, and after impressing her with his (fabricated) knowledge of her favorite "girl bands," she grudgingly agrees to go to a party with him. But by the end of the party, when Kat tries to kiss him, Patrick demurs, apparently suffering from an ethical dilemma. At the same party, Bianca realizes that Joey is a posturing, inauthentic performer, and transfers her affections to the sincerely devoted Cameron.

Kat is once more cold to Patrick. In a heartfelt confession to Bianca, we learn that Kat has been furious at Joey for years: she lost her virginity with him, but he subsequently rejected her, catalyzing the misanthropic identity that Kat has performed throughout the film. Patrick tries to use his bogus knowledge of Kat's tastes to win her over again, confronting her in a bookstore ("Excuse me, have you seen *The Feminine Mystique*? I've lost my copy"), but unsuccessfully. In the lunch line, Michael tries to encourage Patrick with the words, "Sweet love, renew thy force,"[17] but Patrick does not get the reference: "Don't say shit like that to me, people can hear you." Despite his disaffection for Shakespeare, Patrick takes the advice, performing a love song for Kat on the Padua High athletic field. As a result, he achieves the goal for which he was hired: Patrick and Kat go to the Padua Prom together.

The system of keyed performances and fabrications reaches its climax in the Prom scene. Although Cameron and Michael are "keyed" to Patrick's (increasingly sincere) performance, Kat is not. For her, Patrick's (decreasingly fabricated) affection has always been the real thing. Joey is aware that Patrick's affection was initially fabricated (he paid for it), but Joey is not keyed to the secondary fabrication arranged by Cameron, Michael, and Patrick. Furthermore, Patrick, ever since Kat first tried to kiss him, is concerned with the dwindling distance between the "seeming" of his outward performance and the "that within" of his authentic feelings.

All performance frames are threatened when Joey arrives at the Stratford house to pick up Bianca and realizes that she's already gone to the Prom with Cameron: the frames separating the conditions of fabrication collapse. Joey causes a three-way fight at the Prom. He decks Cameron for duping him and upbraids Patrick for taking his money while being aware of Cameron's ploy. Although Bianca steps in and beats up Joey on behalf of Cameron, Kat, and her own new sense of identity, the critical break-frame has already transpired: Kat realizes that her transformation from misanthrope to romantic has taken place as part of a commercial transaction. She becomes furious with Patrick, unwilling to believe that he has experienced a transformation of his own.[18]

"Therefore, they thought it good you hear a play"

The structure of *The Taming of the Shrew* leads its audience through a series of successive losses of performance frames. More specifically, the play progresses to a structure with fewer frames between it and the audience and, in this way, becomes decreasingly distanced from its audience. The most obvious example of this trend within the play is the existence of the Induction, whose characters disappear after the first act of the play. The Induction's plot creates a frame for the five remaining acts. The plot in Padua is then a performance itself, played by the traveling players for Sly and the Page, disguised as his lady. Previously, we suggested that the Induction's purpose is to introduce the themes of metamorphosis, foreshadowing Katherina's transformation. Apart from this content, its form also creates an additional effect: pulling the audience closer to its performance as it progresses. This "closeness" results from the fact that *Shrew*'s audience is likely to forget about the Induction by the end of the fifth act. Although Garber argues that the Induction's frame "performs the important tasks of distancing the later action and of insuring a lightness of tone" (28), after four and a half acts, the audience may be inclined to forget about Sly, and without his presence, the Induction's "distancing" is no longer firm. Haring-Smith's diagnosis is that once Sly is removed, "he is no longer effective for distancing the taming plot. The audience soon forgets him and the play-within-a-play structure evaporates" (Haring-Smith 5). It may be more precise to say that the Induction is simply left behind.

Instead of watching Sly watch the play, the audience shifts roles, and by the end they have assumed the position previously occupied by Sly. In this way, the question of the effect of the performance is subtly transferred from Sly to the audience. Is the audience, by this elision, transformed? Although *10 Things* would seem to have eluded this question by dispensing with an Induction, the resolution of the plot and the final effect of the film depend on two registers of self-conscious performance: a juvenile Shakespearean sonnet that takes the place of Katherina's final speech in *Shrew*, and a meta-cinematic coda to the film that in many ways draws on the framing power of *Shrew*'s Induction to highlight the constructed nature of the foregoing representation.

After Patrick's initial motives are revealed to Kat, she abandons him at the Prom, saying, "You are *so* not who I thought you were." Patrick does not see her again until English class, when Kat volunteers to read her homework assignment, a love sonnet.

> I hate the way you talk to me
> And the way you cut your hair.
> I hate the way you drive my car,
> I hate it when you stare.
> I hate your big dumb combat boots
> And the way you read my mind.
> I hate you so much it makes me sick,
> It even makes me rhyme.
> I hate it—[19] I hate the way you're always right.
> I hate it when you lie.
> I hate it when you make me laugh,
> Even worse when you make me cry.
>   I hate it when you're not around and the fact that you didn't call.
>   But mostly I hate the way I *don't* hate you, not even close, not even
>   a little bit, not even at all.

With the final lines, Kat bursts into tears and runs from the classroom. Patrick meets her in the parking lot, where they kiss and make up. The end.

Though no title for this poem is given in the film, this poetic list must be the "10 Things I Hate About You" of the film's title. The sonnet initially emphasizes the constructed aspects of identity, the very "seemings" by which Kat (and the audience) judge others. But the poem moves on to illustrate the effects these outward appearances have on

Kat, concluding by describing the pain that Kat experienced when the rules of the fabrication were revealed to her. The final lines, however, make clear that a full transformation has occurred: despite the discovery of a disparity between outward performance and inward identity, which the first twelve lines address, the final couplet indicates that Kat's feelings have not changed with that discovery. Kat hates the fact that the performative fabrication had its intended affect, and though she is finally keyed to the rules of Patrick's performance, she cannot reverse the change that has taken place. The fiercely independent Kat is forced to admit that she has been "tamed" by Patrick, even in the face of his duplicity.

The abruptness of the film's conclusion, and Kat's poetic confession of romantic metamorphosis, do nothing to dispel the sense of moral ambiguity that surrounds Patrick's motivations (he got his money, just as did Petruchio), nor does this conclusion empower *10 Things* to redress the patriarchalism and misogyny in *Shrew* (an outspoken, unattached young woman is repeatedly called "bitch"; later, she writes love poetry and starts dating). However much *10 Things* may emphasize the constructed and performative nature of identity and social interaction, there is nothing in the film to "key" the audience to the constructed and performative nature of the film itself. There is not even the trace of an Induction to mitigate the narrow gap between audience and the film: never having once been keyed to the potential for change that performance can effect, there is no reason for an audience to question the resolution of *10 Things*, or challenge the moral dilemmas it presents.

As our preceding analysis shows, there is more at stake in the *Shrew* Induction than a nostalgic attachment to Shakespearean performativity. The Induction is the epicenter of *Shrew*: conceptual, thematic, and performative registers in this play all proceed from the Induction. In dispensing with the Induction, but subsequently following closely the structure of *Shrew*, *10 Things* must labor to highlight the constructed nature of its characters' identities with nothing more than one liners and self-consciously amateurish poetry.

However, those who continue to watch *10 Things* after what would conventionally be considered its ending will have access to a part of the film that belatedly attempts to frame the narrative, responding to the lack of an Induction by highlighting the constructedness of the film after the fact. A sequence of "out-takes" accompanies the credits. Instead of Patrick refusing to kiss Kat after the party, we see Ledger and Stiles goofing around, laughing, and making out. At the party, instead of Michael and Cameron planning Bianca's seduction, we see Michael seducing Cameron. We see Kat and Bianca having a fight in the hallway of Padua High, a fight that never appears in the film, and one that is cut short when Kat and Bianca come round a corner to find Miss Perky and a man wearing only a towel vigorously making out in front of the student lockers; the actors dissolve into laughter.

The out-take scenes serve as a kind of belated Induction, a "Reduction" if you will. This Reduction is a return to key scenes in *10 Things* (and some scenes not in *10 Things*), but with a difference. The return creates the opportunity to redo the scene in a different way, and in each case the redo fails: the scene terminates prematurely either because the actors "go up," or because an off-screen voice calls a halt to the performance. The Reduction reveals the way the film was made. Each out-take reduces the fabrication to practice, separating character from actor, breaking the cinematic frame over and over again. Here, in the Reduction, the audience has access to the meta-performances that constitute the performance-intense *10 Things*, to the fabrication of the fabrications that make up the film.

By emphasizing actor over character, the Reduction provides *10 Things* with the framing function that an Induction would have provided at the outset. If Shakespeare's *Shrew* can be said to open a frame but not close it, then *10 Things* closes with a frame though it did not open with one. Those in the audience who watch the credits have the opportunity to be keyed to a level of meta-fabrication unavailable to other viewers of *10 Things*.

Back to the (Post-Cinema) Future, or, Hamlet takes Manhattan

We return from our flights of Shakespearean–Burtian loser fantasy[20] to the Almereyda *Hamlet*, an example of Shakespearean performativity that strives to modernize the source text without straying too far from the source. This film is one of a minority of current films that, unlike *10 Things*, *Never Been Kissed*, and other distantly Shakespeare-related films, is actually a production of a play by Shakespeare, and announces itself as such.

The Almereyda film stands as an unexpected response to the question posed with no small trepidation by Lynda Boose and Richard Burt in their collection *Shakespeare, The Movie*: "Just what kind of an American *Hamlet* is destined to succeed Mel Gibson's action hero is indeed a topic to puzzle the will" (18). In the years since 1996 when this question was asked in print, the film industry has seen a continued decline in the market dominance of the action film genre, and the continued ascendance of the quirky independent film, a zeitgeist first heralded by the critical and financial success of the low-budget independent film *The Full Monty* (dir. Peter Cattaneo 1996), later to be spectacularly crowned by the shockwave of popularity that followed the release of the ultra-low budget "indie" *The Blair Witch Project* (dir. Eduardo Sanchez et al. 1999). Almereyda himself came to prominence on the strength of an inexpensive black and white vampire indie, *Nadja* (1995), which notably made clever use of a children's toy lens for its most striking visual effect. The response, then, to Boose and Burt's question comes as a response to the cultural moment.

Almereyda characterizes Hamlet as "a poet/filmmaker/perpetual grad student" (22), but it is specifically the second of these three contemporary archetypes of the contemplative hero (as opposed to "Gibson's action hero") that Almereyda puts on screen. Hawke's Hamlet is a melancholy prince of indie art films: in the "Mousetrap" sequence, he attempts to catch the conscience of the king, not with a staging of "The Murder of Gonzago," but with an experimental video presented to Claudius (Kyle MacLachlin), Gertrude (Diane Venora), Ophelia (Stiles), and their entourage in a private screening. Notably, the opening credits of Almereyda's *Hamlet* are identical to the credits for Hamlet's film-within-a-film. In fact, the Almereyda *Hamlet* eliminates the many references to popular theater and the technologies of writing and reading (newly popular in early modern England) that litter Shakespeare's play, and replaces them with references to twentieth/twenty-first century popular media and the technologies of visual recording and reproduction (still relatively new in millennial United States) that are shown in virtually every shot of the film. In other words, despite the fact that Almereyda's film is a radical adaptation, the only liberty that Almereyda has taken with the source text of Shakespeare's *Hamlet* is that he has appropriated a self-consciously theatrical play and made it into a self-consciously cinematic film: a media-savvy adaptation for a mediated era.

Despite the insight driving Almereyda's cultural appropriation of the source text, many scenes in this film are jarringly incongruous. In between virtuoso displays of pop-culture montage, many scenes are lumbering and conventional, lacking the understanding of the source text displayed by other parts of the film. Contemporary (late 1990s/early 2000) Shakespeare films are rife with such lapses: Kenneth Branagh's *Hamlet* (dir. Branagh 1996) presents an Act I, Scene 2 filled with "flashbacks" to events that never occur; the 1999 version of *A Midsummer Night's Dream* is a meditation on the home life of

Nick Bottom (Kevin Kline) and the popularity of the bicycle in nineteenth-century Italy.

Even in a best-case scenario—and the Almereyda *Hamlet* may be one—the understanding of onscreen performance lags far behind the offscreen effort invested in reinventing the context of Shakespeare's play. Almereyda's astute indie deployment of contemporary artifacts and practices in response to congruent artifacts and practices in the source text is really just a more conceptually sophisticated version of what goes on in big-budget studio productions: casting famous names, changing the setting, designing costumes, and riding bicycles consistently takes precedence over figuring out what kind of performance the four-hundred-year-old text needs from contemporary actors.

The public theater of early modern England was located in the marginal places set apart from the mainstream social spaces of London. Geographically, this point is demonstrated by the placement of most public theaters outside the city of London, and/or in its legally flexible "Liberties." The name implied exactly what this space embodied: a freedom from London's social and municipal regulations. In *Becoming Criminal: Transversal Performance and Cultural Dissidence in Early Modern England*, Bryan Reynolds discusses the transversal impact of these theatrical spaces in Southwark and beyond:

> Southwark and its Liberties were spaces of transversality where radical identity becomings transpired, social demarcations were blasted, and where change and its myriad possibilities were triggered and conducted by the transversal theater. The suburb proved so provocative that it also became a battlefield during such demonstrations as the Shrove Tuesday riot of 1617, when a throng of London moralists attempted to demolish a public theater and its neighboring brothels. Yet transversal power did not stop at Southwark's borders. Most strikingly, in addition or subsidiary to criminal praxis, it infected London *as* and *through* the fashion of female-to-male transvestism, which meant not just cross-dressing but also the impersonation of the masculine gender in the "naturalistic sense" characteristic of the public theater. (141)

In the Liberties, citizens pursued unconventional and disreputable pastimes, such as gaming, excessive drinking, bear-baiting, and prostitution.[21] This space was governed by a "play frame," within which the rules were different from the accepted norms. Activities like transvestitism, prostitution, and theater were thus permitted or tolerated in the Liberties, but not in London's conventional spaces.

Numerous Elizabethans and Jacobeans, often Puritans, wrote tracts expressing their disgust with and fear of the theater. The theater was considered deviant and a threat to society because of its transformative nature. In his essay "The Devil's House, 'or worse,'" Reynolds documents the fears associated with the public theater, and argues that the antitheatricalists were right. The theater challenged the notion of God-given identity and, by extension, the social order founded on this notion. The theater exercised its dangerous and alluring powers of permanent metamorphosis (or "identification" with the fiction) for both actor and audience. The period's writings about theater reflect anxiety that the consciousness of the frame will disappear, and the illusion will become truth: for an audience member, the performance becomes "deceiving" fabrication; for a performer, the performance ends and becomes behavior. The early modern English theater was dangerous, as Reynolds shows, because it demonstrated an understanding of identity as transformational.

The contradictions within the period's reflections on the practice of the theater suggest that for Elizabethan and Jacobean theatergoers, consciousness of the theatrical frame is both retained and lost. The awareness of the play itself remains intact. The actor is "insincere," and thus aware of a "seeming" along with a "that within." The reflections of the audience, in the form of the writings about the theater, share this awareness that a play frame separates actor and character. The framing function of the theater itself, however, as Reynolds demonstrates, was not self-contained. The very idea of transformational identity leaked outside the Liberties into the conventional social spaces of London in various forms of criminality. The antitheatricalists themselves display an understanding of the conceptual ideas that they are trying to fight. Their writings reflect a consciousness of the idea that an audience member might learn the negotiability of her or his own subjective boundaries and identify with a performing self, subsequently exercising the power to play another role in society. The audience could learn that identity is not God-given but socially constructed, prompting change in their perceptions of identity, and leading to a conceptual or actual change in their own identities: a transformation through performance. Performance, framed by the practice of theater, held the transversal power to enable such a change of conceptual rules.

The trappings and the suits of woe: a practice of Nudge, Nudge, Wink, Wink, Know What I Mean, Know What I Mean?

Now we want to return, via a circular route, back to acting itself, and acting onscreen in particular. In a statement as true for film acting as for that of theater, Philip Auslander explains how one "reads" acting:

> We arrive at our perception of a performance by implicitly comparing it with other interpretations of the same role (or with the way we feel the role should be played), or with our recollection of the same actor in other roles, or with our knowledge of the stylistic school to which the actor belongs, the actor's private life...our perception of the actor's work derives from this play of differences. (60)

Whether one is aware of it or not, all of these elements are engaged in the interpretation, or "reading," of a film actor's performance. Some audience members bring specialized knowledge to a film that allows them to recognize these individual elements or to reveal them at work. Conversely, some films willingly reveal those constitutive elements that together constitute an actor's "meaning." The theoretical approach to (screen) acting that we playfully choose to call "Nudge, Nudge, Wink, Wink, Know What I Mean, Know What I Mean?"[22] fetishizes this revelatory process by placing on the individual actors the responsibility for keying the audience to the various frames that code and control the post-cinematic representation: by being made aware of the parameters, the audience is made to participate.

Our theory finds correlative support in the Godard masterpiece, *Contempt* (*Le Mépris*). Alain J.-J. Cohen analyses the multiple, interconnected systems at work in the film to reveal "the *levels* at which filmic discourse plays itself out" (118). Cohen's subject is the screening of a film-within-the-film ("FwtF," as Cohen calls it, 115) in *Contempt*, a sequence that most clearly articulates this film's complex "reflexivity": "Aside from the intensity it provokes, just as Freud's *Witz* makes us be at two places at the same time, the film-within-the-film (or the play-within-the-play) illuminates the very principle of composition" (127). And in this case, the compositional principle at work in Godard's film is a complex framing of the way the actor *means*: the famous film director Fritz Lang is an actor in *Contempt*; he plays a character named Fritz Lang who is a famous film director; Jean-Luc Godard, the famous film director, is directing Fritz Lang, but he also plays the character of the assistant director to the character Fritz Lang (played by Fritz Lang, whose films were a great influence on Godard); and in this scene, which Godard has directed, Lang's character screens a film (FwtF), which the Lang character has directed, in which the characters and events of *The Odyssey* are juxtaposed with those in the film that frames them. Reflexivity breeds complicity. The film is only reflexive because its many internal and external references are understood by its audience; these references in turn are only understood because the audience has been keyed to recognize the multiple frames at work in the film, to read intertextually the "levels" of meaning, and, by performing these operations, the audience in fact becomes complicit with the filmmakers in the making of the film.

Notably, Cohen's semiotic analysis of Godard's film-within-the-film begins with a quote from *Hamlet*: "The play's the thing / Wherein I'll catch the conscience / of the King" (115). Though *Contempt* takes full advantage of the film-within-the-film device to demonstrate the many layered reflexivity of composition itself, the film-within-the-film presented in the Hawke/Almereyda *Hamlet* does not: though, as mentioned earlier, the opening credits of Almereyda's film-within-the-film reprise the opening credits of this *Hamlet* itself—creating an interesting *mise en abime* effect—the film-within-the-film subsequently devolves into the merely glib. The key issue here is one of acting, and this issue is not qualitative, it's quantitative: the "Nudge-Wink-Know" approach to performance would demand that, in the role of Hamlet, Ethan Hawke must do more than one thing at a time. Being "in the moment" is not enough: one must also *point out* that one is in the moment; one must be in the moment in a way that is *about* the moment.

This approach would require the director and the performers to do more than Godard and Lang did with *Contempt*. The intense reflexivity of *Contempt*'s film-within-the-film is not accomplished by way of Lang's performance as an actor, but rather by the way Godard deploys Lang's performance. Lee Siegel argues convincingly that a different deployment of the actor is at work in Stanley Kubrick's last film, *Eyes Wide Shut*, where actors are employed, not merely deployed, in the making of meaning. "Just about every critic . . . mocked what they considered to be Cruise and Kidman's stilted performances. They seemed to be acting like actors, everyone complained" (80). Siegel makes it clear that in *Eyes Wide Shut*, Kubrick's celebrity actors are performing not only their characters but also themselves, keying the audience to a consideration of their dual existence both inside and outside the film frame. Siegel continues:

> In a film about life's essential doubleness, Kubrick presents Cruise and Kidman with double lives. They are actors in a film, and they are people we think we know something about. . . . They act with dreamy formality because they exist between dream and reality. Kubrick wants us to watch Cruise and Kidman and think about what people appear to be and who they really are.
>
> Kubrick's genius in *Eyes Wide Shut* is to make us look at the film the way the film looks at life. (80)

This self-aware employment of actors forces the inclusion of the audience (unwilling, unwitting) into the performance with a gesture that is not unlike Peirce's example of the ostention of the drunk, but whereas Peirce's Salvation Army preacher makes the drunk a performer, Nudge-Wink-Know—a theoretical approach to acting we now retroactively apply to *Eyes Wide Shut*—makes the audience an interpreter of performance. By comparison to Pavis's actor-dramaturgs, Nudge-Wink-Know produces "audience-dramaturgs": moviegoers simultaneously engaged with the film and "contributing to... its meaning" (Pavis 79). This dual awareness is at the heart of Nudge-Wink-Know, a multiform theory of acting that has affinities with—yet is simultaneously oppositional to—the effect of acting intended by Brecht.

Brecht's theory of acting is often reduced to its single most famous tenet, the *Verfremdung*: a "defamiliarized illustration... that, while allowing the object to be recognized, at the same time makes it appear unfamiliar" (qtd. in Rouse 234).[23] In order to provoke his audiences to a reevaluation of their quotidian world, Brecht promoted an acting practice in which "the actor has to discard whatever means he or she has learnt of getting the audience to identify itself with the characters he or she plays" (Brecht 192; 193). This, of course, is a theory at odds with conventional Hollywood filmmaking, as well as Nudge-Wink-Know. Nudge-Wink-Know would require identification along with defamiliarization: a practice of film acting that would constantly key the audience (Nudge, Nudge) to an awareness of the film's frames (Wink, Wink), and lead to a practice of viewing that would admit an awareness of identification (Know What I [The Performer] Mean?).

Brecht's attempt to prevent identification stemmed not from a merely dismissive "alienation," but rather from a commitment to social change. As Rouse points out, "The dramaturgical principle most basic to fulfilling this commitment is, in turn, that the theatre must attempt to present society and human nature as changeable" (230). And it is here in the essential transversality of Brecht's vision that Nudge-Wink-Know finds an affinity with the theorist of the *Verfremdung*. For it is in the potential for creating new social codes and manipulating their frame that performance (for theater or film) can generate transversal power. Transversal power incites people to travel into new conceptual realms, to identify and empathize with others, and thus, in effect, to change. Reynolds's assessment of the public theater in early modern England displays this process at work: "Social identities are therefore subject to theatre, and because theatre, as a conceptual conductor of transversality, is boundless, so are social identities" (Reynolds 165). Change can occur because "conductors of transversality" exist, and, in a certain form, performance can be one of them, for better or worse.

Those who are keyed to performance, either as an audience member or a performer, have access to an enabling power to evade control or take control of others and/or themselves. In this latter category is the Induction's Lord, who calls upon the visiting players to "assist him much." In the former is Sly, who proves to be a spectator who "let the world slip," and as a result his identity—conceptual boundaries, perceptions, subjective territory—are transformed. The Reduction of *10 Things* spurs the audience to question the authenticity of the transformation in the film by encouraging an identification not with the characters, but with the actors themselves. By keying the audience to the frames that govern the fabrication, the audience is tacitly informed (albeit after the fact) of the rules that govern the film's transversal power.

We are here predicting/theorizing/witnessing the emergence of an approach to Shakespearean performativity for the post-cinema era, a mode of acting that has the power to fulfil the function of "nudging" and "winking" at the audience, keying the viewer to the rules and the "knowing" that govern the representation-in-progress. What is an example of post-cinema representation? What is "character" on a CD-ROM? What is "performance" on the Internet? Video games and chatrooms may suggest the provisional answers to these questions, but future reconsiderations of acting will increasingly need to accommodate trans-media projects in order to adequately account for the future of post-cinema responses to Shakespeare.

Notes

1. The title for this piece is an assemblage of catchphrases from a scene from the BBC's comedy program *Monty Python's Flying Circus* (that originally aired from 1969 to 1974): "Nudge, Nudge, Know What I Mean, Know What I Mean?" The script can be located online at http://www.mwscomp.com/mpfc/nudg.html.
2. At the time of writing this essay, Kenneth Branagh released a musical version of *Love's Labours Lost* (2000); his *Macbeth* and *As You Like It* were in the works; a high school basketball version of *Othello*, called *O* (2001), and a parody of *Macbeth, Scotland, PA* (2001), have also been released.
3. For those who may not be familiar with the German-derived term, a "dramaturg" is a theater practitioner who serves as a kind of critical and intellectual consultant to a theater company or a theatrical production. Although the term may seem obscure, the role of a dramaturg has become increasingly important in the United States over the last two decades, and has been a vital and active force in Germany for a century or more. Traditionally,

a dramaturg's work is analytical and interpretive, providing the theory that the rest of the company puts into practice. Pavis, however, fuses theory and practice in the "actor-dramaturg." In his notes to Pavis's article, translator (and dramaturg) Jim Carmody makes it very clear: "dramaturg" is not the same as "*dramaturge*," the latter being a French synonym for "playwright." Despite the outdated editorial format of *The New York Times*—which inexplicably insists on using the spelling "dramaturge" when it should use "dramaturg"—a "dramaturge" is a French theater writer whereas a dramaturg is a theater consultant; it is the latter term that Pavis is deploying here. For a discussion of the role of the dramaturg in Western theater, see Bryan Reynolds and D.J. Hopkins' "The Making of Authorships: Transversal Navigation in the Wake of *Hamlet*, Robert Wilson, Wolfgang Wiens, and Shakespace" in this book.

4. The term, "transversally," is used here according to Bryan Reynolds's "transversal theory." For more on "transversal theory," beyond this essay and the introduction to this book, see Bryan Reynolds, "The Devil's House, 'or worse': Transversal Power and Antitheatrical Discourse in Early Modern England," *Theatre Journal* 49.2 (1997): 143–67, and *Becoming Criminal: Transversal Performance and Cultural Dissidence in Early Modern England* (Baltimore: Johns Hopkins University Press, 2002); and Bryan Reynolds and Joseph Fitzpatrick, "The Transversality of Michel de Certeau: Foucault's Panoptic Discourse and the Cartographic Impulse," *Diacritics* 29.3 (1999): 63–80.

5. In his book *Liveness*, Philip Auslander challenges the idea that live performance—be it dance, theater, performance art, or other—can in fact convey any unique impact in a culture in which audience perceptions have been developed by mediated, rather than live, forms. This idea is the logical postmodern correlative to Fredrick Jameson's assertion that "perceptual habits" formed in the era of high modernism cannot make sense of the spaces of postmodernism (Jameson 39). Auslander asserts that the perceptual habits of individuals acclimated to postmodernism's virtual spaces are not attuned to the "actual" spaces of live performance. If we accept Auslander's position, then the goal of the revitalized millennial theater movement is to attempt to restore a sense of the actual in a culture of the virtual.

6. The most recent cinematic adaptation of *Macbeth*, *Scotland, PA* (dir. Billy Morrisette 2002), is a strange conglomeration of these two approaches to adapting Shakespeare. The film's dialogue is almost completely devoid of Shakespeare's words, with the exception of recorded phrases echoing from a self-help tape that soothes Detective MacDuff (Christopher Walken) as he drives from Scotland, PA to Atlantic City and back. However, the opening credits of the film attribute the story to Shakespeare, thereby emphasizing the film's relationship to *Macbeth*. Although *Scotland, PA* advertises itself as a parody of Shakespearean adaptations, and thus is not so much an adaptation of *Macbeth* as an adaptation of recent cinematic adaptations of Shakespeare's plays, the film seeks to "legitimize" itself from the onset by suggesting in its opening credits that, however "unfaithful" the film may be to Shakespeare's play, it is still to some degree "by Shakespeare."

7. This fad irritates us. In the case of *Romeo and Juliet*, a title that would seem to be inextricable from if not synonymous with all that is culturally bound up in the word "Shakespeare," is not the very notion of calling a film *William Shakespeare's Romeo + Juliet* a gross redundancy?
8. *O*, which recasts Othello as a high-school basketball star, was originally scheduled for release in 1999. The film was "canned" following the Columbine shootings, when movie moguls were wary of releasing any film that could be accused of promoting violence in high schools. For a detailed history of the film's shelf-life, see Rebecca Traister's "The Story of *O*, Weinstein Style: High-School *Othello* is Held Up" (*New York Observer*, November 11, 2000). It is interesting to note that the censoring of possibly "subversive" material following the incident in Columbine had extended to Shakespeare, whose plays were, and continue to be, a staple in the high school curriculum in the United States.
9. For a discussion of *Shakespeare in Love* in relation to "Shakespace," see Donald Hedrick and Bryan Reynolds' "Shakespace and Transversal Power" in *Shakespeare Without Class: Misappropriations of Cultural Capital* (New York: Palgrave, 2000): 3–47.
10. In terms of cultural production, *Shakespeare in Love* performs two functions that endear it to the hearts of Shakespeare fans and academic losers alike. First, when Joseph Fiennes appears as Will in *Shakespeare in Love*, a transfilmic link is drawn between Shakespeare and another role that Fiennes played that same year in the film *Elizabeth* (dir. Shekhar Kapur 1998), the Earl of Leicester. The coincidence of casting facilitates that very consummation so devoutly to be wished, gratifying the academic loser's desire for Shakespeare and Queen Elizabeth I to be lovers. Second, *Shakespeare in Love* proposes a vision of the New World (and by extension colonial power) based on Shakespeare. In the final minutes of the film, Will (in Voice Over) is penning the scenario that will become the germ of his next play, *Twelfth Night*. As he describes the shipwreck that befalls a character named for his lost lover, Viola (Gwyneth Paltrow), we see Viola herself suffer a shipwreck while sailing to the New World. We see her wash ashore alone and begin to walk up the beach toward a towering, unbroken forest. The implication would seem to be that Shakespeare's muse and lover comes to shore in the New World, implying a personal connection between Shakespeare and the continent that would eventually give birth to Hollywood. Further, should Viola be pregnant, the relationship between Hollywood and Shakespeare becomes one of direct lineal descent, a birthright implicitly claimed throughout the film (and not unlike that claimed by Davenant at the reopening of the theaters in 1660).
11. This of course is true only of the Folio version; the "bad" Quarto of 1594 provides a return to the frame of the Induction and a resolution to Sly's experience.
12. Of course, *Taming of the Shrew* is not the only play by Shakespeare containing elements that may offend modern audiences. The "anti-Semitic problem" with Shylock in *The Merchant of Venice*, and the "racist problem" with Aaron in *Titus Andronicus*, are but two examples. In order for

Shakespeare to retain his legacy as a humanist in our more politically correct society, critics and adapters alike have had to confront these "problems," which are often "resolved" by interpreting these characters as subversive portrayals intended to question the social prejudices of their own time.

13. In his discussion of "social" and "aesthetic" dramas, Schechner primarily references the following studies by Turner: *Dramas, Fields, and Metaphors* (Ithaca: Cornell University Press, 1974), and *On the Edge of the Bush* (Tuscson: University of Arizona Press, 1985), 291–301. In *Dramas*, Turner defines and explores the idea of social dramas. In *On the Edge*, Turner—expanding on Schechner's "infinity loop" (itself an expansion of Turner's 1974 study)—discusses the interconnectedness of social and aesthetic dramas.
14. Writing of J.L. Austin's notorious antitheatrical bias, Worthen notes: "Austin excluded theater from the category of felicitous performance" (137n2). But even Austin's work implies what Derrida and Butler both subsequently argue: there is a "conventional status" to all utterance and performance.
15. The issue of the *physical body* of a subject does limit the effect of social definitions on identity. Though Judith Butler claims that the body is a completely "historical idea" (271–72), we believe that biological characteristics (such as disability or sex) can affect aspects of identity beyond certain social codings. However, it is not necessary to resolve this debate in this examination since we will be discussing identity in terms of *conceptual* ideas of self, rather than its *physical* aspects.
16. *The Taming of the Shrew* (1.1.155).
17. Shakespeare, Sonnet 56, line 1.
18. Even the geeky Michael Eckman experiences a transformation. Although (or perhaps because) he is the character most obviously coded as a "loser," Michael shows up at the Padua Prom dressed as Shakespeare, where he finds romance with Kat's Bardolotrous best friend (she has a Shakespeare pin-up in her locker).
19. Here Stiles seems to falter in her delivery, and prematurely starts the following line. We have reproduced the seemingly unscripted misstep here in the interest of conveying word for word Stiles's screen performance.
20. Only losers would spend this much time and effort analyzing the relationship between a canonical late sixteenth-century text and its late twentieth-century "kiddie culture" adaptation.
21. See Bryan Reynolds, "The Devil's House," 143–67; John L. McMullan, *The Canting Crew: London's Criminal Underworld 1550–1700* (New Brunswick: Rutgers University Press, 1984); A.L. Beier, *Masterless Men: The Vagrancy Problem in England* (London: Methuen, 1985); and Steven Mullaney, *The Place of the Stage: License, Play, and Power in Renaissance England* (Chicago: University of Chicago Press, 1988).
22. Hereafter abridged to "Nudge-Wink-Know."
23. We have chosen to use John Rouse's translation of this definition from Brecht's "A Short Organum for the Theatre" because Rouse rejects the established term "alienation" in favor of "defamiliarization" (234). We

hope to prevent the frequent misunderstanding that *Verfremdung* implied the existence of an alienating distance between the actors and their audience.

Works cited

10 Things I Hate About You. Dir. Gil Junger. Perf. Julia Stiles and Heath Ledger. Disney, 1999.
A Midsummer Night's Dream. Dir. Michael Hoffman. Perf. Calista Flockhart and Kevin Kline. twentieth Century Fox, 1999.
Alcorn, Marshall W. Jr. and Mark Bracher. "Literature, Psychoanalysis, and the Re-Formation of the Self: A New Direction for Reader-Response Theory." *PMLA* 100.3 (1985): 342–54.
Artaud, Antonin. *The Theatre and Its Double*. Trans. Mary Caroline Richards. New York: Grove Press, 1958.
Auslander, Philip. *Liveness: Performance in a Mediated Culture*. New York: Routledge, 1995.
Barton, John. "Using the Verse: Heightened and Naturalistic Verse." *Playing Shakespeare*. London: Methuen, 1984. 25–46.
Bateson, Gregory. *Steps to an Ecology of Mind*. San Francisco: Chandler Publishing, 1972.
Baumlin, Tita French. "Petruchio the Sophist and Language as Creation in The Taming of the Shrew." *Studies in English Literature 1500–1900* 29.2 (1989): 237–57.
Beier, A.L. *Masterless Men: The Vagrancy Problem in England*. London: Methuen, 1985.
The Blair Witch Project. Dir. Eduardo Sanchez. Artisan, 1999.
Blau, Herbert. *The Audience*. Baltimore: Johns Hopkins University Press, 1990.
———. "Universals of Performance; or Amortizing Play." *By Means of Performance: Intercultural Studies of Theatre and World*. Ed. Richard Schechner and Willa Appel. Cambridge: Cambridge University Press, 1990. 250–72.
Boose, Lynda A. "The Taming of the Shrew, Good Husbandry, and Enclosure." *Shakespeare Reread: The Texts in New Contexts*. Ed. Russ McDonald. Ithaca: Cornell University Press, 1994. 193–225.
Boose, Lynda and Richard Burt, eds. *Shakespeare: The Movie*. New York: Routledge, 1997.
Brecht, Bertolt. *Brecht on Theatre: The Development of an Aesthetic*. Ed. and trans. John Willet. London: Methuen, 1964.
Burt, Richard. *Unspeakable ShaXXXspeares: Queer Theory and American Kiddie Culture*. New York: St. Martin's Press, 1999.
Butler, Judith. "Performative Acts and Gender Constitution: An Essay in Phenomenology and Feminist Theory." *Performing Feminisms: Feminist Critical Theory and Theatre*. Ed. Sue-Ellen Case. Baltimore: Johns Hopkins University Press, 1990. 270–82.
Carlson, Marvin. *Performance: A Critical Introduction*. London: Routledge, 1996.
Case, Sue-Ellen and Janelle Reinelt. Introduction. *The Performance of Power: Theatrical Discourse and Politics*. Ed. Sue-Ellen Case and Janelle Reinelt. Iowa City: University of Iowa Press, 1991. 9–29.

Cohen, Alain J.-J. "Goddard/Lang/Goddard—The Film-Within-the-Film: Finite Regress and Other Semiotic Strategies." *American Journal of Semiotics* 9.4 (1992): 115–29.
Contempt (*Le Mépris*). Dir. Jean-Luc Goddard. Perf. Brigitte Bardot and Michel Piccoli. Nelson Entertainment, 1964.
de Lauretis, Teresa. "Imaging." *Alice Doesn't: Feminism, Semiotics, Cinema*. Bloomington: Indiana University Press, 1984. 37–69.
Diamond, Elin. "Rethinking Identification: Kennedy, Freud, Brecht." *Kenyon Review* 15.2 (1993): 86–99.
Dolan, Jill. "Geographies of Learning: Theatre Studies, Performance, and the 'Performative.'" *Theatre Journal* 45 (1993): 417–41.
Dusinberre, Juliet. "The Taming of the Shrew: Women, Acting, and Power." *Studies in the Literary Imagination* 26.1 (1993): 67–83.
Eco, Umberto. "Semiotics of Theatrical Performance." *The Drama Review* 21.1 (1977): 107–17.
Elizabeth. Dir. Shekhar Kapur. Perf. Cate Blanchett, Geoffrey Rush, and Joseph Fiennes. Polygram, 1998.
Eyes Wide Shut. Dir. Stanley Kubrick. Perf. Tom Cruise and Nicole Kidman. Warner Brothers, 1999.
Freud, Sigmund. *Group Psychology and the Analysis of the Ego*. 5th ed. Trans. James Strachey. New York: Liveright Publishing Corporation, 1949.
Friedberg, Anne. "A Denial of Difference: Theories of Cinematic Identification." *Psychoanalysis and Cinema*. Ed. Ann E. Kaplan. New York: Routledge, 1990. 36–45.
Fuller, Graham. "Shooting Shakespeare." *Madison* (2000): 78–83.
Garber, Marjorie B. *Dream in Shakespeare: From Metaphor to Metamorphosis*. New Haven: Yale University Press, 1974.
Goffman, Erving. *Frame Analysis: An Essay on the Organization of Experience*. Cambridge: Harvard University Press, 1974.
Greenblatt, Stephen. *Renaissance Self-Fashioning: From More to Shakespeare*. Chicago: University of Chicago Press, 1980.
Hamlet. Dir. Kenneth Branagh. Perf. Kenneth Branagh and Kate Winslet. Turner, 1996.
Hamlet. Dir. Michael Almereyda. Perf. Ethan Hawke and Julia Stiles. Miramax Films, 2000.
Haring-Smith, Tori. *From Farce To Metadrama: A Stage History of The Taming of The Shrew, 1594–1983*. Westport: Greenwood Press, 1985.
Hedrick, Donald and Bryan Reynolds. "Shakespace and Transversal Power." *Shakespeare Without Class: Misappropriations of Cultural Capital*. Ed. Donald Hedrick and Bryan Reynolds. New York: Palgrave, 2000. 3–47.
Hodgdon, Barbara. "Katherina Bound; or Play(K)ating the Strictures of Everyday Life." *Publications of the Modern Language Association of America* 107.3 (1992): 538–53.
Hopkins, D.J. and Bryan Reynolds. "The Making of Authorships: Transversal Navigation in the Wake of *Hamlet*, Robert Wilson, Wolfgang Wiens, and Shakespace." *Performing Transversely: Reimagining Shakespeare in the Critical Future*. New York: Palgrave, 2002: 265–86.

Jameson, Frederick. *Postmodernism: Or, The Cultural Logic of Late Capitalism.* Durham: Duke University Press, 1991.
Kernan, Alvin B. "Shakespeare's Stage Audiences: The Playwright's Reflections and Control of Audience Response." *Shakespeare's Craft: Eight Lectures.* Ed. Philip H. Highfill, Jr. Carbondale: George Washington University, 1982. 138–55.
Kristeva, Julia. "Identification and the Real." Trans. Shaun Whiteside. *Literary Theory Today.* Ed. Peter Collier and Helga Geyer-Ryan. Ithaca: Cornell University Press, 1990. 167–76.
Londré, Felicia Hardison. "Confronting Shakespeare's 'Political Incorrectness' in Production: Contemporary American Audiences and the New 'Problem Plays.'" *Staging Difference: Cultural Pluralism in American Theatre and Drama.* Ed. Marc Maufort. New York: Peter Lang, 1995. 85–95.
Marcus, Leah. "The Shakespearean Editor as Shrew-Tamer." *English Literary Renaissance* 22.2 (1992): 177–200.
McMullan, John L. *The Canting Crew: London's Criminal Underworld 1550–1700.* New Brunswick: Rutgers University Press, 1984.
Mullaney, Steven. "Civic Rites, City Sites: The Place of the Stage." *Staging The Renaissance: Reinterpretations of Elizabethan and Jacobean Drama.* Ed. David Scott Kastan and Peter Stallybrass. New York: Routledge, 1991. 17–26.
———. *The Place of the Stage: License, Play, and Power in Renaissance England.* Chicago: University of Chicago Press, 1988.
Myerhoff, Barbara. "The Transformation of Consciousness in Ritual Performances: Some Thoughts and Questions." *By Means of Performance: Intercultural Studies of Theatre and Ritual.* Ed. Richard Schechner and Willa Appel. Cambridge: Cambridge University Press, 1990. 245–49.
Never Been Kissed. Screenplay by Abby Kohn and Marc Silverstein. Dir. Raja Gosnell. Perf. Drew Barrymore. Twentieth Century Fox, 1999.
Novy, Marianne L. *Love's Argument: Gender Relations in Shakespeare.* Chapel Hill: University of North Carolina Press, 1984.
O. Screenplay by Brad Kaaya. Dir. Tim Blake Nelson. Perf. Josh Hartnet, Mekhi Phifer, and Julia Stiles. Lions Gate, 2001.
Orlin, Lena Cowen. "The Performance of Things in The Taming of the Shrew." *Yearbook of English Studies* 23 (1993): 166–88.
Patterson, Annabel. "Framing The Taming." *Shakespeare and Cultural Traditions: The Selected Proceedings of the International Shakespeare Association World Congress.* Ed. Telsuo Kishi, Roger Pringle, and Stanley Wells. London: Associated University Presses, 1994. 304–13.
Pavis, Peatrice. "Premature Synthesis: Temporary Closure for End-of-Century Inventory." *TheatreForum* 17 (2000): 74–81.
Reynolds, Bryan. *Becoming Criminal: Transversal Performance and Cultural Dissidence in Early Modern England.* Baltimore: Johns Hopkins University Press, 2002.
———. "The Devil's House, 'or worse': Transversal Power and Antitheatrical Discourse in Early Modern England." *Theatre Journal* 49 (1997): 143–67.
Romeo Must Die. Dir. Andrzej Bartkowiak. Perf. Aaliyah and Jet Li. Warner Bros., 2000.

Rouse, John. "Brecht and the Contradictory Actor." *Acting (Re)Considered: Theories and Practices*. Ed. Philip B. Zarrilli. London: Routledge, 1995. 228–41.
Schechner, Richard. *Performance Theory*. Rev. ed. New York: Routledge, 1988.
———. *Between Theater & Anthropology*. Philadelphia: University of Pennsylvania Press, 1985.
Scotland, PA. Dir. Billy Morrissette. Perf. James LeGros, Maura Tierney, and Christopher Walken. Lot 47 Films, 2001.
Shakespeare in Love. Dir. John Madden. Perf. Joseph Fienes and Gwenyth Paltrow. Miramax, 1998.
Shakespeare, William. *The Taming of the Shrew*. Ed. Brian Morris. London: Routledge, 1981.
———. *The Riverside Shakespeare*. Ed. G. Blakemore Evans. Boston: Houghton Mifflin Co., 1974.
Shapiro, Michael. "Framing the Taming: Metatheatrical Awareness of Female Impersonation in The Taming of the Shrew." *Yearbook of English Studies* 23 (1993): 143–66.
Shurgot, Michael. "From Fiction to Reality: Character and Stagecraft in The Taming of the Shrew." *Theatre Journal* 33.3 (1981): 327–40.
Siegel, Lee. "What the Critics Failed to See in Kubrick's Last Film (Review)." *Harper's Magazine*. Vol. 299, No. 1793. October 1999: 79–81.
Stanislavksi, Constantin. *An Actor Prepares*. Trans. Elizabeth Reynolds Hapgood. New York: Routledge, 1964.
The Blair Witch Project. Dir. Edward. Sanchez. Artisan, 1999.
The Full Monty. Dir. Peter Cattaneo. Twentieth Century Fox, 1997.
Thompson, Ann. "Introduction." *The Taming of the Shrew*. By William Shakespeare. Ed. Ann Thompson. Cambridge: Cambridge University Press, 1984. 1–41.
Traister, Rebecca. "The Story of O, Weinstein Style: High-School *Othello* is Held Up." *New York Observer* November 13, 2000: F1.
Turner, Victor. *Drama, Fields, and Metaphors*. Ithaca: Cornell University Press, 1974.
———. *On the Edge of the Bush*. Tuscon: University of Arizona Press, 1985.
Weller, Barry. "Induction and Inference: Theater, Transformation, and the Construction of Identity in The Taming of the Shrew." *Creative Imitation: New Essays on Renaissance Literature in Honor of Thomas M. Greene*. Ed. David Quint, Margaret W. Ferguson, G.W. Pigman III, and Wayne A. Rebhorn. Binghamton: Medieval and Renaissance Texts and Studies, 1992. 297-329.
———. "Identity and Representation in Shakespeare." *ELH* 49.2 (1982): 339–62.
Williams, Gwyn. *Person and Persona: Studies in Shakespeare*. Cardiff: University of Wales Press, 1981.
Worthen, William B. "Shakespearean Performativity." *Shakespeare and Modern Theatre*. Ed. Michael Bristol, Kathleen McLuskie, and Chistopher Holmes. New York: Routledge, 2001. 117–41.
———. "Is it Not Monstrous: The Demonic Dialectic of Renaissance Acting." *The Idea of the Actor; Drama and the Ethics of Performance*. Princeton: Princeton University Press, 1984. 10–69.

CHAPTER 7
"A LITTLE TOUCH OF HARRY IN THE NIGHT": TRANSLUCENCY AND PROJECTIVE TRANSVERSALITY IN THE SEXUAL AND NATIONAL POLITICS OF *HENRY V*

Donald Hedrick & Bryan Reynolds

The multiple roles of Henry V have become a critical commonplace, whether he is viewed positively or negatively as savvy prince or as calculating politician. The specific interrelation of the roles and the mechanisms of their production have been less attended to, and are, because of their paradoxical interfaces, particularly inviting of a transversal analysis. In employing "transversal theory" for this purpose, we find that the play foregrounds a mechanism we call the "principal of translucency," by which one signifier or identity is incompletely concealed within another, a disguise apparently flawed or inadequate, producing a particular kind of mixed coding for audiences or spectators, who experience transversality in response.[1] While historically nonspecific as both a social and theatrical trope, this principle of translucency is specifically historical as a feature of early modern social performativity theorized in terms of courtly spectacle by Baldessare Castiglione. It is a visual means whereby transversal territory may be open for wonder, whereby the subjective dislocations of transversality may be approximated visually for an audience's understanding and enticement. Shakespeare, as we hope to demonstrate, illustrates an employment of transversality onstage and offstage, including what we call a "projective transversality," by which one effects, or attempts to affect, transversality in the other rather than in oneself.

The principle of translucency operates when a social or theatrical performance incorporates inconsistent and contradictory codes that in effect present mixed or simultaneous messages. However subtly, the codes must be identifiable and featured. Because the coding forces the audience to become active readers of the performance, rather than merely passive voyeurs, in order for the coding to produce reactions, it must be accessible and interpretable. In other words, if there are many codes at work or if the codes are obscure, then the audience will be too confused and detached, and the principle's power will be lost. Translucency cannot occur when the coding is blurred and amorphous. Each code displayed must be distinct and recognizable, even while it is a component of a larger picture. In most cases, translucency is most effective when there are just two or three featured codes. The more codes, the greater the potential for obfuscation. If skillfully executed, translucency can promote a transversal performance. For this reason, the transversal performer often deliberately pursues translucency. Should the principle of translucency be achieved, the spectator, and often the performer (particularly in social performance) experiences disempowerment, but also fascination relative to the transversal power generated by the performance event.

The spectator or audience member is manipulated by the specific rhetoric and atmosphere of the event, by the event's power that moves the spectator transversally outside of his or her own subjective territory. But the transversal territory into which the spectator moves is not necessarily that of a subjective territory of an other, but rather a creative space. The artfulness with which the principle of translucency is realized produces the conditions for both the desire for artistic development and its pursuit on the part of the spectator. Movement into transversal territory is prompted by the coalescence of the contradictory coding during the interpretive process. Thus, the event's transversal power emerges through the resonance and wonder it manufactures. The spectator is jarred from the parameters of conventional discourse and reason, exposed to the multilayeredness always present in all performances but rarely so apparent, and is attracted to the new found pellucidity. The principle of translucency is therefore a complex occurrence by which the spectator sees one or two or more things through others while at the same time seeing the others themselves. Faced with a radical situation, rather than wanting to deny or flee from the experience, the spectator wants to assimilate into the transversal territory of the performance and emulate the performer of such alluring translucency. These wants are demonstrated either by submission to or convergence with both performance and performer. Convergence takes the form of counter, complimentary, or more intimate rhetoric and engagement. The spectator wants to be in cahoots with the transversal performer.

Blind Cupid's curse, or transversal vision?

In a short, charged scene often cut from performances of *Henry V*, presumably because of an additional damaging effect it might have on the critical problematic of the heroic characterization of Henry, the visual or translucent dimension of transversality is italicized.[2] After what is often called the "wooing scene," Henry and Burgundy share a lighthearted, locker room dialogue specifically about the sex that Henry will have with Catherine. The banter is punctuated by references to "love," perhaps implying some sense of mild protest from Henry that his relationship with Catherine is not just about domination and conquest. Nevertheless, the resonance of domination and conquest fully overrides the topic of love, as the men, through a dense series of ostensible double-entendres, the meaning of which is sometimes virtually unrecoverable, speak about how submissive Catherine will be in bed, and, more lewdly, how she will be brought around to a willingness to do absolutely anything, whatever that might entail. The scene warrants quoting at length:

> BURGUNDY: God save your majesty! My royal cousin, Teach you our Princess English?
> KING: I would have her learn, my fair cousin, how perfectly I love her, and that is good English.
> BURGUNDY: Is she not apt?
> KING: Our tongue is rough, coz, and my condition is not smooth, so that having neither the voice nor the heart of flattery about me I cannot so conjure up the spirit of love in her that he will appear in his true likeness.
> BURGUNDY: Pardon the frankness of my mirth if I answer you for that. If you would conjure in her, you must make a circle; if conjure up love in her in his true likeness, he must appear naked and blind. Can you blame her then, being a maid yet rosed over with the virgin crimson of modesty, if she deny the appearance of a naked blind boy in her naked seeing self? It were, my lord, a hard condition for a maid to consign to.
> KING: Yet they do wink and yield, as love is blind and enforces.
> BURGUNDY: They are then excused, my lord, when they see not what they do.
> KING: Then good my lord, teach your cousin to consent winking.
> BURGUNDY: I will wink on her to consent, my lord, if you will teach her to know my meaning. For maids well summered and warm kept are like flies at Bartholomew-tide, blind, though they have their eyes and then they will endure handling, which before would not abide looking on.

> KING: This moral ties me over to time and a hot summer; and so I shall catch the fly, your cousin, in the latter end, and she must be blind too.
> BURGUNDY: As love is, my lord, before that it loves.
> KING: It is so: and you may some of you thank love for my blindness, who cannot see many a fair French city for one fair French maid that stands in my way.
> FRENCH KING: Yes, my lord, you see them perspectively, the cities turned into a maid; for they are all girdled with maiden walls that no war hath entered. (5.2.278–318)

For the purposes of exploring visual dimensions of transversality, what is striking in this moment is how the dialogue moves, from the conventional *topos* of love's blindness to something altogether different, even visually pornographic in its implications, revolving around the key concept of "winking." As Andrew Gurr observes, the word "winking" is used here in both its Elizabethan sense of closing the eyes, as well as in the current sense, apparently then emergent, of the leering blink. Both, as it turns out, have ideological and theoretical implications illustrating transversality.

When Burgundy asks if Henry is "teaching our Princess English," and questions whether she is "apt," he employs a "cluing-in technique" for social interaction and performance framing. Borrowing signature catchphrases from the BBC's comedy program "Monty Python's Flying Circus" (that originally aired from 1969 to 1974), D.J. Hopkins, Catherine Ingman, and Bryan Reynolds call this technique "Nudge, Nudge, Wink, Wink, Know What I Mean, Know What I Mean?" Nudge, Nudge, Wink, Wink, Know What I Mean, Know What I Mean? is a technique by which a particular audience is made privy to ("keyed into," as Erving Goffman puts it [24]) a meaning implied by a performance, whether offstage in everyday conversation or theatrically onstage, in order to foster a non-explicit information flow and unspoken bond between the performer and that audience (see chapter 6 of this book, on *The Taming of the Shrew*). In response to Burgundy's entreaty, Henry seems a bit apologetic, claiming more romantically that he wants her to learn how "perfectly I love her, and that is good English," though inadvertently acknowledging that this love has an analogy in domination that is both linguistic and cultural. Self-deprecatingly, as he is at times in the wooing scene with Catherine, Henry notes his inability at wooing from his "rough tongue," "not smooth" condition, and failure to perform courtly flattery. But Burgundy returns the conversational direction to the bawdy, turning the image of blind cupid into that of the blind phallus to "make a circle." Thus, Burgundy presumes leeringly to see through Catherine's necessary pose of modesty ("rosed over with the virgin crimson of modesty") as a female translucency, one that would make it a "hard condition to consign to" for her to immodestly allow the "appearance of a naked blind boy in her naked seeing self." Henry's conversational attempt to return love to the realm of the nonsensual—"Yet they do wink and yield, as love is blind and enforces"—has the immediate effect of reintroducing power and domination into the remarks ("enforces").

From this point, the visual references of blindness take another direction entirely; Burgundy goes on to suggest that blindness excuses one from responsibility ("They are then excused, my lord, when they see not what they do"). Henry takes the hint, requesting Burgundy to encourage Catherine's blindness by "winking," that is, by having her close her eyes during sex, thereby self-blinding; he says: "Then good my lord, teach your cousin to consent winking." Burgundy then builds on the image of Catherine closing her eyes during sex to suggest that in doing so she will be like the infamous flies at Bartholomew-tide who let themselves "endure handling, which before would not abide looking on." The metaphor of blindness thus transforms from romantic passion to the masculine fantasy of total power during sex, the English fantasy of total domination of the French, total surrender by the other, and an infinite variety of sexually submissive acts. That thought in turn seems to produce Henry's next suggestion about a different form of sex-without-looking, namely, when performed not frontally, but from behind: ". . . and so I shall catch the fly, your cousin, in the latter end, and she must be blind too." The coarseness of the banter would seem to be difficult to ameliorate by regarding it simply as a reflection of Catherine's increasing sexual awareness (Wilcox 67), for it naturally invites us to consider what it would mean from *her* perspective, in her reading of the King.

Sex-without-seeing, from the perspective of this male fantasy, indicates a trajectory of male and national domination. But a transversal reading suggests a different possibility altogether—a key to the scene, if not to the entire play: by closing one's eyes, one "disappears" the other, or even transforms the other into someone else. Instead of transversality as a becoming-other of one's own subjectivity, becoming what you are not, transversality might be now thought in terms of transforming the other outside himself or herself, a projective transversality or "Renaissance other-fashioning." What we are suggesting is that transversality and translucency may act as mechanisms with an entirely different outcome or purpose for Catherine than for Henry in his performance of himself. The "naked seeing self" of the modest maiden, in Burgundy's remark, constitutes an image of body, and identity, as made up entirely of eye. Again, the scene's self-reflectiveness goes far beyond its function merely to serve as "sexual titillation" for the audience, as Jean Howard and Phyllis Rackin argue (206).

The scene proposes a masculine, joint French–English expectation for Catherine to give in to the power of love, or the will of the King, in the relationship to follow their marriage. Is this on all fours, however, with the representation of Catherine in the earlier scenes of the play, particularly in the famous (often infamous) scene of Henry "wooing" her for her love? Is the closing up of sight, the blinding of the "naked seeing self," not potentially resistant or transgressive of domination, the entire body, made up entirely of eye, thus closing itself off? Is Catherine as submissive in those earlier scenes as we might expect from Henry and Burgundy's anticipatory banter? Moreover, is Henry's translucent disguise as the noble interior wearing the demeanor and habit of the crude, savage soldier, as fully effective as he would have it, or is there a dimension of translucency remaining to be understood, namely, failure at translucent effect?

"I cannot tell vat is *baiser en* English": sexual and linguistic transversality

From Catherine's perspective, a perspective made available to the audience chiefly through the means of the actress portraying her, the main point of the scene is not that of the usual critical questions about it (How does Henry "woo" her?), but rather that of *reading the King*. Her scant dialogue in the scene, whether it is played coyly or coolly depending either on our understanding of her character or else on contemporary tastes involving gender relations, continually interrogates Henry's appearance to her. The play thus presumes a code-savvy audience for translucent performance, one capable of judging codes embedded in codes. In brief, Catherine's questions are continually suspicious of his motives, as she first suggests that he is mocking her and her linguistic disability: "Your majesty shall mock at me; I cannot speak your England" (102). She brands him as a typically deceitful man ("*les langue des hommes sont pleines de tromoperies*" [116]), and even suggests that his motives may be antagonistic ("Is it possible dat I sould love de enemy of France?" [169]). Her decided resistance would disqualify a reading of her as simply "bubbly and girlish" (Wilcox 63), on the one hand, or simply as a "symbol of subjugation" (Ellis-Fermor 137), on the other.

Henry's performance itself involves the deliberate layering constituting translucency. In its broadest outline, in fact, reading the King calls forth the same aesthetic strategy of translucency as described by Castiglione, namely, the prince appearing as a rustic but with emblems of his nobility showing through the gaps. What has been termed the principle of translucency is thus operating here, a principle described in fact as one operative in Renaissance court entertainments. Castiglione describes the practice of Continental masquing whereby even a prince may "take up counterfeiting of false visages," since it is "no noveltie at all to any man for a prince to bee a prince." Specifically, he describes "disgracing" disguises in which the concealment is only *partial*, such as a young man's disguising himself in an old man's attire, but in such a way that his garments are not "a hindrance to him to shew his nimblenesse of person," or a soldier might be disguised as a rustic but at the same time ride a splendidly martial steed (Castiglione 99–100).

Henry's bluff rhetorical posture is, like one of Castiglione's "disgracing" disguises appropriate for the prince in masque, that of the English rustic, literalized as he tells Catherine that he is glad her English is bad enough that she *cannot* read him accurately, "for if thou couldst thou wouldst find me such a plain king that thou wouldst think I had sold my farm to buy my crown" (124–27). The manliness he portrays rhetorically and physically, in fact, is actually one of multiple layers of masculinity, for he describes how he would win her with rural athleticism: "if I could win a lady at leapfrog," "if I might buffet for my love," but with deliberate hints of noble masculine activities appended: "by vaulting into my saddle with my armour on my back" (138).[3] This translucent double-coding, one masculinity through the other, and one social class through the other, is Henry's deliberate, general performance strategy of "mystery and wonder" (Porter 150) throughout the scene. While the scene has been described as a "masterful stage managing of the entire encounter," and Henry's performance accordingly as "flawless" (Howard 150–51), we want to argue that this rather reflects Henry's desire more than his success, and, as we will show, an attempted transversality that overreaches itself.

Of course, the possibility of a proliferation of masculinities, complicating the translucency, complicates as well this potential reading of the English King by Catherine. This proliferation may, moreover, include unintentional layers, strategies without strategists, reflecting the ruthlessness we have already seen in war, from Harfleur to the killing of the prisoners of war. Killing looks and killing appearances are, to be sure, the theme that the French Queen herself brings up as she expresses her hope that the "murdering basilisks" and the "venom of such looks" will be transformed, lose their "quality," and change "griefs and quarrels into love" (17–20). Civilization versus barbarism is also the theme of the Duke of Burgundy's lament that France, "the best garden of the world," is declined into overgrown weeds of "darnel, hemlock and rank fumitory . . . hateful docks, rough thistles, kecksies, burrs," her "vine . . . unpruned dies," and the vineyards "defective in their natures, grow to wildness" (36–55). The French children now having grown "like savages"; "savagery" in this portrait, however, is viewed, Hobbes-like, as natural, with the tools of civilized agriculture designed to "deracinate such savagery" (47).

Henry's translucent masculinity, then, partakes in one of its codings as savage and barbarian, the layer revealed in his threats of rape at Harfleur (3.3.90–115), carried on through the more innocent-seeming scene of his performance before Catherine. Although her recognition of translucency would involve the conventional Castiglione model of the nobility through the peasant, the process of moving through such identities would have no ulterior governing mechanism, potentially revealing the savagery if not psychopathology of masculinity, the image of the rapist alongside that of the seducer as in a Renaissance perspective drawing. While the most degrading of images is made available to Catherine in the scene, translucent decoding allows them to remain unstable. Thus, when at the start of the scene she is announced by Henry to be "our capital demand, comprised / Within the fore-rank of our articles," Catherine may be reminded of the even more degrading circumstance of his already having rejected her in the negotiations at Harfleur, when we hear that the offer of her together with "to dowry, / Some petty and unprofitable dukedoms" (3.1.29–32) has been rejected, curtly dismissed by the Chorus with the phrase, "the offer likes not" (3.0.32). But she may equally read this as Henry's tactic for having a private conference with her, an opportunity to persuade her to love, his ruthlessness carried out in the service of her.

The specifically sexual politics of the scene, that is, its topic of future sex as distinct from the merely gender politics, are carried out by the same translucency of multiple codings, the most striking of which appears to be that the King becomes sexually aroused during the scene. This critically ignored or underattended moment (but see Hedrick, forthcoming, in *PMLA*) occurs in a rather obvious sequence of stages of arousal mid-scene. First, the topic of sex is brought up directly by Henry, as he suggests that Catherine will make a "good soldier-breeder," and that between them they will "compound a boy, half French, half

English, that shall go to Constantinople and take the Turk by the beard? Shall we not?" (203–07). Unsure as to whether he is more excited about proposing sex or military conquest, her coy resistance to the idea, perhaps relying on feigned linguistic lack of understanding ("I do not know dat") seems to drive him to rhetorical excess, as he constructs his own "false French" to praise her extravagantly as *"la plus belle Catherine du monde, mon trés cher et divin déesse?"* (214–15). In this arousal sequence, he suddenly declares his love: "By mine honour, in true English, I love thee, Kate" (218–20). Following this "confession," he pointedly adds, "by which honour I dare not swear thou lovest me, yet my blood begins to flatter me that thou dost" (219–21). The semi-autonomy of his blood, like his rapist soldiers at Harfleur, who are represented as not within his control, subversively flatters him; the reference to his blood operates as both lewd joke and dramatic conflict for the scene's persuasions and adventures in propriety and cultural difference. If, as it seems, the King actually reveals an erection onstage, translucency takes on the sexual dimension to which the scene has been tending all along, as the "crude" Henry is now visible through the gallant one, a moment of potentially comic awkwardness of looks and glances, an "empowerment" that at the same time socially disadvantages the King.

Given the overt sexual discourse and interest of the scene, one might have expected that translucent performance would occur in the manner of Harfleur, as an overwhelming display of transversal power whose reception is one of being overloaded or overwhelmed, defeated by proliferating codes. But Henry's domination and conquest in this final scene is hardly monolithic and certainly not complete, even though he creates a transversal space outside his own subjectivity. One might argue, in fact, that the very linguistic instability of the scene creates its own space of possibility, the linkage of Catherine's bad English to Henry's bad French, the implication of which is that they share a dissident or unclassifiable space of badness, where neither's language exerts the kind of control through its speaker that the speaker might intend. Language learning, as it were, acts to enforce not only humility but also a temporary release from their respective subjective territories; it's an anti-subjective force in the sense of loosening the construction of subjectivity. Of course, as has been often observed, the earlier language-learning scene between Catherine and her waiting woman is distinctly oriented toward the sexual, not only in the crosslinguistic punning by which English words sound like French sexual slang to Catherine, but also by the implicit forecasting of Catherine's focus on learning the English words for body parts, presumably a practical knowledge in anticipation of both eventual union with the English King as well as the defeat of her nation.

The intersection of bad French and bad English produces a transversal space of social creativity or experimental alterity, a space of marital negotiation rather than patriarchal dominance, pointedly brought out when Henry persuades her to violate her social custom of not kissing before marriage with the trump line, "We are the makers of manners, Kate" (268–69). The image of transversal power presented here, therefore, is not akin to the display at Harfleur, producing concession and defeat, a process of being overwhelmed by multiple codings, but rather one of creative social space appropriate to transversal power: "O, Kate, nice customs curtsy to great kings" (266). While customs are feminized, Kate is by implication masculinized, given honorary masculinity of the sort shared sexually in equally "compounding" a boy in bed. Henry, moreover, expresses the "barometer" of his erection in terms of translucent reading. It is not the conventional reading, as we might expect, that he interprets it as a sign of his desire for her. Instead, he regards it as evidence of *her* desire for him, that his blood flatters him that she loves him, "notwithstanding the poor and untempering effect of my visage" (221–22). In effect, his anti-subjectivity takes place as viewing her desire within his own physiological response. In pop-Lacanian fashion, he sees in her the mirror of his own plenitude. In the psychology of the moment, this fullness allows him to imagine or declare, as he does a few lines earlier, his love for her.

The sexualized reading of the moment is underscored further with the following thoughts of Henry, directly about sex again, but this time about what it was like in bed between *his* father and mother, as Shakespeare imagines English succession as a succession of moments in beds through the ages. Again in a mode of self-deprecation, Henry presents a scene not of idealized sexuality, nor of mutuality by which the sexual space is shared, but of another bad transversal space like that in which language is neither French nor English, an almost comical scene of bad sex whereby the "dominator" is preoccupied with war rather than sex: "Now beshrew my father's ambition! He was thinking of civil wars when he got me: therefore was I created with a stubborn outside, with an aspect of iron, that when I come to woo ladies I fright them" (223–26). Here, the sexual subtext of the play, attained by linking the random sexual threads and allusions of the scene, would focus on the future, namely, what will happen in bed between Henry and Catherine. And, as fitting a transversal space of becoming, what exactly will happen is not predetermined by Henry's position of dominance, the political and military outcome, or conventional, "nice customs" of cultural expectations. And yet domination, as long as it is not considered reductively as what the scene is about, must always be present translucently, as it were, *in potentia*.

In fact, the transversal space of bad English/bad French, combined with the hint about "making manners," feuls yet another of the scene's fascinations for both audience and Catherine. Simply put, we do not know, nor do either of the protagonists of the scene know, what exactly will happen when in bed together. If we are to follow the folk logic of the scene as expressed by Henry, that is, that the style of sex produces the character or appearance of the children begot, the moment of conception becomes historically charged, more politically weighty than one might expect from such gossipy business, and through the scene's "foreplay," we glimpse the character of the future monarch. Knowing this, however, and knowing from history and from the play's epilogue that the product was in fact the weak and inept Henry VI, who "lost France and made his England bleed" (Epil. 12), Henry's "core" masculinity becomes questionable. Whether produced civilly or savagely—perhaps overdetermined in his fantasy of having sex with Catherine from behind—the "product" of Henry's masculinity leads audiences to consider just how far beyond his subjective territory he may go. In effect, Catherine is responsible for the outcome of Henry's glorious project, in that it is her body that will ultimately produce— or, as it happens, fail to produce—the kind of conquering son that Henry requires imaginatively for his performance. While, as it has been noticed, it becomes the place of the woman to secure man's place in history (Rackin 107), and her consent is a "gift" to his patriarchal legitimation (Tennenhouse 69), our transversal reading emphasizes that her role goes further: she threatens to control history itself, writing an alternative future with her body as counter-historiographer.

"That's more than we know": winking patriotism at Harry le Roy

Through his elaborations of transversality in relation to sexual practice and close-eyed sex, Shakespeare makes a remarkable analogue between Catherine and, as it happens, the British troops in their underdog status at Agincourt. The scene of "a little touch of Harry in the night" (4.1.), more directly of course, represents the kind of prince-in-disguise that Castiglione promotes in his aesthetics of courtly entertainment, though with the slight variation that has the prince disguised not as a rustic, but as an inferior only some steps down, another noble, yet not royalty. While not recognized as the King, therefore, his performance before the troops might acknowledge his superiority in sensibility and reasoning, as he makes arguments, rather than rousing speeches, to try to bolster morale in what are obviously demoralized ground troops the night before the battle. In his project in the scene, as we shall see, he tries to produce the same sort of willingness that his men might be "handled" anyway whatsoever that he later fantasizes with Catherine, a willingness to do anything for him, in this case to die for him.

When Henry borrows the cloak of Sir Thomas Erpingham, and half-lies to the confronting foot-soldier that he serves "under Sir Thomas Erpingham" (92), we find that he has indeed in translucent fashion mixed social codes of rhetoric, costume, and social space: by speaking much more eloquently and pulling out all the stops of rhetorical persuasion; by the costume of the nobility; and by inhabiting the part of the camp where the infantry lodges. His disguises are adventures in interpellation, as he responds to Williams's initial challenge, "Who goes there?" with "A friend" (90), presuming in advance the subject position that he will try to inhabit. The debate that follows, in which he argues with these soldiers about the King's responsibility, or lack of it, for the deaths to follow, becomes a debate about the "subject" of the King, that is, about subject position itself. The idea is addressed explicitly by Williams's argument that if the men do not die well in war, it is a "black matter for the King that led them to it—who to disobey were against all proportion of subjection" (137–38). For Henry "proportion of subjection" is indeed the key issue for enforcing domination, for having his way, which he attempts both through force and through rhetoric.

One of the most curious moments of the scene is his comment about morale at the beginning, when he concedes to the men that the outcome of the battle looks bleak. While we might expect that the argument would be that they should be as brave as the King is, what Henry says is quite different and unexpected. First, he concedes, in effect, that because the King is "but a man, as I am," appearing "naked" like other men when without his clothes, and with similar "affections," which stoop as low as do those of the common man. But what follows logically is quite surprising, since he goes on to argue that this is the reason the men must not show their fear *to him*, since the fear will be reflected and in turn demoralize the troops. Here, Henry is aware of the transversal situation of projecting the other into a subject position: they find themselves the source of the *King's* subjectivity. Of course, the argument in its entirety replicates the rhetorical posture of Henry throughout the scene, namely, that he is continually abjuring responsibilities (it is the men's fault if he is scared; it is their responsibility if they die badly or well, just as it is not the merchant's responsibility if the son he sends out to sea miscarries on the boat that sinks). Again, in terms of the subject, Henry significantly concludes, "Every subject's duty is the King's, but every subject's soul is his own" (164–65). The King's radical individuality here imagines a subjectivity that forecloses the possibility of transversality, a subjectivity without any porous or flexible boundaries by which one might get "outside" oneself, if only temporarily.

As with Catherine, however, the King is not as fully persuasive as the saturation bombing of his rhetorical assault would have it. His extremely lengthy rhetorical analogy to prove his non-responsibility, encompassing some thirty-three lines of prose, is followed by a divided response. The more naive Bates parrots the King's position: " 'Tis certain, every man that dies ill, the ill upon his own head. The King is not to answer it" (173–74). Like the falcon who "bates" or flutters out of control while on the leash, Bates is played as supremely malleable, the kind of ideal that Henry will later imagine for Catherine sexually in his fantasy of her pliability. But, to Henry's assertion that the King had told him he would not be ransomed, Williams responds far more cynically and resistant: "Ay, he said so, to make us fight cheerfully" (179), whereas once their throats are cut they will not know what he does. In this Williams echoes his remarkable response to Henry's assertion of the King's "cause being just and his quarrel honourable" with the straightforward but ultimately subversive skepticism, "That's more than we know" (123).

With this line, Williams directly acknowledges not only patriotism but its blindness, the inherent position of the subject to sovereign power, relying on the legitimacy of that power without any access to modes of proof or verification. In effect, Henry has by his rhetorical assault created a double bind for himself. Whereas on the one hand he would invite his hearers to think transversally as if they were in the position of the King (thinking that this will help him out), on the other hand, if they should actually do so, as Williams does in his provocative challenges, the King is placed in a very dangerous and disadvantaged situation, since they will *see through* his rhetoric and his disguise, recognizing the self-interest and manipulation that motivate them. To consider the matter of, in Williams's words, "proportion of subjection," raises the key issue for these soldiers of the extent to which one resides in assigned subjective territory, and the extent to which one might be able to go outside it transversally. To see through the disguise, to regard it as the translucency involved in the court masque, with Williams constituting a "subversive voice" (Rackin 247) in the scene, is thus acknowledging Shakespeare's demystification of patriotism itself (Holderness 139).

As in the scene with Catherine, the King's persuasion rests on a dialect of open-eyed versus shut-eyed, and the contradictory way in which he would, in a sense, have it both ways at once. Henry in effect not only experiences a transversality in getting outside the subjectivity of a king for a time being, but he also "projects," as it were, transversality onto others, by his own arguments inviting them to go outside themselves (and to take on, for instance, the subjectivity of the King). The King requires a sense of consent for hegemony to work properly, both from his troops in battle and from Catherine in bed, thus linking the visual thematic of "winking" to the terms of transversality through translucency. If "to die" is taken in its Renaissance context as sexual climax, the concealed pun of this military history is not unlike the "dying" Shakespeare puns on in *Antony and Cleopatra*. He desires to have them do absolutely anything he wants, going against whatever impulses and instincts have governed their "manners" prior to this. The same fantasy governs both the excesses of sexual submission implied in the jokes previously analyzed with Burgundy, as well as the excesses of war that Henry describes himself unleashing when he threatens and promises not to "half-achieve" it: that if they do not surrender immediately, he will not be able to restrain his soldiers from "mowing like grass / Your fresh fair virgins and your flow'ring infants," from subjecting "your pure maidens [to] fall into the thand / Of hot and forcing violation," from defiling their "shrill-shrieking daughters," dashing the "reverend heads" of their fathers against the city walls, and spitting their "naked infants ... upon pikes" (3.3.90–115).

Thus Henry's transversality in this scene implies a terroristic imagination of total force requiring total submission, just as the final scene, as we have seen, implies a pornographic imagination of total force requiring total submission. The critical question of whether or not Henry means what he says in this threat is, in transversal terms, somewhat beside the point, for the mechanism of translucency allows him to show various selves, in various contradictory codes, without necessarily privileging one over another. Henry shows ruthlessness to the English conspirators when he exposes them in the earlier scene by first pretending to be compassionate, just as he will show ruthlessness through "compassion" to the enemy by having executed his former friend Bardolph for stealing a French church's relic, or by having the French prisoners-of-war executed in response to the outrage committed on his camp and the boys attending his luggage in it. In each case, the debate about his "real" character is produced by the very presence of the contradictory codes and translucency in his performance, suggesting that it is in the contradictoriness itself, not in its resolution, where his power and charisma reside.

A "little touch of Harry" in us all?

Throughout this analysis, we have concentrated on the way that Henry as actor-sovereign engages in transversal movement through different subjective spaces in order to consolidate, or rather to attempt to consolidate, his control; we have shown how he relies on the spectators of his performance to inhabit the sovereign's subjectivity. We have further suggested that such an identification is risky, for it opens up the observer's subjective territory by means of identification with the King, giving the observer either a newly acquired sovereign power, or awareness of the manipulative means by which that sovereignty is secured and extended. Inviting them to "close their eyes"—on the one hand to accept blindness while on the other hand to knowingly "wink"—Henry seeks the absolute submission of his troops, and later of Catherine. Watching Henry's onstage transformations, we move outside our usual subjectivity territory as we learn to resist having the joke or the trick played on *us*. As if to confirm the connection between subjugator and subjectified, Shakespeare has Henry, as he speaks to the ground troops, refer to the soldiers among them who would "beguile virgins," thus projecting his own later stance toward Catherine onto the soldiers he would reject or condemn. But in so doing, he invites them outside themselves, and, in effect, to replace himself as king with their own subjective rendering of him, to become their own "makers of manners." Therefore, translucency, as we have argued, is a chief visual mechanism for transversal power, and even for projecting transversality onto others, making others, as it were, experience movements across the conceptual and emotional boundaries of subjective territories, becoming-other themselves.

We have seen the transversal power of translucency illustrated in the strong case of Henry's own practices as monarch. But is translucency a transversal mechanism for the theater audience as well? The Renaissance principle of translucency has become the mechanism for transversality for the audience, but, as Bryan Reynolds has argued,[4] the outcome is not as simple as might have been imagined. Unlike the Castiglione example of the practice, the seeings-through of *Henry V* do not stop at the viewing of nobility through baseness, the initial premise of the "little touch of Harry in the night" scene. Instead, the situation becomes more unstable than that, with the possibility of seeing the baser, more ruthless Henry through the noble one, as well as the actor-commoner through the multiple class layers of Henry's disguise. While Henry in fact fails at fully transversalizing his onstage audience of either Catherine or the troops, in the process of opening himself up subjectively, he allows a partial entry into the subjectivity of the King, just as his "blinding" of Catherine in the imagined sex scene opens up for her the possibility of "erasing" the King. So, too, does the spectator of the play discover the possibility of sovereignty, and the erasure of the stage sovereign, in viewing Henry's manipulating skills and ruthlessness.

Henry's failed transversality, therefore, opens up a space for the audience of imaginative sovereignty and rule, whether over Henry, the adjacent spectators or neighbors, or the monarch herself. In effect, the empowerment allows the spectator to transversally enter into the subject position of an actual noble, the Earl of Essex, Elizabeth's favorite and the figure of an attempted *coup d'état* against her a few years following the performance. While it was Shakespeare's *Richard II* that the rebels hired the Lord Chamberlain's men to perform on the eve of their failed attempt, it may rather have been *Henry V*, through its transversality, that opened up a wedge into the subjective territory otherwise organized by the state, a becoming "out of all proportion of subjection." Since *Henry V* is Shakespeare's only history play alluding to a contemporary figure of the political realm, it becomes even more significant that this figure is the Earl of Essex, identified by the Chorus as being like King Henry in Essex's glorious return from Ireland. The transversality of sovereignty from king to rebel-in-becoming is thus made a condition of possibility by the particular visual workings of translucency of the play; hence its considerable power and danger is unleashed for, and onto, the audience.

Notes

1. For more on "transversal theory," beyond the introduction to this book, see Bryan Reynolds, "The Devil's House, 'or worse': Transversal Power and Antitheatrical Discourse in Early Modern England," *Theatre Journal* 49.2 (1997): 143–67, and *Becoming Criminal: Transversal Performance and Cultural Dissidence in Early Modern England* (Baltimore: Johns Hopkins University Press, 2002); and Bryan Reynolds and Joseph Fitzpatrick, "The Transversality of Michel de Certeau: Foucault's Panoptic Discourse and the Cartographic Impulse" *Diacritics* 29.3 (1999): 63–80.

2. Our interpretation of the play was inspired by Donald K. Hedrick's forthcoming essay in the *PMLA* special issue of "Imagining History." That essay, "Advantage, Affect, and History, Shakespeare's *Henry V*," focuses particularly on the concept of maximally "affective labor" linking theories of historiography to ideas about love through the play's motif of "advantage." For the classic essay of the divided view of the King, see Rabkin.
3. For a theoretical discussion of Elizabethan stage masculinity as the production of plural masculinities in an entertainment economy, see the forthcoming essay by Donald K. Hedrick, "Male Surplus Value," in the special issue on "Affect" in *Renaissance Drama* (2002).
4. For more on the transversal power of early modern England's public theater, see Bryan Reynolds, "The Devil's House," 143–67 and Bryan Reynolds, *Becoming Criminal* 125–55.

Works cited

Altieri, Joan. "Romance in *Henry V*." *Studies in English Literature* 21 (1981): 223–40.

Castiglione, Baldessare. *The Book of the Courtier*. Trans. Sir Thomas Hoby (1561). New York: 1928; rpt. London: 1937.

Ellis-Fermor, Una. "Shakespeare's Political Plays." In *Twentieth-Century Interpretations of "Henry V."* Ed. Ronald Berman. Englewood Cliffs, NJ: Prentice-Hall, 1968.

Goffman, Erving. *Frame Analysis: An Essay on the Organization of Experience*. Cambridge: Harvard University Press, 1974.

Hedrick, Donald. "Advantage, Affect, History, *Henry V*." *PMLA* (2003).

———. "War is Mud." In *Shakespeare, The Movie: Popularizing the Plays on Film, TV, and Video*. Ed. Lynda E. Boose and Richard Burt. London and New York: Routledge, 1997.

———. "The Masquing Principle in Marston's *The Malcontent*." *English Literary Renaissance* 8 (1978): 24–42.

Hedrick, Donald and Bryan Reynolds. "Shakespace and Transversal Power." In *Shakespeare Without Class: Misappropriations of Cultural Capital*. New York: Palgrave, 2000.

Holderness, Graham. *Shakespeare's History*. New York: St. Martin's Press, 1985.

Howard, Jean. *The Stage and Social Struggle in Early Modern England*. New York and London: Routledge, 1994.

Howard, Jean E. and Phyllis Rackin. *Engendering a Nation: A Feminist Account of Shakespeare's English Histories*. London and New York: Routledge, 1997.

Porter, Joseph. *The Drama of Speech Acts: Shakespeare's Lancastrian Tetralogy*. Berkeley: University of California, 1979.

Rackin, Phyllis. *Stages of History: Shakespeare's English Chronicles*. Ithaca and London: Cornell University Press, 1990.

Shakespeare, William. *The Norton Shakespeare, Based on the Oxford Edition*. Ed. Stephen Greenblatt. New York and London: W.W. Norton, 1997.

Shakespeare, William. *King Henry V*. Ed. Andrew Gurr. Cambridge and New York: Cambridge University Press, 1992.

Tennenhouse, Leonard. *Power on Display: The Politics of Shakespeare's Genres*. New York and London: Methuen, 1986.

Wilcox, Lance. "Katharine of France as Victim and Bride." *Shakespeare Studies* 17 (1985): 61–76.

CHAPTER 8
INSPRITEFUL ARIELS:
TRANSVERSAL TEMPESTS

Bryan Reynolds & Ayanna Thompson

That afternoon, at the end of a year of classes, the venerable old teacher, who by allusion to the wise magician of Shakespeare's *Tempest* was often called Prospero, was bidding his young disciples farewell, gathering them about him one last time.... An exquisite bronze of *The Tempest's* Ariel, like the presiding spirit of that serene atmosphere, dominated the room. It was the teacher's custom to sit beside that statue...

Shakespeare's Ariel symbolizes the noble, soaring aspect of the human spirit. He represents the superiority of reason and feeling over the base impulses of irrationality. He is generous enthusiasm, elevated and unselfish motivation in all actions, spirituality in culture, vivacity and grace in intelligence. Ariel is the ideal toward which human selection ascends, the force that wields life's eternal chisel, effacing from aspiring mankind the clinging vestiges of Caliban, the play's symbol of brutal sensuality.

The regal statue represented the "airy spirit" at the very moment when Prospero's magic sets him free, the instant he is about to take wing and vanish in a flash of light. Wings unfolded; gossamer, floating robes damascened by the caress of sunlight on bronze; wide brow uplifted; lips half-parted in a serene smile—everything in Ariel's pose perfectly anticipated the graceful beginnings of flight. Happily, the inspired artist who formed his image in solid sculpture had also preserved his angelic appearance and ideal airiness.

José Enrique Rodó's *Ariel*

The Tempest has been a homage to art, an autobiographical farewell to theater, a colonialist celebration of English expansion, an exhortation on early modern (mis)understandings of racial differences, and an expression of patriarchal power, to name just some of the diverse interpretations that have marked its progress through "Shakespace."[1] Debates over the play's meaning have spanned topics ranging from geographical determinations to identifying various characters as humans, symbols, or monsters.[2] Consistent with the play's hermeneutic history is the now popular assertion that the play in fact invites conflicting responses. Defending and complicating new historicist readings of the play, for example, Paul Brown reads *The Tempest* as an "ambivalent text," which struggles "to produce a coherent discourse adequate to the complex requirements of British colonialism in its initial phase."[3] Stephen Orgel has also championed the notion of the text's inherent ambivalence. Declaring, "Historically, there has been a consistent tendency to ignore [the play's] ambivalences, sweeten and sentimentalize it," Orgel goes on to use the terms "ambivalent" and "ambiguous" no fewer than fifteen times to describe the play.[4] Mixed meanings have come to characterize *The Tempest*'s ideological coordinates, for which it serves as a magnetic sociopolitical conductor, attracting its opposites and propelling those with strong similarities into perhaps more confusing territories. The difficult task at hand is how to assess the influential power and directional inclination of its ambivalent forces, forces that neither necessarily subvert "coherent discourse" nor necessarily lead to indeterminacy.

In *The Tempest*, it is possible to identify and delineate below the surface of ambivalence an ever-developing source of indeterminacy, but not essentially, philosophically speaking, indeterminacy itself. In our view, *The Tempest* pursues what we call "incoherent discourse," a discourse roaming through transversal territory, rather than an emergent "coherent discourse" that Brown sees through his dissective-cohesive reading of the play. Central to the play's transversality is Ariel, the character that most exemplifies the play's subtextual indeterminacy, and works to produce both incoherent and coherent discourses within and beyond the play.[5] The sylph's gender, sexuality, humanity, birth, origin, and future existence all evade exact terms within the play, and its nebulous presence has gone on to inform a variety of discourses through Shakespearean criticism, adaptation, and production, as well as through self-referentiality in other modes of artistic expression. Ariel has worked transversally to expand Shakespace into territories where no other Shakespearean character has gone.

In contrast, although many critics have raised questions about Caliban's nature, "the thing of darkness," as Prospero calls him, has a clearly identified gender, sexuality, familial lineage, and physicality

(5.1.275).[6] It should come as no surprise, then, that Caliban, the figure in *The Tempest* with so many determined aspects, has received so much more critical attention than his indeterminate counterpart.[7] Ariel seems too flimsy and whimsical to grasp and hold up to the light of critical examination. But a look at the sprite's disparate incarnations in rewritings and appropriations of *The Tempest* enables an exploration of its nature and the use of ambivalence and indeterminacy in Shakespeare's original work. Ariel's "affective presence" (its combined material, symbolic, and imaginary existence) has served as a nexus for transversal intercourse, becoming the inspiration for both transversal movements and efforts to promote state power and consolidate subjective territories.[8]

The Tempest has inspired readings and adaptations that try to define and render determinate the play's ambiguities, often strategically in the explicit interest of appropriating the text ideologically for specific cultural and political purposes. The characters, actors, and audience participate in a performance that is something other than what is commonly expected from Shakespeare's play, especially given its consistent Western performance history in regard to traditional casting (European, white) and setting (Mediterranean, island). In effect, characters, actors, audience, and whatever else is associated with the play usually move transversally, becoming what they are not, even though the ultimate destination may not be a territory outside the boundaries of official culture. Thus, we want to focus on two functions of adaptation in regard to *The Tempest*, although there is often overlap between them: (1) adaptation that works to create/maintain/enforce an official culture, methodologically as well as ideologically; and (2) adaptation that attempts, by appropriating an official text (in this case, *The Tempest*), to raise questions concerning official constructs and to instigate movement beyond established social, cultural, and political determinations (which are informed by issues concerning materiality, species, race, gender, sexuality, and class). Like a gambler with dice, Shakespeare tossed his *Tempest*, and through criticism, adaptation, production, and allusion the gale continues to span and generate Shakespace. With this essay, we hope to provide some navigation.

Restoring the tempest-tossed

As many critics have noted, William Davenant and John Dryden's adaptation, *The Tempest, or The Enchanted Island* (1667), was the most popular stage production in England during the late seventeenth and early eighteenth centuries.[9] It inspired an operatic revision with the same name in 1674, a burlesque in 1675 from a rival theater company entitled *The Mock-Tempest: Or The Enchanted Castle*, and another operatic revision in 1756, *The Tempest, An Opera*. The Davenant/Dryden adaptation was so popular that Shakespeare's original was not staged again until 1838, when the nineteenth-century veneration of the Bard overtook the popularity of the revised text. Retaining only around thirty percent of the original dialogue from Shakespeare's play, Davenant and Dryden liberally changed the plot of *The Tempest*. Miranda has a sister, Dorinda, who shares in her innocence and isolation; Prospero keeps a foster man-child, Hippolito, in complete isolation from his two daughters; Caliban has a monster/sister/companion named after their mother, Sycorax, whom he tries to marry to Trincalo; and Ariel is a clearly gendered, male spirit, who flies off to his beloved Milcha at the end of the play. Davenant and Dryden's revisions refocus the attention of *The Tempest* on a series of paired unions in order to explore the tensions between innocence and experience, and in doing so they also make explicit and determinate and more relatable (by adding conventional human traits—gender, heterosexual love-interests, siblings, family) many of the ambivalent and indeterminate aspects of the original.

The exploration narratives, to which Shakespeare's *Tempest* seems to allude ever so faintly (critics who have looked for Shakespeare's use of exploration source narratives have made much of a few words like "Bermoothes" and "new world"), are made much more explicit in the Davenant/Dryden *Tempest*.[10] While Shakespeare's Miranda does not understand why a tempest is wracking the ship on the sea, she does nevertheless understand that it is a ship. Davenant and Dryden's Miranda and Dorinda, however, do not even understand this. With this change, Davenant and Dryden encourage their audience to step outside of themselves and imagine not knowing what a ship is, not being European, as possibly a way of reinforcing some of the common knowledge that allows Europeans to differentiate themselves from others. Dorinda reports to her sister:

> This floating Ram did bear his Horns above;
> All ty'd with Ribbands, russling the wind,
> Sometimes he nodded down his head a while,
> And then the Waves did heave him to the Moon;
> He clamb'ring to the top of all the Billows,
> And then again he curtsy'd down so low,
> I could not see him: till, at last, all side long
> With a great crack his belly burst in pieces. (1.2.304–11)[11]

Echoing lines that had been recorded as the American Indians' first reaction to seeing large ships, Davenant and Dryden make explicit the connections between the guileless girls and naive Indians. In fact, they employ similar language to that used by Dryden in an earlier play recounting the conquest of America by the Spaniards, *The Indian Emperour, or the Conquest of Mexico by the Spaniards* (1665). In *The Indian Emperour*, an Indian reports his first glimpse of ships on the sea:

> The object, I could first distinctly view,
> Was tall straight trees, which on the waters flew;
> Wings on their sides, instead of leaves, did grow,
> Which gathered all the breath the winds could blow:
> And at their roots grew floating palaces,
> Whose outblowed bellies cut the yielding seas. (1.2.107–12)[12]

Because of the popularity of Dryden's *Indian Emperour*, the two plays may have been produced within months of each other. Although staged by different companies in different theaters, many of the audience members would have seen both plays (including Samuel Pepys, who wrote about both in his diary). In other words, Davenant and Dryden's rewriting of this narrative of first encounter would not have been missed by many. Linking Miranda and Dorinda with the native inhabitants of America situates the play more firmly in the New World, and shows Davenant and Dryden altering the source text to create a "coherent discourse" by erasing the play's ambiguities, thereby transforming the indeterminate text (and its subtext) into a coherent means for reinforcement of dominant English subjective territory. The linking of Miranda and Dorinda creates a "coherent" conceptual space for the audience, an identifiable context that reduces the indeterminacy that could allow space for questioning the official subjective spatial and cultural divisions that parameterize their own, domestic world.

Likewise, the fear of unlicensed and uncontrolled sexual desire, which Prospero expresses in Shakespeare's *Tempest*, is transformed into a more explicit fear of miscegenation in the Davenant/Dryden rewrite. While in Shakespeare's original Caliban's desire to have "peopled else / This isle with Calibans" presents an underlying threat, it is unclear whether the threat is that of bestiality, miscegenation, or simple insubordination (1.2.353–54). Caliban's monstrous status is not clearly defined, and therefore identifying this threat proves difficult. While Davenant and Dryden's Caliban remains ambiguously identified, racialized language and the fear of miscegenation are employed more consistently throughout their revised play. The addition of Caliban's sister

Sycorax, and the new plot to marry her to one of the drunken, rebellious sailors, Trincalo, creates a space for this new language to exist. These additions also work to make Caliban more relatable, having a family and family interests not unlike Europeans. His beingness is thereby more coherently presented, providing him with an identifiable subjective territory. This may have made his racialized otherness more threatening, since infiltration into official culture could be more easily accomplished given the similarities in social structuring and values.

Nevertheless, upon first seeing Sycorax, Trincalo mockingly declares that "She's monstrous fair indeed" (3.3.6), and proceeds to refer to her as "Blobber-lips" (3.3.12), and as "Queen Slobber-Chops" (3.3.130). As Kim Hall has convincingly argued, the early modern use of "fair" was often employed as a "term in praise of female beauty that is closely associated with white [and] becomes a term referring to skin color."[13] In this instance, Trincalo mocks Sycorax's *unfair* appearance, and locates her difference in specific physical terms. He employs familiar racialized language in which a certain feature, in this case her lips, is described as being abnormally large or disproportionate. This focus on an easily located physical trait allows for a belief in the essential differences between races, even if social differences are less discernible. Thus, Sycorax's desire to reproduce "twenty sycoraxes and . . . twenty calibans" with Trincalo reads more like the threat of miscegenation than anything in Shakespeare's original (3.3.41–42).

Along with these crystallizations of Shakespeare's ambiguities, Davenant and Dryden alter Ariel's role significantly, and it is in this alteration that it is possible to see the adapters' larger aim. As many twentieth-century critics have noted, the ambiguities of Shakespeare's original play pose several staging challenges.[14] Where Shakespeare left several areas open to directorial interpretation, Davenant and Dryden sought to clarify these interpretative challenges. Ariel is no longer an ambiguously gendered sylph in the Davenant/Dryden *Tempest*, but rather becomes clearly gendered, and helps to conclude the happy resolution of the play by dancing a "saraband" with his beloved Milcha. It is speculated that Ariel was a breeches role on the Restoration stage.[15] Dryden's prologue alludes to the fact that the largest divergence from Shakespeare's text is the transformation of a female actress into a male role:

> But, if for Shakespeare we your grace implore,
> We for our Theatre shall want more:
> Who by our dearth of Youths are forc'd t'employ

> One of our Women to present a Boy.
> And that's a transformation you will say
> Exceeding all the Magick of the play.
> Let none expect in the last Act to find,
> Her Sex transform'd from man to Woman-kind.
> Whate'er she was before the Play began,
> All you shall see of her is perfect man. (27–36)

Accepting that Ariel is the breeches role discussed, then it is clear that the sprite has been transformed not only from a female actress to a male character, but also from a genderless sylph to a "perfect man." Despite this opportunity for Shakespeare's character to move transversally outside of itself, Ariel's role in Davenant and Dryden's play is greatly diminished: in response to Ariel's transversal power, this change works to constrain Ariel's affective presence. In addition to the cross-gender casting of Ariel, many of Ariel's songs and interlude scenes with Prospero have been cut in order to make room for the new plot twists. But do these changes adequately erase the ambiguity and thus the power of Shakespeare's character? It seems that alluding to the stage convention of having a woman play a boy in the prologue would call attention to the "gender discrepancy," that might go unnoticed if unannounced. In performance, would the character have been so clearly male, or would the actor's opposite gender, expressively stated to the audience, suggest androgyny? Could the dual genders (actor and role) have created what transversal theory recognizes as a "translucent effect": when contradictory coding in a performance forces the audience to engage more intensely with less predictable results in its pursuit of meaning?

What is enlarged, however, is Ariel's role in saving the day. As one critic notes, Ariel prevents the Davenant/Dryden *Tempest* from becoming a revenge tragedy, and allows it to become a comic romance.[16] By miraculously saving Hippolito's life, which was fleeting after his attempt to duel with the more worldly and experienced Ferdinand, Ariel transforms the tragic turn of the play into one of resolution and union. The fact that Ariel accomplishes this task without Prospero's guidance or knowledge points to a newfound independence of mind and spirit, and thus a more powerful affective presence both within and beyond Shakespeare's original. And the fact that this occurs in a play in which *he* has been granted a clearly defined (male) gender is significant, suggesting that such independence is only possible for a male.

The ambiguities that Shakespeare's *Tempest* leaves unresolved—most notably Ariel's past and future—are writ large by Davenant and Dryden. The simple addition of a female counter-sprite, Milcha, gives Ariel the history that was so conspicuously absent in Shakespeare. Ariel informs the audience of his past history when he declares that Milcha "twice seven years hath waited for my freedom" (5.2.257). While reading the Davenant/Dryden *Tempest* as a play that celebrates the loyal subject's salvation of an unyielding monarchy is compelling, the gendered nature of Ariel must not be overlooked as part and parcel of this political reading.[17] An ethereal Ariel with no past or identifiable future cannot help to foster another generation of loyal subjects. A solitary figure, Shakespeare's Ariel gains freedom in the end of the play: freedom from authorial control and identification. Davenant and Dryden's gendered Ariel, with his female partner, offers the hope that generation and procreation occur even for spirits, and in turn hope for a more coherent future. Why else have male and female fairies? Loyalty and citizenship are refigured as imaginative acts that can be nurtured and taught to future generations. This hope, of course, counters the potential horror that is induced by Trincalo and Sycorax's marriage. Their miscegenous union could produce not only future generations of racially undetermined children, but also more rebels and dissenters.

Revolutionizing the tempest-tossed

Almost exactly three hundred years later, another adaptation of *The Tempest*, Aimé Césaire's *Une Tempête* (1968), sought to crystallize the ambiguities of Shakespeare's text in a distinctly different way. Césaire's cast of characters locates some of these alterations:

CHARACTERS
As in Shakespeare

Two Alterations:	Ariel, a mulatto slave
	Caliban, a black slave
An Addition:	Eshu, a black devil-god[18]

Césaire firmly places *Une Tempête* within colonialist and imperialist discourses. Ariel and Caliban are specifically designated as slaves before the play begins, and Prospero becomes a self-professed imperialist in the first lines of the play. (See photo 8.1.) Like Dryden three centuries earlier, Césaire sought to make explicit what he saw as shadowy implications in Shakespeare's original play. But as a Martinican growing up in the grip of British colonial rule, in which the Bard was required civilized and civilizing reading, Césaire's relationship with Shakespeare's text is fraught and complex.[19] As one critic has noted, Césaire believed that it

TRANSVERSAL TEMPESTS / 197

Photo 8.1 Aimé Césaire's *Une Tempête* at The Gate in London, September–October 1998. Césaire's specifies that Caliban is a "black slave" while Ariel is a "mulatto slave" (photo by Pau Ros).

was important to "protagonize the colonized . . . to ensure that their voices were heard loud and clear from within the citadel of Western culture."[20] Césaire has said:

> The Martinican malaise is the malaise of a people that no longer feels responsible for its destiny and has no more than a minor part in a drama of which it should be the protagonist.[21]

With this specific political agenda in mind, Césaire moved both Caliban and Ariel to the center stage in order to "protagonize" their existences.

In *Une Tempête*, Caliban and Ariel are set apart by race (black versus mulatto), service (house slave versus field slave), and outlook (integrationist versus separatist). Although they both despise Prospero's reign and desire freedom, their tactics are remarkably different. Césaire deliberately created the two characters to echo the ideological differences between black separatists like Malcolm X and integrationists like Martin Luther King, Jr.[22] But despite the fact that both characters receive sympathetic treatment by Césaire, it becomes increasingly clear that he imagines Caliban's ideological position as the one that will eventually gain the greater good. Caliban decides to stay on the island with Prospero because he desires to make Prospero "impale" himself on his own "mission":

> Prospero, you're a great magician:
> you're an old hand at deception.
> And you lied to me so much,
> about the world, about myself,
> that you ended up by imposing on me
> an image of myself:
> underdeveloped, in your words, undercompetent
> that's how you made me see myself!
> And I hate that image . . . and it's false!
> But now I know you, you old cancer,
> And I also know myself!
> And I know that one day
> my bare fist, just that,
> will be enough to crush your world!
> The old world is crumbling down! (64–65)

In the end, Césaire even grants the audience a glimpse of a mad and defeated Prospero, as a joyous Caliban takes the last lines of the play: "FREEDOM HI-DAY, FREEDOM HI-DAY!" (68).

Ariel's desires, however, are painted in much more ambiguous terms, which is consistent with him being mulatto, but not with his being male. Unlike the Davenant/Dryden version, where there still is possibility for ambiguity (since the role is to be played by a woman dressed as a man), Césaire has eliminated the gender ambiguity, while at the same time erasing the character's inhuman status by emphasizing his racial background. When Prospero grants him his freedom, he describes his future plans in the following manner:

> There, where the Cecropia gloves its impatient hands with silver,
> Where the ferns free the stubborn black stumps
> from their scored bodies with a green cry—
> There where the intoxicating berry ripens the visit
> of the wild ring-dove
> through the throat of the musical bird
> I shall let fall
> one by one,
> each more pleasing than the last
> four notes so sweet that the last
> will give rise to a yearning
> in the heart of the most forgetful slaves

> yearning for freedom!
> . . .
> Or on some stony plane
> perched on an agave stalk
> I shall be the thrush that launches
> its mocking cry
> to the benighted field-hand
> "Dig, nigger! Dig, nigger!"
> and the lightened agave will
> straighten from my flight,
> a solemn flag. (60)

To which Prospero responds, "That is a very unsettling agenda!" And so it is in more ways than one. While Césaire's Ariel begins echoing Shakespeare's original words ("Where the bee sucks . . . "), his plan is to remind slaves of their "yearning for freedom." This disturbs the "totalitarian" Prospero, but the next part of Ariel's plan contains all of the ambiguity that the mulatto slave symbolizes in Césaire's text, which in turn echoes Shakespeare's text. Ariel wants to be the mocking thrush, whose cry resembles a taunt to field hands, so that he can fly away to leave the slaves to peace. But as it becomes clear in the convoluted syntax used to explain this desire, Ariel must be that taunting bird in order to foster the flight. Césaire's mulatto Ariel cannot conceive of a world in which slavery does not exist; even in the dream world of his future existence, inequities of power are inevitable.

Césaire, like so many Caribbean and African writers in the mid-twentieth century, explicitly named Caliban as his cultural, political, and literary progenitor:[23]

> I was trying to "de-mystify" the tale. To me Prospero is the complete totalitarian. I am always surprised when others consider him the wise man who "forgives." What is most obvious, even in Shakespeare's version, is the man's absolute will to power. Prospero is the man of cold reason, the man of methodical conquest . . . . Caliban can still participate in a world of marvels, whereas his master can merely "create" them through his acquired knowledge. At the same time, Caliban is also a rebel—the positive hero, in a Hegelian sense. The slave is always more important than his master—for it is the slave who makes history.[24]

In this view, Ariel is much less important, and much less interesting, because of the need and desire to identify with a "rebel" figure.

Of course, Césaire was not alone in establishing a "Calibanic lineage."[25] Roberto Fernández Retamar's essay "Caliban: Notes Toward a Discussion of Culture in Our America" (1971) sought to theorize what Césaire performed artistically. Writing in Cuba during the Cuban revolution, Retamar locates Caliban as an important symbol of rebellion and revolution. Unlike Césaire, however, Retamar does not read Shakespeare's text as ambiguous; he reads Shakespeare's *Tempest* as a text explicitly about imperialism:

> This is something that we, the *mestizo* inhabitants of these same islands where Caliban lived, see with particular clarity: Prospero invaded the islands, killed our ancestors, enslaved Caliban, and taught him his language to make himself understood. What else can Caliban do but use that same language—today he has no other—to curse him, to wish that the "red plague" would fall on him? I know no other metaphor more expressive of our cultural situation, of our reality. From Tupac Amaru . . . [to] Frantz Fanon—what is our history, what is our culture, if not the history and culture of Caliban?[26]

The appropriation, which Retamar envisions as an empowering one, sets Caliban apart from Ariel. Ariel, whom Retamar repeatedly defines as an "intellectual," must unite with the rebellious Caliban in order to make the revolution complete. To demonstrate the possibility of this union, Retamar figures Ernesto Che Guevara as an example of an Ariel turned Caliban. Quoting from a speech of Guevara's at the University of Las Villas, Retamar seizes on Guevara's former status "as a member of the middle class, as a doctor with middle class perspectives" in order to show that the intellectuals (Ariels) must become "black, mulatto, a worker, a peasant," a Caliban:

> He [Guevara] proposed to Ariel, through his own most luminous and sublime example if ever there was one, that he seek from Caliban the honor of a place in his rebellious and glorious ranks (45).

Appropriating Caliban as a symbol for rebelliously demanded freedom, Retamar wants to transform all Ariels into this more militant figure.

Retamar was not only responding to Shakespeare's *Tempest*, but also to the famous Uruguayan essayist José Enrique Rodó, who at the turn of the twentieth century sought to make Ariel the symbol for Latin American intellectual prowess. In *Ariel* (1900), Rodó writes that it is the duty of the State to provide the means for human perfection, declaring that if in a democracy "all are granted initial equality, subsequent inequality will be justified, since it will bear the sanction either of the mysterious selection of Nature or of meritorious effort of will."[27] Written as a farewell address to a group of students by an aging professor, *Ariel* seeks to distinguish between the "Calibanistic" North American "dominance of mediocre individualism" and an "Arielistic," Latin American "ability to find effective ways to encourage the emergence and sovereignty of true human superiorities" (58, 59).

When, nearly seventy years later, Retamar wrote, "Our symbol then is not Ariel, as Rodó thought, but rather Caliban," he ostensibly rejected Ariel's (and Rodó's) pacifism in favor of the more active and rebellious Caliban (14). And yet, the desire to appropriate a "Calibanic lineage" may have more to do with the unambivalent aspects of Ariel's character than has been readily admitted by critics and artists alike. The conclusion to Rodó's *Ariel* pinpoints these aspects clearly:

> In making this plea, I have taken my inspiration from the gentle and serene image of Ariel . . . He is the immortal protagonist: his presence inspired the earliest feeble efforts of rationalism in the first prehistoric man when for the first time he bowed his dark brow to chip at rock or trace a crude image on the bones of the reindeer; his wings fanned the sacred bonfire that the primitive Aryan, progenitor of civilized peoples, friend of light, ignited in the mysterious jungles of the Ganges in order to forge with his divine fire the scepter of human majesty . . . Often I am transported by the dream that one day may be reality: that the Andes, soaring high above our America, may be carved to form the pedestal for this statue, the immutable alter for its veneration. (99–100)

Rodó's Ariel is unambiguously linked with the "Aryan, progenitor of civilized peoples," the beacon out of primitive and barbaric darkness. The sprite, in other words, far from being an ambiguous and indefinable figure, is clearly racialized and gendered (male). Both Retamar and Rodó, affected by Ariel's and Caliban's presence, as well as Shakespeare's, have taken Shakespeare's characters off the page/stage and into the "real" political sphere, transforming them from literary/theatrical figures into "literal" characters. This is especially the case for Retamar, who referred to Che Guevara "as an example of an Ariel turned Caliban" (14), thereby invoking multiple presences, both fictional and real, "middle class" and revolutionary.

Inspriteful Shakespace

Rodó was not the first man to notice Ariel's fair complexion. Artistic renderings of *The Tempest* from the eighteenth century on depicted the fairy as fair. (See photos 8.2, 8.3, and 8.4.) Whiteness is the assumption for the spirit of the air. While it is certain that white actors played both Ariel and Caliban on the Jacobean stage, Caliban's monstrosity made his

Photo 8.2 William Hogarth, *Scene from "The Tempest,"* ca. 1735–1740. Notice that while Caliban is depicted as monstrous and of an indeterminate race, Ariel is clearly the white sprite who hovers above (Nostell Priory).

race impossible to define with clarity. (See photo 8.5.) In both Hogarth's and Buchel's renderings of Caliban, the demonic and monstrous characteristics attributed to him make determining his race impossible and, by extension, irrelevant: neither the devil nor the prehistoric man has a determinable race. Sprites and spirits, on the other hand, do not seem to pose the same interpretative problem.

Focusing on the "ambivalent" and "ambiguous" aspects of *The Tempest*, Stephen Orgel discusses the gendered history of Ariel on stage. He claims that by the eighteenth century, Ariel "had become exclusively a woman's role, usually taken by a singer who was also a dancer, and so it remained until the 1930s" (70). But what Orgel fails to comment upon is Ariel's unfailing racialized representation on stage. In an interesting study of interracial casting in Shakespeare's play, Errol Hill declares that "Caliban was one of the first nonblack roles offered to black actors."[28] (See photo 8.6.) Hill further notes that a nonwhite actor did not play Ariel until 1970.[29] For over three hundred and fifty years, then, Ariel's whiteness was unchallenged in stage productions, as well as artistic renderings.

Photo 8.3 Henry Fuseli's *Ariel*, ca. 1800–1810. Fuseli creates an artistic rendering of Ariel's song, "Where the bee sucks." Notice that Fuseli depicts Ariel ambiguously (Folger Shakespeare Library).

While a handful of recent critics have focused on Ariel's position within *The Tempest*, the play seems to set a trap for these critical inquiries. If one assumes or asserts that Ariel represents a human type, then one woefully goes against the text, which repeatedly denies the sprite's humanity.[30] And if one assumes or asserts that Ariel merely represents artistic endeavors, then one woefully ignores the historical

Photo 8.4 F. Miller's *Ariel*, ca. 1850. Miller makes Ariel's gender more masculine, but his whiteness is still maintained ("Passages from the Poets," *The Art Journal* 12[1850]: 278).

and political implications imbedded within the text.[31] This trap does not arise in the same fashion for critical inquiries into Caliban's character because his gender and gendered desires are so human and identifiable.

Despite Caliban's gendered identity, he is the figure in *The Tempest* whose racial identity cannot be defined definitively. The appropriation of a "Calibanic lineage," then, seeks to make determinate what initially existed indeterminately. Perhaps this hardening occurs because "glimpses" of a racial identity can be seen in Shakespeare's text, but the creation of a clearly racialized Caliban within the various African and

TRANSVERSAL TEMPESTS / 205

Photo 8.5 Charles Buchel's charcoal drawing of Sir Herbert Beerbohm as Caliban, ca. 1904. First published in *The Tatler*. Caliban's monstrousness prevents him from being clearly racialized (Folger Shakespeare Library).

Caribbean appropriations of *The Tempest* makes essential what had been more flexible. In other words, Caliban's ambiguous racial status within Shakespeare's *Tempest* may provide a more powerful tool to combat beliefs in racial essentialism than the clearly determined racialized Caliban created by Césaire, Retamar, and others. Reading their works against Davenant and Dryden's seventeenth-century appropriation of *The Tempest* makes the problems of these racialized crystallizations clear.

Photo 8.6 Canada Lee as Caliban in Margaret Webster's 1945 production of *The Tempest* in New York. Canada Lee is thought to be the first black actor to portray Caliban, and it is thought that this represents one of the first nonblack Shakespearean roles assigned to a black actor (photo by Eileen Darby in *Theatre Arts Magazine*, February 1945: 88).

Davenant and Dryden employ the same hardening of terms to make the threat of miscegenation more apparent in *The Tempest, or the Enchanted Island*, but their essentialism is the same as Césaire's and Retamar's.

On the other hand, there have been few modern African and Caribbean appropriations of Ariel precisely because the spirit's racial identity is not truly undetermined in the text. While Ariel's gender, sexuality, humanity, birth, origin, and future existence evade determinacy in Shakespeare's original play, *its* race is oddly inscribed within it; sprites are not raceless, but assumed to be the default color of the culture that imagines its own (white) homogeneity. But the successful appropriation of Ariel, as a symbol of racial and political freedom, offers the ability to deconstruct the assumption that indeterminacy (or more to the point, unspecificity) equates whiteness. Appropriating Ariel's presumed racial indeterminacy and actually maintaining it creatively would highlight the problems of essentialism. Perhaps the more powerful political and literary move would be to question indeterminacy in general. Ariel's indeterminate status could be turned into a transversal power if one acknowledges the freedom of ambiguity and ambivalence.

The question remains as to how one could render Ariel's indeterminacy onstage, presenting Shakespeare's sprite as an ambivalent creature outside the socially constructed categories of race, sex, and gender in order to challenge the boundaries of those very categories. In practice, how could Ariel, played by an actress or actor, be nonprescribed, stripped of the assumptions that have been made historically so as to resolve the character's textually ambivalent features? If embodied by a performer, whose own form would inform the role, providing a platform on which audiences could read the character and resolve for themselves its ambiguities, would Ariel's indeterminacy not be erased? How, then, could Ariel be transversally reproduced in the theater, an art form based on the capacity of audiences to readily identify performers as specific, *not* unspecified, characters?

It is our contention that by embracing Ariel's textually determined feature, its otherworldliness, the character could retain its transversally empowering ambiguity, even if played within the confines of the human form. In effect, Ariel's only link to the human world *is* its creator, Shakespeare, and perhaps because Shakespeare can be considered Ariel's author/father, the character has been, historically, brought down to Earth, transformed from an ethereal being of the air into a tangible form that could be easily read and experienced by spectators as human. However, if, by capitalizing on the character's otherworldliness as a sprite, Ariel were dehumanized, its indeterminacy could be powerfully fueled by transversal power. By reproducing Ariel as a non-gendered and nonracialized physical presence—such as a structurally nonhuman form, like a metal-based robot,[32] or an unspecified, *in*human (within the human) form, such as a non-gendered nude whose body was painted in a nonracially coded color, say orange—Ariel could retain its indeterminable "otherness." On the other hand, by embracing the character's ethereal state, Ariel could, through the use of an indeterminately gendered voice-over, be rendered as an unspecified present-absence—an omnipotent power without visible form that is present yet seemingly absent.

These means of dehumanizing Shakespeare's sprite would free Ariel from the bounds of meaning socially prescribed onto it, as well as onto audiences, and would allow Ariel's inspriteful indeterminacy to be harnessed in order to inspire transversal tempests both off and on the stage.

Notes

1. For a definition of "Shakespace," see Don Hedrick and Bryan Reynolds, "Shakespace and Transversal Power," in *Shakespeare Without Class: Misappropriations of Cultural Capital*, ed. Donald Hedrick and Bryan Reynolds (New York: St. Martin's Press, 2000), 3–47. For a more traditionalist reading of *The Tempest*, see Frank Kermode's influential introduction to the play: *The Tempest*, ed. Frank Kermode (Cambridge: Harvard University Press, 1954), i–lxxxviii. George Will took up this traditionalist mantle when new historicism began to gain national attention in the 1990s: George Will, "Literary Politics: *The Tempest*? It's really 'about' imperialism. Emily Dickinson's poetry? Masturbation," *Newsweek*, April 22, 1991: 72. Interest in the colonial aspects of the play was started by Stephen Greenblatt's new historicist essay, "Learning to Curse: Aspects of Linguistic Colonialism in the Sixteenth Century," in *First Images of the Americas: The Impact of the New World on the Old*, ed. Fredi Chiappelli (Berkeley: University of California Press, 1976), 561–80. Feminist readings of the play have also grown out of this late twentieth-century discomfort with traditionalist views. See, e.g., Ann Thompson, " 'Miranda, Where's Your Sister?': Reading Shakespeare's *The Tempest*," in *Feminist Criticism: Theory and Practice*, ed. Susan Sellers (Toronto: University of Toronto Press, 1991), 45–55.
2. Interestingly, the critical debates about the colonial aspects of *The Tempest* have often focused on the play's indeterminate geographical location. Deborah Willis, "Shakespeare's *Tempest* and the Discourse of Colonialism," in *Studies in English Literature 1500–1900* 29 (1989): 277–89. Jerry Brotton, " 'This Tunis, sir, Was Carthage': Contesting Colonialism in *The Tempest*," *Post-Colonial Shakespeares*, ed. Ania Loomba and Martin Orkin (New York: Routledge, 1998), 23–42. David Scott Kastan, *Shakespeare After Theory* (New York: Routledge, 1999), especially his chapter, " 'The Duke of Milan / And His Brave Son': Old Histories and New in *The Tempest*."
3. Paul Brown, " 'This Thing of Darkness I Acknowledge Mine': *The Tempest* and the Discourse of Colonialism," in *Political Shakespeare: New Essays in Cultural Materialism*, ed. Jonathan Dollimore and Alan Sinfield (Manchester: Manchester University Press, 1985), 48.
4. Stephen Orgel, "Introduction," *The Tempest*, ed. Stephen Orgel (Oxford: Clarendon Press, 1987), 10.
5. For more on "transversal territory," "transversality," and "transversal theory" in general, beyond the introduction to this book and other chapters, see Bryan Reynolds, "The Devil's House, 'or worse': Transversal Power and Antitheatrical Discourse in Early Modern England," *Theatre Journal* 49.2 (1997): 143–67, and *Becoming Criminal: Transversal Performance and*

Cultural Dissidence in Early Modern England (Baltimore: Johns Hopkins University Press, 2002); and Bryan Reynolds and Joseph Fitzpatrick, "The Transversality of Michel de Certeau: Foucault's Panoptic Discourse and the Cartographic Impulse" *Diacritics* 29.3 (1999): 63–80.

6. William Shakespeare, *The Tempest* in *The Tempest: A Case Study in Critical Controversy*, ed. Gerald Graff and James Phalen (Boston: Bedford/St. Martins, 2000). All quotations from this play come from this edition.

7. Critical interest in Caliban has been consistently strong since Greenblatt's "Learning to Curse" essay was first published in 1976. For a few examples, see: Trevor Griffiths, "'This Island's Mine': Caliban and Colonialism," *Yearbook of English Studies* 13 (1983): 159–80; Meredith Ann Skura, "Discourse and the Individual: the Case of Colonialism in *The Tempest*," *Shakespeare Quarterly* 40 (1989): 42–69; and Virginia and Alden Vaughan, *Shakespeare's Caliban: A Cultural History* (Cambridge: Cambridge University Press, 1991).

8. For more on "affective presence," see Bryan Reynolds, *Becoming Criminal*, 6, 46,128; for a detailed account of "subjective territory," see the articles dealing with transversal theory cited in note 5.

9. For a full treatment of the new theatrical resources utilized by Restoration adaptation of *The Tempest*, see George Guffey's introductory materials to *After the Tempest* (Los Angeles: William Andrews Clark Memorial Library, 1969), i–xxiv; Maximilian Novak and George Guffey, eds., *The Works of John Dryden*, Vol. X (Berkeley: University of California Press, 1970), 319–43; and Montague Summers, *Shakespeare Adaptations* (London: J. Cape, 1922).

10. Painstaking research has been done to locate possible sources for *The Tempest*. Much of this work, connecting Michel Montaigne's "Of the Cannibals," William Strachey's "True Repertory of the Wrack," and Sylvester Jourdain's "A Discovery of Bermudas," can be traced in: Charles Frey, "*The Tempest* and the New World," *Shakespeare Quarterly* 30 (1979): 29–41; Peter Hulme, "Hurricanes in the Caribbees: The Constitution of the Discourse of English Colonialism," in *1642: Literature and Power in the Seventeenth Century*, ed. Francis Barker (Colchester: University of Essex, 1981), 55–83; and Stephen Greenblatt, "Martial Law and the Land of Cockaigne," *Shakespearean Negotiations: The Circulation of Social Energy in Renaissance England* (Berkeley: University of California Press, 1988), 129–63.

11. John Dryden and William Davenant, *The Tempest, or the Enchanted Island* (1670). *The Works of John Dryden*, Vol. X, ed. Maximillian Novak and George Guffey. All quotations from this play come from this edition.

12. John Dryden, *The Indian Emperour, or Conquest of Mexico by the Spaniards* (1665). *The Works of John Dryden*, Vol. IX, ed. John Loftis and Vinton Dearing (Berkeley: University of California Press, 1966).

13. Kim Hall, "'These bastard signs of fair': Literary Whiteness in Shakespeare's Sonnets," in *Post-Colonial Shakespeares*, ed. Ania Loomba and Martin Orkin (New York: Routledge, 2000), 68. Kim Hall's other works also touch on this subject: "'I rather would wish to be a blackamoor': Race, Rank, and Beauty in Lady Mary Wroth's *Urania*," in *Women, 'Race' and*

Writing, ed. Margo Hendricks and Patricia Parker (New York: Routledge, 1994), 178–94, and *Things of Darkness: Economies of Race and Gender in Early Modern England* (Ithaca: Cornell University Press, 1995).

14. For an interesting examination of the staging and interpretative challenges that face any production of *The Tempest*, see Stanley Wells's article: "Problems of Stagecraft in *The Tempest*," *New Theatre Quarterly* 10 (1994): 348–57. Interestingly, Wells does not address the interpretative challenges that face staging Caliban and Ariel.
15. For a thorough discussion of the breeches roles in Davenant and Dryden's *The Tempest, or The Enchanted Island* see Joseph Roach's article: "The Enchanted Island: Vicarious Tourism in Restoration Adaptations of *The Tempest*," in "*The Tempest*" *and Its Travels*, ed. Peter Hulme and William Sherman (Philadelphia: University of Pennsylvania Press, 2000), 60–70. Roach speculates that either Hippolito or Ariel was the breeches role mentioned in the prologue, and he offers a fascinating reading of the Restoration habit of visiting backstage, including the dressing room, as an aspect of vicarious tourism/voyeurism.
16. Katharine Eisaman Maus, "Arcadia Lost: Politics and Revision in the Restoration *Tempest*," *Renaissance Drama* 13 (1982): 198.
17. Maus, for example, offers a fascinating reading of Ariel as the loyal and imaginative subject who helps to preserve political stability. Maus argues that this figure is typical for Dryden, who liked to figure his own writing as an imaginative act that provides a similar type of political preservation.
18. Aimé Césaire, *A Tempest*, trans. Richard Miller (New York: Ubu Repertory Theater Publications, 1992). All citations from this play come from this edition.
19. For a brief biography of Césaire's life and writing, see Rob Nixon, "Caribbean and African Appropriations of *The Tempest*," *Critical Inquiry* 13 (1987): 557–78.
20. Lucy Rix, "Maintaining a State of Emergence/y: Aimé Césaire's *Une Tempête*," "*The Tempest*" *and Its Travels*, ed. Peter Hulme and William Sherman (Philadelphia: University of Pennsylvania Press, 2000), 236.
21. Aimé Césaire, quoted in Richard D.E. Burton, *Assimilation or Independence? Prospects for Martinique* (Montreal: Center for Developing-Area Studies, McGill University, 1978), 1.
22. Judith Holland Sarnecki, "Mastering the Masters: Aimé Césaire's Creolization of Shakespeare's *The Tempest*," *The French Review* 74 (2000): 283–84.
23. Some other Caribbean and African appropriations of *The Tempest* include: George Lamming's *Pleasures of Exile* (1960); John Pepper Clark's "The Legacy of Caliban" (1968); Edward Braithwaite's *Islands* (1969); David Wallace's *Do You Love Me Master?* (1977); and Ngugi wa Thiong'o's "Towards a National Culture" (1983).
24. Aimé Césaire, quoted in S. Belhassen, "Aimé Césaire's *A Tempest*," in *Radical Perspectives in the Arts*, ed. Lee Baxandall (Harmondsworth: Penguin, 1972), 171–184.
25. "Calibanic lineage" is a term that Rob Nixon uses to trace the Caribbean and African appropriations of *The Tempest*. For other discussions of

literary/theatrical appropriations see: Thomas Cartelli, "Prospero in Africa: *The Tempest* as Colonialist Text and Pretext," in *Shakespeare Reproduced: The Text in History and Ideology*, ed. Jean Howard and Marion O'Connor (New York: Routledge, 1990), 99–115; and "On the World Stage," "*The Tempest*" *and Its Travels*, ed. Peter Hulme and William Sherman (Philadelphia: University of Pennsylvania Press, 2000), 147–67.

26. Roberto Fernández Retamar, "Caliban: Notes Toward a Discussion of Culture in Our America," *Caliban and Other Essays*, trans. Edward Baker (Minneapolis: University of Minnesota Press, 1989), 14.
27. José Enrique Rodó, *Ariel*, trans. Margaret Sayers Peden (Austin: University of Texas Press, 1988), 66.
28. Errol Hill, "Caliban and Ariel: A Study in Black and White in American Productions of *The Tempest* from 1945–1981," *Theatre History Studies* 4 (1984), 2.
29. Jonathan Miller's Mermaid Theatre Production in London (1970) starred Norman Beaton as Ariel. John Barton's Royal Shakespeare Theatre Production in Stratford-upon-Avon (1970) starred Ben Kingsley as Ariel. These productions were influenced by Aimé Césaire's rewrite of *The Tempest* in 1968.
30. Even though critics who assume/assert that Ariel represents a human type reach radically different conclusions, nevertheless they must sidestep the text's declarations of Ariel's nonhuman status. Andrew Gurr argues that Ariel represents the fabled industrious servant set against the idle and failed servant type, Caliban. In making this argument, Gurr declares that Ariel should not be viewed as "the light spirit of fancy," but "as a servant of sorts." Derek Cohen, on the other hand, declares that while both Ariel and Caliban are slave-types, Ariel represents the slave whose social death has been achieved. In making this argument, Cohen declares that "reading [Ariel and Caliban] as non-human is an established practice which tends to remove from the reader the responsibility of and connection to motive." Andrew Gurr, "Industrious Ariel and Idle Caliban," in *Travel and Drama in Shakespeare's Time*, ed. Jean-Pierre Maquerlot and Michelle Willems (Cambridge: Cambridge University Press, 1996): 193–208. Derek Cohen, "The Culture of Slavery: Caliban and Ariel," *Dalhousie Review* 76 (1996): 153–75.
31. Critics who find postcolonial readings of *The Tempest* misguided have made such arguments. See, for example, Alvin Kernan, *The Playwright as Magician* (New Haven: Yale University Press, 1979), and Jonathan Bate, "Caliban and Ariel Write Back," *Shakespeare and Race*, ed. Catherine Alexander and Stanley Wells (Cambridge: Cambridge University Press, 2000): 165–76.
32. In *Forbidden Planet* (dir. Fred McLeod Wilcox, 1956), the sci-fi adaptation of *The Tempest*, the character of Ariel, played by Robby the Robot (Marvin Miller), was radically dehumanized and rendered into a machine. While this is, to our knowledge, the farthest a version of *The Tempest* has come to capturing Ariel's nonhuman quality, the character was still linked to the terrestrial world: Robby the Robot was man-made (created by the film's

Prospero character, Dr. Edward Morbius [Walter Pidgeon]), and, in name and voice, a clearly gendered male robot.

Works Cited

Bate, Jonathan. "Caliban and Ariel Write Back." *Shakespeare and Race*. Ed. Catherine Alexander and Stanley Wells. Cambridge: Cambridge University Press, 2000. 165–76.

Belhassen, S. "Aimé Césaire's *A Tempest*." *Radical Perspectives in the Arts*. Ed. Lee Baxandall. Harmondsworth: Penguin, 1972. 175–77.

Brotton, Jerry. " 'This Tunis, sir, Was Carthage': Contesting Colonialism in *The Tempest*." *Post-Colonial Shakespeares*. Ed. Ania Loomba and Martin Orkin. New York: Routledge, 1998. 23–42.

Brown, Paul. " 'This Thing of Darkness I Acknowledge Mine': *The Tempest* and the Discourse of Colonialism." *Political Shakespeare: New Essays in Cultural Materialism*. Ed. Jonathan Dollimore and Alan Sinfield. Manchester: Manchester University Press, 1985. 48–71.

Burton, D.E. *Assimilation or Independence? Prospects for Martinique*. Montreal: Center for Developing-Area Studies, McGill University, 1978.

Cartelli, Thomas. "Prospero in Africa: *The Tempest* as Colonialist Text and Pretext." *Shakespeare Reproduced: The Text in History and Ideology*. Ed. Jean Howard and Marion O'Connor. New York: Routledge, 1990. 99–115.

Césaire, Aimé. *A Tempest*. Trans. Richard Miller. New York: Ubu Repertory Theater Publications, 1992.

Cohen, Derek. "The Culture of Slavery: Caliban and Ariel." *Dalhousie Review* 76 (1996): 153–75.

Dryden, John. *The Indian Emperour, or The Conquest of Mexico by the Spaniards* (1665). *The Works of John Dryden*. Vol. IX. Ed. John Loftis and Vinton Dearing. Berkeley: University of California Press, 1966. 1–112.

Dryden, John and William Davenant. *The Tempest, or the Enchanted Island* (1670). *The Works of John Dryden*. Vol. X. Ed. Maximillian Novak and George Guffey. Berkeley: University of California Press, 1970. 1–103.

Forbidden Planet. Dir. Fred McLeod Wilcox. Perf. Walter Pidgeon, Anne Francis, Leslie Nielson, and Marvin Miller. Warner Brothers, 1956.

Frey, Charles. "*The Tempest* and the New World." *Shakespeare Quarterly* 30 (1979): 29–41.

———. "Martial Law in the Land of Cockaigne." *Shakespearean Negotiations: The Circulation of Social Energy in Renaissance England*. Berkeley: University of California Press, 1988. 129–63.

Greenblatt, Stephen. "Learning to Curse: Aspects of Linguistic Colonialism in the Sixteenth Century." *First Images of the Americas: The Impact of the New World on the Old*. Ed. Fred Chiappelli. Berkeley: University of California Press, 1976. 561–80.

Griffiths, Trevor. " 'This Island's Mine': Caliban and Colonialism." *Yearbook of English Studies* 13 (1983): 159–80.

Guffey, George. Introduction. *After The Tempest*. Los Angeles: William Andrews Clark Memorial Library, 1969. i–xxiv.

Gurr, Andrew. "Industrious Ariel and idle Caliban." *Travel and Drama in Shakespeare's Time*. Ed. Jean-Pierre Maqerlot and Michelle Willems. Cambridge: Cambridge University Press, 1996. 193–208.

———. " 'These bastard signs of fair': Literary Whiteness in Shakespeare's Sonnets." *Post-Colonial Shakespeares*. Ed. Ania Loomba and Martin Orkin. New York: Routledge, 2000. 64–83.

———. *Things of Darkness: Economies of Race and Gender in Early Modern England*. Ithaca: Cornell University Press, 1995.

Hall, Kim. " 'I rather would wish to be a blackamoor': Race, Rank, and Beauty in Lady Wroth's *Urania*." *Women, 'Race' and Writing*. Ed. Margo Hendricks and Patricia Parker. New York: Routledge, 1994. 178–94.

Hedrick, Donald and Bryan Reynolds. "Shakespace and Transversal Power." *Shakespeare Without Class: Misappropriations of Cultural Capital*. Ed. Donald Hedrick and Bryan Reynolds. New York: St. Martin's Press, 2000. 3–47.

Hill, Errol. "Caliban and Ariel: A Study in Black and White in American Productions of *The Tempest* from 1945–1981." *Theatre History Studies* 4 (1984): 1–10.

Hulme, Peter. "Hurricanes in the Caribbees: The Constitution of the Discourse of English Colonialism." *1642: Literature and Power in the Seventeenth Century*. Ed. Francis Barker. Colchester: Univesity of Essex, 1981. 55–83.

Hulme, Peter and William Sherman, "On the World Stage." *"The Tempest" and Its Travels*. Ed. Peter Hulme and William Sherman. Philadelphia: University of Pennsylvania Press, 2000. 147–67.

Kastan, David Scott. *Shakespeare After Theory*. New York: Routledge, 1999.

Kermode, Frank. Introduction. *The Tempest*. Cambridge: Harvard University Press, 1954. i–lxxxviii.

Kernan, Alvin. *The Playwright as Magician*. New Haven: Yale University Press, 1979.

Maus, Katharine. "Arcadia Lost: Politics and Revision in the Restoration *Tempest*." *Renaissance Drama* 13 (1982): 189–209.

Nixon, Rob. "Caribbean and African Appropriations of *The Tempest*." *Critical Enquiry* 13 (1987): 557–78.

Novak, Maximillian and George Guffey. "Notes on *The Tempest*." *The Works of John Dryden*. Vol. X. Ed. Maximillian Novak and George Guffey. Berkeley: University of California Press, 1970. 319–43.

Orgel, Stephen. Introduction. *The Tempest*. Oxford: Clarendon Press, 1987. 1–87.

Retamar, Roberto Fernández. *Caliban and Other Essays*. Trans. Edward Baker. Minneapolis: University of Minnesota Press, 1989.

Reynolds, Bryan. *Becoming Criminal: Transversal Performance and Cultural Dissidence in Early Modern England*. Baltimore: Johns Hopkins University Press, 2002.

———. "The Devil's House, 'or worse': Transversal Power and Antitheatrical Discourse in Early Modern England." *Theatre Journal* 49.2 (1997): 143–67.

———. "Venetian Ideology or Transversal Power?: Iago's Motives and the Means by which Othello Falls." *Critical Essays on Othello*. Ed. Philip Kolin. New York: Routledge, 2002. 203–19.

Reynolds, Bryan and Joseph Fitzpatrick. "The Transversality of Michel de Certeau: Foucault's Panoptic Discourse and the Cartographic Impulse." *Diacritics* 29.3 (1999): 63–80.
Reynolds, Bryan and D.J. Hopkins. "The Making of Authorships: Transversal Navigation in the Wake of *Hamlet*, Robert Wilson, Wolfgang Wiens, and Shakespace." *Shakespeare After Mass Media*. Ed. Richard Burt. New York: Palgrave, 2002. 265–86.
Rix, Lucy. "Maintaining the State of Emergence/y: Aimé Césaire's *Une Tempête*." *"The Tempest" and Its Travels*. Ed. Peter Hulme and William Sherman. Philadelphia: University of Pennsylvania Press, 2000. 236–49.
Roach, Joseph. "The Enchanted Island: Vicarious Tourism in Restoration Adaptations of *The Tempest*." *"The Tempest" and Its Travels*. Ed. Peter Hulme and William Sherman. Philadelphia: University of Pennsylvania Press, 2000. 60–70.
Rodó, José Enrique. *Ariel*. Trans. Margaret Sayers Peden. Austin: University of Texas Press, 1988.
Sarnecki, Judith Holland. "Mastering the Masters: Aimé Césaire's Creolization of Shakespeare's *The Tempest*." *The French Review* 74 (2000): 276–86.
Shakespeare, William. *The Tempest. The Tempest: A Case Study in Critical Controversy*. Ed. Gerald Graff and James Phalen. Boston: Bedford/St. Martin's Press, 2000: 10–88.
Skura, Meredith Ann. "Discourse and the Individual: the Case of Colonialism in *The Tempest*." *Shakespeare Quarterly* 40 (1989): 42–69.
Summers, Montague. *Shakespeare Adaptations*. London: J. Cape, 1922.
Thompson, Ann. " 'Miranda, Where's Your Sister?': Reading Shakespeare's *The Tempest*." *Feminist Criticism: Theory and Practice*. Ed. Susan Sellers. Toronto: University of Toronto Press, 1991. 45–55.
Vaughan, Virginia and Alden. *Shakespeare's Caliban: A Cultural History*. Cambridge: Cambridge University Press, 1991.
Wells, Stanley. "Problems of Stagecraft in *The Tempest*." *New Theatre Quarterly* 10 (1994): 348–57.
Will, George. "Literary Politics: *The Tempest*? It's really 'about' imperialism. Emily Dickinson's poetry? Masturbation." *Newsweek* April 22, 1991: 72.
Willis, Deborah. "Shakespeare's *Tempest* and the Discourse of Colonialism." *Studies in English Literature 1500–1900* 29 (1989): 277–89.

Chapter 9
"For Such a Sight Will Blind a Father's Eye": The Spectacle of Suffering in Taymor's *Titus*

Courtney Lehmann, Bryan Reynolds, & Lisa Starks

Disrupting the cinematic narrative of Julie Taymor's *Titus* (1999), an adaptation of Shakespeare's *Titus Andronicus*, is a nightmare sequence in which the newly crowned empress, Tamora, stands face-to-face with her enemy, Titus. Images of dismembered limbs engulfed in flames appear in the background, sailing forward, until they inundate the screen behind the characters' silhouetted profiles.[1] Invoking the murders of Tamora's eldest son Alarbus and Titus's youngest son Mutius, the flying, burning body parts symbolize the powerfulness, unpleasantness, and mysteriousness of the creative process, harkening back, via homage to the horror of Seneca, to the ritualistic, religious roots of theater in the Greek festival of Dionysus. In *Titus*, especially in such magnificent scenes as this, Taymor combines a slew of culturally rich metaphors, in effect demonstrating what transversal theory calls "investigative-expansive wherewithal" and the "principle of translucency." (See photo 9.1.)

Our term, "investigative-expansive wherewithal" ("i.e. wherewithal"), is derivative of what Bryan Reynolds and James Intriligator call the "investigative-expansive mode of analysis" ("i.e. mode"), which is a research methodology that works to break down any subject matter under a given investigation into variables that are then partitioned and examined in relation to other influences, both abstract and empirical, beyond what is immediately observable and outside of the scope of any proffered hypothesis. The purpose is to situate the analysis in respect to the environments of which it is a part, and in light of those that act upon it. Open to anything, the investigative-expansive approach continually adapts to changes in circumstances and atmosphere, whether conceptual,

Photo 9.1 Titus and Tamora

emotional, or physical, reparameterizing as the investigation progresses. In typical postmodern fashion, the investigative-expansive approach resists predetermination and circumscription, and does not aspire to fantasies of definitive answers, totalization, or universality.[2]

I.e. wherewithal shares the i.e. mode's investment in expansive thinking and exploration, seeking to draw from various cultural and philosophical sources when pursuing artistic expression. Although i.e. wherewithal is adventurous, evolutionary, and malleable, it also relates to attitude. Implementation of i.e. wherewithal necessitates a readiness to coalesce cultural tropes, appropriating icons, rituals, and aesthetics from other cultures and ethnicities (those "other" to the artist or context in which the art is displayed or performed), and to deliver them to often resistant and uninitiated audiences. In other words, in our view, i.e. wherewithal requires transversal thought, courage, and *chutzpah*. These are traits Taymor exhibits in *Titus*, which shine delightfully and powerfully through her employment of what Donald Hedrick and Bryan Reynolds have termed the "principle of translucency."[3]

The principle of translucency can function through diverse media, but it is especially effective when operative through social, theatrical, and cinematic performances. It involves the incorporation of contradictory or incoherent codes in a context where a discernible, typically singular or overarching meaning is anticipated, such as in the theater or cinema. The articulation of such inconsistent coding works to present varied and multiple messages at the same time. The codes, however, must be apparent and identifiable. Inasmuch as the coding forces the audience to become active readers of the performance, rather than merely passive voyeurs, in order for the coding to produce active responses, it must be accessible and reasonably comprehensible. If there are many codes at work or if the codes are unintelligible, then the audience will become bewildered and alienated, and the principle's potency and potential will not be realized. Translucency cannot occur when the coding is nebulous. Each code presented must be distinct and recognizable, even while it continues to participate as a variable in an expression that is more complex, that is to say, part of a bigger picture. The codes themselves must not be complexly displayed, such that they are convoluted or that there are too many for the audience to track and differentiate. The principle of translucency is most effective when there are just a few featured codes. If skillfully executed, the principle of translucency can promote a transversal performance, one that instigates transversal movements: feelings, thoughts, and actions alternative to those that work to circumscribe and maintain a particular "subjective territory," the socioculturally prescribed, hence ideologically informed, conceptual, and emotional spatial range from which a given subject perceives and experiences the world.[4]

Supercharged by the principle of translucency, the audience member or spectator is manipulated by the performance event's power, and consequently moved transversally outside of his or her own subjective territory. The transversal territory into which the audience member moves is not necessarily that of a subjective territory of an other, but rather potentially a creative space. The artfulness with which the principle of translucency is manifested generates the conditions for both the desire and quest for artistic development on the part of the audience. The performance event's transversal power emerges through the resonance and wonder it manufactures. According to Hedrick and Reynolds:

> The spectator is jarred from the parameters of conventional discourse and reason, exposed to the multilayeredness always present in all performances but rarely so apparent, and is attracted to the new found pellucidity. The principle of translucency is therefore a complex occurrence by which the spectator sees one or two or more things through others while at the same time seeing the others themselves. Faced with a radical situation, rather than wanting to deny or flee from the experience, the spectator wants to assimilate into the transversal territory of the performance and emulate the performer of such alluring translucency.[5]

As with many collaborative mediums, it is unclear who the actual performers ultimately are in the case of Taymor's film *Titus*, whether they are the characters, actors, scene designers, the director, and so on. Yet, through the principle of translucency, *Titus* creates a transversal space wherein spectators are invited to expand and reconfigure the coordinates of their own subjective territories. Insofar as they are transformed into active viewers, working to glean meaning from the challenge of deciphering the codes, the spectators are also held accountable, like the characters in the film, for the reconfigurations that they produce. In its employment of transversal techniques, *Titus* becomes a performance of unusual radicalism, venturing through the principle of translucency beyond the expected. As we will show, the agency with which the audience becomes endowed solicits a particular emotional state, abjection, as a means by which to pass into transversal territory.

We are not the first critics to notice Taymor's transversality. In *Playing with Fire*, the most comprehensive analysis of Taymor's work to date, Eileen Blumenthal points to Taymor's investigative-expansive wherewithal and her use of the principle of translucency: "Harvesting existing art in order to cross-breed new work" is described by Blumenthal as "one key to Taymor's creativity" (8). As an artist, Taymor thus "conceives new theatrical organisms, by combining traits from the most disparate sources to bring original hybrids to life" (52, 57). This i.e. wherewithal quality has led to descriptions of her work that defy lexical consistency, itself a reflection of the plurality of techniques, cultures, genres, and media that constitutes Taymor's productions. Taymor has called herself a "theater-maker" and a "mixed-media artist" (qtd. in Schwartz, Skipitares, and Taymor 468), while others have referred to her work in terms of the dynamics of "crossing," characterizing Taymor's art as "cross-bred," "cross-pollinated," or "cross-cultural" (see esp. Blumenthal and Taymor 7, 8, 52). Although such descriptors emphasize the diversity and scope of Taymor's work, they always seem to fall short in describing its full quality and impact on audiences, challenging critics to generate a theoretical lexicon that is commensurate in complexity with Taymor's aesthetic. To this end, we argue that Taymor's work, from her early theatrical productions in Indonesia to her work in Japan and on Broadway to her cinematic narratives, realizes what we call a "transversal aesthetic," one that goes beyond the mixing of techniques and crossing of boundaries to spur a revolutionary theatrical–cinematic experience, in effect spurring transversal movements. In *Titus*, Taymor exploits the traditional filmic medium to transform the voyeuristic audience into active agents in an onscreen spectacle of suffering.

Taymor's transversal aesthetic

On the level of character, the principle of translucency as implemented by Taymor begins with the actor's iconic physical embodiment of a concept, also known as the "ideograph." Herbert Blau, who developed the concept of the ideograph and was a major influence in Taymor's theatrical background, describes the ideograph as "tightly formed, consolidated, volatile moments of apprehended energy" (qtd. in Blumenthal and Taymor 12). The minimal physical movement serves to signify the primary facet of the character, while becoming, as Blau puts it, a "vortex" from which "Everything outside . . . is whirling in a kind of indeterminate way, but as you get toward the center it becomes dense, and it focuses the mind with sufficient intensity that it explodes outward" (qtd. in Blumenthal and Taymor 12). For Taymor, the ideograph serves as a device to deflect the spectator's interest from the person of the performer to the multifaceted significations of the gestures that register within the language of theater. For this reason, Blumenthal calls the ideograph "a theatrical sign language that facilitate[s] the layering and counter-pointing of subtexts" (12). For Taymor, it also allows for an emotional intensity quite distinct from that elicited by naturalistic acting. Providing a way of "getting at psychology through physicality" (*Bravo Profiles*), the ideograph runs counter to realism, which emphasizes the emotions of the individual through primarily naturalistic representation; the ideograph also poses an alternative to some forms of postmodern theater, which valorize the intellect at the expense of the body (see Blumenthal and Taymor 52).

Taymor invokes Antonin Artaud's theory of the Theater of Cruelty, which imagines a theater that does not permit the audience the opportunity to suspend disbelief willingly or to become mere "Peeping Toms" passively observing an imitation of events that are usually distressing; this theater is instead "a believable reality which gives the heart and the senses that kind of concrete bite which all true sensation requires" (85).[6] Unlike Artaud's Theater of Cruelty, nevertheless, Taymor's theater seeks to inspire empathy as a means to greater audience involvement, as opposed to just wanting to stimulate their senses with, say, representations of horrific acts and trauma. Applying Blau's ideograph in combination with some of Artaud's ideas, Taymor pushes toward a gestural theater, wherein the body and voice signify through plural, often unconventional ways. When she does highlight spoken language, whether it be through music, poetry, or standard dialogue, she combines the spoken language with abstract physicality, making the words seem at once familiar and strange, and the speakers' identities appear multidimensional, even contradictory.

From the abstract foundation of the ideograph, Taymor may then add complex layers of incongruous modes and identities by combining masks and puppets with human actors. Through this overlay of manifold characterizations, Taymor often exploits the principle of translucency: the mask or puppet transparently envelopes other identities, producing multiple, simultaneous codes of characterization that force collisions between illusionary and emblematic styles of representation. As Jeffrey Horowitz, Artistic Director of the Theater for a New Audience, describes this effect, "you know you're looking at a puppet, at a mask. And while you know you're looking at that, you're thinking of something totally different" (*Bravo Profiles*). The result is to broaden the spectator's perception of character and their attendant conception of human subjectivity. Rather than concealing the person underneath, the mask allows for the release of alternative aspects of the actor's personality, as well as for the cathexis of the spectator, all the while presenting an iconic rendering of character on the surface.[7] The mask, therefore, enables transformation: it changes person into puppet and, in combination with human and nonhuman characters, creates an effect of alienation that distances even as it produces an intense emotional response in the viewer (Taymor qtd. in Schwartz, Skipitares, and Taymor 469–70).

The implementation of puppets itself plays on concepts of identity, blurring borders between the nonhuman and human, inanimate and living. This effect is beautifully illustrated in Taymor's designs for the theater piece *King Stag*, where the main character appears first as an actor wearing a mask, then as a figure wrapped in material and rattan, and finally as a Bunraku doll (Blumenthal and Taymor 25). Taymor extended this idea further in *Juan Darien: A Carnival Mass* (New York, 1988 and 1990; Edinburgh, Lille, Montreal, Jerusalem, and San Francisco, 1990–1991; New York, 1996–1997), which she directed, designed, and also wrote in collaboration with composer Elliot Goldenthal. *Juan Darien* explores the boundary between what is and is not categorized as "human" through experimentation with puppets and masked actors. As Blumenthal explains: "Taymor took advantage of the ability of puppets to tease at that border. The audience saw Juan change from a little Bunraku-technique jaguar, to a small wooden baby, to a Bunraku puppet boy—all interacting with the masked actress playing Juan's mother" (32). *King Stag* and *Juan Darien* revealed ways by which Taymor and Goldenthal's use of puppets and masks, along with a unique mix of dissimilar techniques, media, modes of representation, and genres, can give a theatrical production the opportunity, as Taymor puts it, to "move through different levels of reality" (qtd. in Schwartz, Skipitares, and Taymor 461); the intersection and amalgamation of representational modes enabled the Goldenthal–Taymor production itself, and possibly its audience, to enter a transversal space.

One of the chief ways by which Taymor exercises her i.e. wherewithal and achieves her multifaceted theater is through the assimilation of diverse cultural influences, including techniques derived from Indonesia and Japan, Eastern and Western Europe, as well as South and North America. These techniques are drawn from an equally wide range of time periods, encompassing, for example, ancient ritual, eighteenth-century neoclassicism, and contemporary performance art. From these various theatrical traditions, Taymor adopts "shifts in scale" and an "eclectic array of theatrical techniques" in staging (Guatam Dasgupta qtd. in Schwartz, Skipitares, and Taymor 460), which she combines with contrasting religious, mythological, and philosophical perspectives, as invoked by the play, opera, or film. She adopts, however, not by simply appropriating these cultural *styles*, but rather by borrowing the *technique*, its primary form, and working it into a style-in-process that is suited to the new project at hand (Taymor qtd. in Schwartz, Skipirates, and Taymor 469).[8]

Taymor does not just borrow across cultural, national, and historical borders, however; she also trespasses those of media and genre to create the translucent effect of her transversal aesthetic. Indeed, mixing media, particularly that of stage and screen, has always been a vibrant quality of Taymor's work, another example of her ethos of constitutive crossings. In her theatrical productions, Taymor often shifts scales and angles of perspective to generate a cinematic look, a technique that is especially evident in productions like *Way of Snow* and *Juan Darien*. To further play on the border that distinguishes the two media, Taymor frequently employs techniques that blur the distinction between the two, as in shadow puppetry, which is a kind of "live cinema" (Bluementhal and Taymor 8–9). In her films, in addition to importing stage theatricality, through design, choreography, and a combination of stylistic and naturalistic modes of representation, Taymor transcribes many aspects of the stage itself into the cinematic, as in the use of curtains in *Fool's Fire* and staged tableaux in *Titus*.[9] The i.e. wherewithal of Taymor's work that distinguishes it from most Western cinema and theater is nowhere better exemplified than in her cinematic foray into Shakespace and the potentiality of its posthuman occupation.

"Inhuman traitors"

Titus not only assays an oppositional aesthetic of transversality, but also investigates the possibly enabling relationship between abjection and identity as delineated by Julia Kristeva. Indeed, in conjunction, the transversal aesthetics employed in *Titus* resonate in provocative ways with both the promise and perversion associated with the posthuman. In her "Cyborg Manifesto," Donna Haraway explores the cyborg as a figure whose rich topography, comprising "transgressed boundaries" and "potent fusions," offers unique opportunities to "contest for meanings, as well as for other forms of power and pleasure in technologically mediated societies" (154). The crux of deriving an oppositional poetics *or* politics from the cyborg, however, is that the cybernetic organism is itself the offspring, if not the *telos*, of patriarchal capitalism, militarism, biotechnological advances, and, of course, the sovereign individual abstracted from all dependency—and responsibility. Therefore, Haraway focuses on the *systems* environment of the cyborg, wherein the prevailing network of C^3I (the command-control-communication-intelligence technology employed to police borders in a potentially libratory nexus of information flow) becomes a locus of opportunity for "recoding communication and intelligence to subvert command and control" ("Cyborg Manifesto" 175). Yet, if Haraway's "socialist-feminist" vision of the cyborg remains an unrealized, indeed "ironic dream of a common language for women in the integrated circuit" ("Cyborg Manifesto" 149), then Taymor's film both materializes and extends this empowering circuitry by awarding the pleasure, and the horror, of boundary subversion to the spectator. Dispersing the cultural grid of gender and identity across tantalizing collisions of human, animal, and machine, Taymor's film redefines the subjective territory of spectatorship as a perilous negotiation of, and ultimately a choice between, alliance-building and terrorism, accountability and complicity, radicalizing the act of consumption itself as a gateway, via Shakespeare, to a "possible world" of "connection that exceeds domination" (Haraway, "Actors are Cyborg" 23).

Though *Titus* is in many respects a brutally faithful representation of Shakespeare's play, perhaps the most striking difference between the play and the film is Taymor's augmentation of the role of Young Lucius, the child whose gaze is implicitly aligned with that of the film spectator. The film begins with the boy sitting at the kitchen table, wearing a crude paper-bag mask and playing increasingly aggressive war games with his action figures. A manic, *noir*-style lighting effect is created from an unseen television screen that casts an intermittent blue glow on the scene of the toy carnage that Lucius sets about creating. Cleverly employing basic household items as virtual weapons of mass destruction, Lucius splatters his action figures with ketchup and cheese, stabbing and plundering them with kitchen utensils and plates. A striking example of contradictory coding, the more Lucius mimics the unguided rage of battery-operated toys, the more his plastic action figures assume human qualities, uncannily appearing not only to attempt enfeebled escapes from their child assailant, but also to writhe in agony under his unrelenting blows. It is this bizarre translucency of human and machine forms that ushers the viewer into the transversal space of the Roman coliseum, a place wherein an entourage of stiff-legged gladiators march toward Young Lucius like so many toy soldiers seeking revenge against his child's play. The gladiators' eyes, partially concealed by blue smears of battle paint, recall both the torn eye-holes of the boy's paper-bag mask and the silver-blue glow of the TV, as if to mark these machine-like warriors as the progeny of the opening scene—only now the relations of power have changed, and their gaze is trained searchingly, intensely, on us. Posing in stark contrast to these robotic warriors, nimble Roman "chariots" comprising high-tech combinations of Hummers, horses, and humans amble across the coliseum floor, linking the military prowess of ancient Rome to the cyborg technology of postmodern warfare, wherein remote-control, "smart" weaponry demonstrates the extent to which "our machines are disturbingly lively, and we ourselves frighteningly inert" (Haraway, "Cyborg Manifesto" 152). In Taymor's film, however, the inertia of such transversal dislocations in time and space does not last long, for as Young Lucius is held up like a trophy to the thunderous applause of an invisible crowd, we get the sinking feeling that the fun and games are over, but the toys are *us*.

It is no coincidence that the audience enters and exits the film-proper through the Roman coliseum, for this "archetypal theater of cruelty," as Taymor describes it, marks the threshold between violence and entertainment.[10] The two on-location shooting sites for *Titus*, the remains of Mussolini's government center and a Roman coliseum in Croatia, localize this theatrical metaphor in twentieth-century tragedies of ethnic cleansing, born out of catastrophic commitments to sacrosanct borders and singular identities. Pitted against these bastions of C^3I, or, what Haraway calls the "informatics of domination," are the artificial sets of the crossroads and the swamp, two places that, as prosthetic landscapes, implicitly resist the naturalizing imperatives of patriarchal command and control. But there is nothing transcendent about these sites of subversion in Taymor's film. At the crossroads location, for example, Taymor inverts the god's-eye perspective of the camera in order to position the spectator at ground level where Titus lies, facedown, with his legs and arms splayed in a posture of abject submission. Mimicking the swastika-like motif emblazoned on the floor of the emperor's headquarters, Titus converts his body into a grid in a desperate tribute to the phallocratic directives of Roman C^3I, even as the gladiators he once presided over proceed to march, quite literally, over *him*. But as his emergent articulation of solidarity with the feminized elements of "the earth" and "the sea" implies,[11] Titus is becoming more porous and, therefore, potentially open to new affiliations of power; in this respect, then, the crossroads location might be read as an emblem of transversality, a province of unforeseen coalitions forged between "problematic selves and unexpected others" in the wake of self-devastation (Haraway, "Actors are Cyborg" 24). Yet, if Titus is the character who occupies the affective center of Shakespeare's tragedy, it is because he learns how to strategize around his own vulnerabilities and, in the process, dismantle the existing informatics of domination—albeit too late to prevent his own destruction. In Taymor's film, it is the *female* characters who are invested with this propensity for "hacking," for it is their security breaches and design flaws that contain the code for a theater of *accountability*, a locus of alliances born of mutual abjection, based on relations of affinity, not identity. The fact that such connections are forged only in the aftermath of murder and rape is brutal confirmation of the fact that "the Gods" do indeed "delight in tragedies"; what Taymor's film seems more intent on proving, however, is that so do we.[12]

It is in the film's other principal artificial location, the swamp, that Taymor assays an oppositional mode of representation, subjecting the viewer to a barrage of visual pleasure and horror that, ultimately, challenges the audience to "take back the look." In his provocative and complex critique of *Titus*, Richard Burt claims that one of the consequences of the film's richly allusive, dense cinematographic texture—its preoccupation with its own "look"—is that Taymor loses sight of the extent to which her "attempt to critique violence via Fascism and the Holocaust are inevitably in tension with her reinscription of the horror genre" (92). Yet *Titus* is a horror film with a difference, for it uses the ontological foundation of horror—the spectacle of the female body subject to rape, mutilation, and murder—as a starting point for a "look" all its own, one that employs abjection, paradoxically, as a means of going through and beyond victimization. The swamp scene begins as a familiar horror episode featuring two pubescent psychopaths whose adolescent sexual confusion, exacerbated by their castrating mother, compels them to transform Lavinia into a vision of "what men would be if they had no penises—bereft of sexuality, helpless, incapable" (Lurie 166). When they tear off Lavinia's clothing, leaving her trembling in her white bustier and petticoat, we are reminded of the "beast in the boudoir" motif that is the classic topos of Victorian horror—the white, virginal bed clothes underscoring the contrast between the victim's innocence and the "beastie boyz" poised to defile her.[13] In characterizing Lavinia's assailants as video game junkies, however, Taymor dresses up this Gothic theme with contemporary coding that underscores the film's ongoing preoccupation with violence as entertainment, for Chiron and Demetrius are clearly presented as boys whose digital mastery of virtual beings is inseparable from their desire to decimate, even as they inseminate, real bodies. Indeed, the culture of video games, as Haraway observes, is a potent incubator for the production of "[h]igh-tech, gendered-imaginations . . . imaginations that can contemplate destruction of the planet and a sci-fi escape from its consequences" ("Cyborg Manifesto" 168).

In *Titus*, the obvious artificiality of the swamp location is evocative of a sci-fi fantasy scape; yet it also implies that the attendant view of women as an extension of "nature" and, therefore, something to be dominated by men, is also a man-made "science-fiction." Rather than deconstructing this fiction, however, Taymor plays it to the hilt, elevating the ensuing scene of murder, rape, and dismemberment to the status of a spectator sport. Using a dollying camera to re-create the effect of coliseum-style spectatorship in-the-round, Taymor's lens tracks Chiron and Demetrius as they circle Bassianus and Lavinia in an increasingly frenzied manner, cruelly playing with their prey as a prelude to the kill. As a thunderous rock music track drowns out all traces of dialogue, the fictitious world of video games yields to the fantasy milieu of music videos—another virtual reality that Chiron and Demetrius call home, as suggested by their spiky, yuppie-punk, bleach-blonde hair and loud leather pants. And, like rock stars, they flee the scene, leaving us feeling oddly accountable for the "look" that comes back to us from Lavinia, as she is left alone, her body the living emblem of a gaping wound.

Thus we arrive by familiar steps at the primal scene of horror, abjection, the point at which we would traditionally be compelled to avert our eyes.[14] "There are excellent reasons," Linda Williams explains in her article, "When the Woman Looks," for refusing to look at disfigured figures like Lavinia, especially since such scenes seem to punish the female spectator in particular, who "is often asked to bear witness to her own powerlessness in the face of rape, mutilation, and murder" (15). But Williams concludes her analysis of spectatorship and the horror film by *imploring* us to look; for to close our eyes as a gesture of sympathy and resistance may, in fact, incur a greater evil, which lies in assuming that such films "have maintained this sympathy [for the female victim] while simply escalating the doses of violence and sex" (32). In fact, Williams notes, the trend is toward a disturbing disjunction between sympathy and escalating violence that perversely renders female victimization more acceptable. By contrast, *Titus* pries our eyes *open*, compelling us to look, and to look further, by exploring a politics of horror that patently rejects the stasis of sympathy in an effort to induce accountability. In order to arrive at this point, however, Taymor must seduce, even temporarily inoculate, the spectator to what the film has in store with the help of stunning production values and a mise-en-scene that comes dangerously close to aestheticizing violence. It is little wonder, then, that *Titus* has justifiably evinced high levels of critical discomfort. Burt's argument, for example, takes Taymor to task for "toning down" the violence and, in the process, subverting her critique of Fascism by attempting to redeem the play "as a work of art" (92). Nevertheless, it is precisely this combination—Taymor's refusal to rupture the suturing effects of cinematic identification and her appeal to familiar visual tropes culled from art history and popular culture—that *sustains* the look, compelling us *not* to avert our eyes, but rather to go through the horror, as in Artaud's Theater of Cruelty, as though we were experiencing, and producing, it ourselves. This is the spectatorial arc that we identify as "radical accountability," an act of bearing witness to the unbearable that renders us complicit in the horror, inducing abjection, paradoxically,[15] as a transversal passage to a new life.

"Will it consume me? Let me see it then"

In contrast to dominant cinematic modes, *Titus* goes a step beyond graphic realism by involving its audience in an investigation of the act of viewing itself, revealing the abjection underpinning the horror that both fascinates and repels audiences.[16] In the process, Taymor transforms the horror film genre. Eliding the boundaries between the real/surreal, inside/outside, human/animal, Taymor's *Titus* plays on the border in-between, continually invoking a sense of disorder and disjunction, as the audience enters into a potentially threatening transversal space. In psychoanalytic terms, this effect of horror is derived from the role of abjection in subjectivity and art. In her book *Powers of Horror*, Kristeva theorizes the abject as that from which the subject must detach itself in order to form a separate identity; the subject projects the abject onto the maternal body, thereby developing a feeling of repulsion toward that body and also toward that which recalls the corporeal within itself. Because the abject rejects the division that underlies the subject's fragile sense of identity, the abject generates the effect of translucency in its own right, marking the boundary of the "in-between, the ambiguous, the composite" (Kristeva 4). Alternately fascinated and repelled by the abject, the subject nevertheless is drawn continually toward it, toward a transversal space wherein "meaning collapses" (Kristeva 2). Thus, the abject forces the subject to face the boundaries of its subjective territory, the established conceptual and emotional lines that separate what is and what is not the "self."[17] Emerging from the inner loss of separation, the abject cannot be wholly repressed. Sublimated, the abject often returns in a variety of forms (taboos, rituals, visual art, and literature), surfacing particularly in the topos of the grotesque, uncanny, or humor of the absurd.[18]

Nowhere is the abject more apparent than in horror films. The themes predominant in the horror genre (food loathing, the corpse, the decaying or polluted body, bodily fluids and wastes, the half human/animal or human/nonhuman monster) document the uncanny power of abjection both to terrorize and to titillate the spectator. According to Barbara Creed, the classic horror film centers on the dual pleasure and terror elicited by these composite images, exploiting the anxieties they produce in order to diminish the power of the abject on viewers, who are, in the end, reassured of its compartmentalization and distance ("Monstrous-Feminine" 14–16). Despite their transgressive potential, then, these films ultimately reaffirm their viewers' sense of order and identity. Alternatively, in *Titus* and in much of her art in general, Taymor dismisses such consolation by exploring abjection as a horrifying, but potentially empowering, aesthetic and political landscape, refusing to vanquish its gravitational pull on viewers by drawing them into its transversal territory. As Shakespeare's bloody revenge play *Titus Andronicus* contains all the elements of modern horror, along with a keen sense of the absurd, it poses the ideal opportunity for Taymor to exploit abjection as the very basis of her transversal aesthetics, eventually made manifest by her use of the principle of translucency.

A cursory glance at *Titus* reveals Taymor's wide-ranging appropriation of cultural referents, from popular images such as the Zapruder film-style political parade to the *haut couture* vision of Titus reclining in a tub fashioned after Jacques-Louis David's painting of "The Death of Marat" to the overall Daliesque mise-en-scene, all of which serve as a visual shorthand for bridging the gap between early modern play and postmodern film. However, the most complex and powerful sites of visual accretion are the film's female characters, Lavinia and Tamora, through whom Taymor explores disturbingly enticing, abject couplings of "the human, animal, and the divine" in order to force the audience to confront the otherwise "unfathomable layers of a violent event" (Taymor 183). Taymor's brutally tender, post-mutilation tableau of Lavinia is the first of these experiments in transversal visual poetics and, arguably, the most horrific. In both this scene and the ensuing Penny Arcade Nightmare, Lavinia's posture explicitly recalls the famous image of Marilyn Monroe coyly poised over a subway ventilation grate, a comic motif that Taymor uses to disengage the audience's flight response. Yet, in the very instant that we register this connection to Lavinia's delicate figure awkwardly balancing on a tree stump, Taymor wrests the image of Monroe from popular memory by filtering it through the defamiliarizing topos of abjection which, as Kristeva explains, refuses to "respect borders, positions, rules," for it is that which "disturbs identity, system, order" (4). Like the swamp location that is at once "earth" and "sea," then, Lavinia's body becomes a borderland in its own right, for it is no longer legible in terms of the virgin/whore dichotomy central to the play's articulation of C^3I. Consequently, we can only watch with an uncomprehending combination of horror and fascination as she proceeds to make exquisitely unsuccessful attempts to swat down the billows of her bloodstained dress with the tree branches that have replaced her severed hands. As the camera draws toward her face, Lavinia's full, alluringly parted lips recall Monroe's sexy, pouting expression, inviting the audience in closer still; however, when we realize that her lips are swollen not with desire, but with beating, our proximity to her wounded body suddenly becomes undesirable. For as Lavinia utters a silent scream, expelling a surreal stream of blood directly at the camera, Taymor forces us to *hear with our eyes*, baptizing the spectator in trauma, and hurling our own prurient look back at us. When this scene is replayed in the Penny Arcade Nightmare, a doe's head and hooves are superimposed on Lavinia's body, which is now positioned on a column, and lunging tigers threaten her on both sides. Taymor hereby converts her Degas-style rendering of Lavinia as a ruined prima ballerina into an image that is stunningly evocative of the pain-infused, yet empowering self-portraits of Frida Kahlo. Through this shifting tableau, then, Taymor not only implies Lavinia's rejection of victimization by ascribing agency to her act of looking back but, also, in refusing to let us look away, renders us accountable both for her devastation and, potentially, her regeneration. (See photo 9.2.)

Taymor's allusion to Kahlo's signature explorations of the wounded female body grafted onto human, animal, and machine forms points to her own interest in prosthetic identity,[19] or "ghostings," which she describes as the experience of getting into "a concrete exterior" and getting out of oneself (Blumenthal and Taymor 36–37),[20] a phenomenon we refer to as transversal movement. Essential to this process is the principle of translucency, which refuses to conceal the apparatus of connection between self and other, and instead accentuates the seams or scars that mark the point at which one subjective territory collides with another. Kahlo's self-portraits, such as "Roots," "The Wounded Deer," and "The Column"—featuring, respectively, Kahlo's body in contorted combinations with tree, animal, and architectural parts—are not merely invoked in Taymor's tableau of Lavinia as part-tree, part-deer, but also suggest the very basis of the film's preoccupation with transmutation through fusion.[21] Rather than fostering disassociation and disgust, such catalytic collisions, according to Taymor, generate a sense of mutual accountability at the point of contact, the place of prosthetic connection. "Sure, I could hide those people in a costume with stilts,"

Photo 9.2 Lavinia

she explains of her construction of the giraffes in her musical, *The Lion King*: "But then no one would feel anything. The fact that as a spectator you're very aware of the human being . . . and you see the straps linking the actor to the stilts, that there's no attempt to mask the stilts and make them animal-like shapes—that's why people cry [when the giraffes enter]" (Blumenthal and Taymor 52). The stilts and straps, in this case, are the prosthetic limbs and sinews that join the human and animal forms into something more than the sum of their parts—possibly into something that transports the audience into a transversal space where its members experience "becomings-other-identities" (such as becomings-animal), a transcending of subjective territories in the process of transforming into other things.[22] In *Titus*, Young Lucius is the figure that connects the spectator to this space of potential alliances and metamorphoses. In an interpolated scene, Young Lucius confronts Lavinia with a pair of prosthetic hands from a nearby doll shop, a gesture that signals his own transformation from the destructive figure of the film's opening frames to an agent of interanimation between people and playthings alike, as girls and dolls—like boys and their toys—merge unexpectedly. The crucial difference between these bookend scenes, nonetheless, is that here prosthesis is not represented as a "cyborg orgy," but rather as a powerful mode of re-membering, indeed, a "critical, deconstructive relationality" that reveals unforeseen connections and associations in the wake of devastation.[23]

Taymor thus converts Shakespeare's nihilistic preoccupation with dismemberment into a regenerative emphasis on prosthesis and, as we shall see, "hacking." According to Andrew Ross, hacking performs the vital social function of exposing "security deficiencies and design flaws" (113); as an expression of "guerilla know how," it "is essential to the task of maintaining fronts of cultural resistance and stocks of oppositional knowledge as a hedge against a technofascist future" (114). In the context of Taymor's technocratic vision of "ancient Rome," Tamora becomes the site of a provocative variation on this theme, mobilizing *her own* security deficiencies and design flaws in order to gain access to and, subsequently, recode the Roman informatics of domination. In effect, Tamora is the vehicle through which Taymor explores the ways in which Renaissance women, like the cyborgs in Ridley Scott's *Bladerunner*, become unruly "replicants," deriving a subversive order of difference from the very practice of imitation. As a Goth, an identity associated in the English Renaissance imagination with the idea of the nomadic, barbaric other, and—post-Reformation—with racial purity and longing, Tamora suggests the ideal prototype of the hacker, for she represents an outsider culture whose survival is contingent upon strategic interfacing with insiders, as well as "unexpected others." Distinct from Lavinia, then, whose self-transformations ultimately signal her ongoing violation, Tamora denies the sanctity of her subjective territory from her very first appearance within Roman borders, turning abjection into opportunism by presenting her body as transversal terrain wherein "any component can be interfaced with any other if the proper standard, the proper code, can be constructed for processing signals in a common language" (Haraway, "Cyborg Manifesto" 163). Merging lexicons, that of the Goths with that of the Romans, Tamora complicates her coding, subsequently becoming more alluring in her insider/outsider status.

When in Rome, do as the Goths do

In Taymor's film, Tamora's body is continually subject to visual and ontological recoding as a reflection of her changing coalitions with insiders and outsiders alike. When we first encounter the Queen of the Goths, her body is dwarfed by a bear fur that immediately distinguishes her as a "savage" Other;[24] her appearance from the neck up, however, poses stark contrast to her grossly attired body, revealing blonde hair tightly braided in rows and adorned with golden shards of metal, signifying the dual-inflection of her identity as both primitive Goth remnant and sophisticated Gothic revenant. Assuming a posture of submission within the Roman camp, Tamora transforms her sense of breached security as Goth, mother, and woman into a position of strength when Saturninus, intrigued by her apparent vulnerability as well as her exotic foreign "hue," impishly singles her out as a desirable new component for his command console.[25] Accordingly, Tamora's striking reappearance in a regal gladiatorial breastplate signals her new insider status, confirming her sudden ascent from prisoner to empress. Central to Tamora's subversive revenge plot to "hack into" and reprogram her Roman "host" is her commitment to a politics of affinity rather than identity; like the elusive self-portraits of Frida Kahlo, Tamora's articulation of partial, permanently unclosed selves implicitly resists the dismantling imperatives of C^3I, precisely because they stake no claim to completion in the first place. Moreover, Tamora is a master of data encryption; for in the very instant that her body is remapped according to her alliance with the Roman Emperor, Tamora "pollutes" this image by unveiling the serpent-like tattoos adorning her arms, which betoken her outsider alliance with the wily Moor, Aaron, a relationship from which she derives not only personal pleasure, but also collective revenge against the Andronicus clan. Thus, what makes Tamora's infiltration of the Roman power structure so successful is that while the coding of her body operates, like Lavinia's, according to the principle of translucency, it defies pattern recognition; despite the transparent prosthetic mechanisms of connection that bind her to both insiders and outsiders, Tamora is, essentially, unreadable—her differential replications are always one step ahead of the C^3I enlisted to control them. While Tamora is a character that is clearly *marked* by difference in the culture of *Titus Andronicus*, she refuses to be *fixed* by it.

Both Shakespeare's play and Taymor's film stage the critique and collapse of C^3I only to punish those who participate in the diaspora, a fate that, as we shall see, not even the audience can escape. For although Tamora uses the very mechanisms of her technocratic captors to generate a prosthetic identity that is "technologically advanced rather than erased," the challenge of the hacker is to "utilize the resources of 'high-tech facilitated social relations' toward the elimination of fixture in racial, sexual, and class identities, without losing sight of the ways in which the same technologies embody patriarchal-capitalist 'informatics of domination' and repression" (Lennon 72–73). Tamora can, therefore, take us only so far into the wilderness of transversality, retreating at the crossroads wherein "vulnerability," as Haraway puts it, becomes "a window onto life" ("Biopolitics" 225). Indeed, Taymor's description of Tamora as the death-dealing Goddess of Revenge is no subtle indication that the once-liberating connections forged in the wake of her abjection have, by the end of the film, been reconfigured as the very matrices of domination:

> At first the audience should believe that the event is a figment of Titus' tortured mind. Seated on a giant lion is the Goddess Revenge, her crown of daggers reminding us of the Statue of Liberty, while the strip of gauze that hides her eyes hearkens to Blind Justice. Where hands should be she wears two coned gauntlets. Her huge pendulous breasts form her shield, and from the nipple of one of them a plastic tube is attached. This end of the tube feeds smoke into the mouth of Murder, a lounging minister to Revenge who sports a tiger head as a hat. To Revenge's right shoulder perches Rape, her other minister, his head enveloped with the outstretched wings of an owl, his naked body dressed only with a little girl's training bra and panties. (*Illustrated Screenplay* 185)

Situated in nightmarish maternal trinity with her sons, Chiron and Demetrius, Tamora emerges as an allegorical cross between biological warfare and biological welfare, as the coalition she forges with Rape and Murder creates an image of reproductive technology gone horribly awry. With disturbing currency, Taymor's allusion to the Statue of Liberty as the inspiration for her vision of Revenge points to the larger, geo-political consequences of converting vulnerability—embodied in Lady Liberty's embrace of the "huddled masses"—into an excuse for acting unilaterally once insider, indeed super-power, status is achieved. Consequently, in a grotesque illustration of what we have been calling radical accountability, Tamora is forced to consume her own children, an implosion that turns her insider status violently against her.

In *Titus*, Shakespeare's apocalyptic banquet scene concludes by implying that Tamora's children live on—in us—the consumers of the spectacle. As Taymor observes, although the coliseum is empty at the beginning of the film, "This time the bleachers are filled with spectators. Watching. They are silent. They are we" (*Illustrated Screenplay* 185). Taymor's message is simple: we are what we eat, that is, what we consume. But the situation is more complicated than this. When an audience is exposed to transversal aesthetics, as Reynolds argues of early modern England's public theater, "Transversal power is thereby transmitted visually: seeing is infectious and instigates becomings. It is also transmitted aurally and physically in the theater *as* and *through* sound waves and resonance" (*Becoming Criminal* 146). Indeed, the subtle addition of spectators to Shakespeare's play has the powerful effect of positioning *us* as not only "the birds of prey" to which Tamora's body is thrown at the end of the film, but also the vulnerable receivers of everything performed. Our instructions proceed from Lucius, who stipulates that since Tamora's life was "devoid of pity, [. . .] let birds on her take pity" (5.3.198–99). Although "to pity" in this case means to consume, *Titus* also gestures obliquely toward the idea of consumption as a route to connection: indeed, as a kind of privilege that can also be a form of punishment, consumption can be an aperture through the fragile membranes of our subjective territories into unknown and unpredictable space–time: the future–present of radical accountability.

Young Lucius, the character who carries Aaron's baby out of the coliseum in the film's concluding tableau, has learned this lesson about radical accountability the hard way, for unlike the spectators who remain stranded in the coliseum, Young Lucius now knows that our machines and screens *are* us: their codes reflect our drives, their programs our processes, their networks our destinies. They are, as Haraway concludes, nothing less than "an aspect of our embodiment" ("Cyborg Manifesto" 180). Thus, if Young Lucius gets to take up the living while we are left to take up the dead, or something in between, existing in the traces and phantasmagoria of a bygone engagement with Taymor in Shakespace, it is because we have yet to prove that we can consume with the care necessary to *preserve* life, and not just imitate it.

"Behold the child"

In the closing sequence of *Titus*, a final, bizarre scene of apocalypse and promise, Taymor returns, oddly enough, to the horror genre. Yet she does so with a difference, positioning Young Lucius as what we term the "Final *Boy*" who, in this case, may be able to teach us how to handle-with-care the virtual so that we may be entrusted with the real. Through the figure of Young Lucius, Taymor assays one last act of transformation, seeking to revise the role of the spectator from within the parameters of the horror genre itself.[26] To be sure, *Titus* ends with Young Lucius taking on the role of the "Final Girl," who functions as a "stand-in" for the spectator in the subgenre of the slasher film.[27]

The Final Girl, as Carol Clover explains, becomes the main character of the slasher film. She is the one who discovers the other (usually female) victims, comprehends the situation and even the psychological state of the murderer, confronts the horrors of death itself, and, at last, survives until she is rescued or until she herself vanquishes the killer (45). Although this survivor is female, she has traditionally masculine attributes, and she fulfills the male spectator's expectations of the slasher narrative, which dramatizes the male's transition from "feminine" childhood to "masculine" adulthood through the Final Girl character (Clover 50). In this sense, Clover notes, "The Final Girl is, on reflection, a congenial double for the adolescent male" (51). Because he is a she, the male spectator can identify with the traditionally feminine responses to terror and experience transgressive masochistic pleasure from the heroine's perspective. Nevertheless, as "she" is a tomboy, more traditionally masculine than feminine in her behavior, the Final Girl is not woman enough to pose the dread of difference for those male viewers (Clover 51). While it is true that slasher films, especially lower budget ones, more willingly and easily play with gender boundaries than the classic horror film counterpart, which tends to work within variations of conservative notions of gender, the slasher film—like the classic horror film—fulfills this masculine narrative. "If the slasher film is 'on the face of it' a genre with at least a strong female presence," Clover argues, "it is in these figurative readings a thoroughly male exercise, one that finally has very little to do with femaleness and very much to do with phallocentrism. Figuratively seen, the Final Girl is a male surrogate in things Oedipal, a homoerotic stand-in, the audience incorporate; to the extent she means 'girl' at all, it is only for purposes of signifying male lack, and even that meaning is nullified in the final scenes" (53). Thus, despite the composite gender she suggests, the Final Girl functions as "an agreed-upon fiction" that brings the male spectator into a cinematic narrative in which traditional gender boundaries are collapsed and sexual norms are challenged (61-64).

Taking up the convention of the Final Girl with investigative-expansive wherewithal, Taymor transforms Lucius's son of Shakespeare's play into the "Final Boy" of her revisionary and hence revolutionary horror film.[28] The masculine narrative underwriting traditional horror films, in which a feminine boy becomes a man in the figure of the Final Girl, is transformed into its opposite in *Titus*, wherein a masculine boy assumes the effeminized persona of the Final Boy. Taymor exploits the well-known and therefore anticipated scenario of the Final Girl, using preconceptions about this character, however latent and non-prompted, to structure the audience's point of view, while at the same time contradictorily coding what eventually emerges, surprisingly, as the Final Boy.

Taymor frames the film with Young Lucius, so that the events unfold before his sensitive gaze in a self-conscious foregrounding of spectatorship. She also emphasizes his growth in this ultimately "non-male" narrative by focusing on his desire to please his father and grandfather in echoing and emulating their masculine efforts at phallic revenge: Taymor has the boy, rather than his uncle, murder the innocent fly and comment on it deserving death for being black like the Moor; she also includes him in the scene of Titus's attack with arrows.[29] Nevertheless, despite these characteristically male-coded gestures, Young Lucius responds ultimately in unexpected ways to these scenes in which he is expected to fulfill masculine expectations. Instead of becoming the revenging warrior of the next generation, the "masculine" boy, first seen playing war at the kitchen table, confronts the horrors of death and the price of revenge and, in the final analysis, becomes both the "feminine" healer toward Lavinia and the nurturing mother figure to Aaron and Tamora's baby, clearly rejecting the world he has inherited. Like the Final Girl, Young Lucius has survived the carnage, but not because he has enlisted his phallic prowess against the horrors that confront the male subject in its tenuous hold on masculine identity. Rather, he has survived, and has ensured the baby's survival, because he has confronted, in fact crossed over, the abyss of abjection by exploiting his own vulnerabilities, and taking on the maternal role himself. In a sense, then, Young Lucius's tale of transversal suffering becomes, through his becomings-woman, a story of transgendered healing. (See photo 9.3.)

It is not quite so apparent, however, where Young Lucius as "Final Boy" is headed as he walks toward a sunrise that is obviously artificial, illuminating, in the process, an obscure horizon. Yet the film's coda does signal the *possibility* of a new beginning, wherever that might lead

Photo 9.3 Final boy

(Taymor, "Director's Commentary"). Commenting on this ending, Taymor notes that she filmed it "without . . . giving an answer or going into horrific, pathetic cliché." But, she explains, "after such a dark story," she liked "the idea of this child getting towards the exit, moving out of this coliseum towards a bleak and barren landscape (oh, some water in the distance, potentially) . . . towards, maybe, a sunrise—maybe . . ." ("Director's Commentary"). Too ambiguous to provide neat closure, the final scene is by no means a conventional Hollywood happy ending. Like the rest of the film, it plays uneasily on the borderlines of the real/nonreal, refusing to give the audience a full release from the unnerving, transversal effect of the film. The spectator, like the Final Boy, Young Lucius, occupies the transversal space evoked in this nightmare world, without providing any easy reassurance or simple way out. Instead, we are left with only a possibility of moving *toward* the exit, a gateway made available through the contributory process of perpetrating, suffering, and surviving the horrors and the fun-house mirrored reflections of ourselves translucently displayed on screen.

Notes

1. In her first film, a television adaptation of Edgar Alan Poe's "Hop-Frog," renamed *Fool's Fire* (1992), Taymor foregrounds the climactic flames that destroy the corrupt, repulsive king and his court (embodied by gigantic and grotesque puppets), who had cruelly subjected the dwarf Hop-Frog and his love Trippetta to a life of subjugation and abject humiliation. This metaphor of fire links Taymor's more recent film productions to her earlier stage work as a "theater-maker" in Indonesia at the Bengrel ("repair shop") Theater, where she collaborated with W.S. Rendra, the playwright, director, and novelist, who first encouraged Taymor to direct. This experimental theater emphasized the sacred, communal, and living creative energy of theater, views that greatly influenced Taymor's future work in Indonesia and beyond. For Taymor then and now, fire figures as the "transforming agent" that illuminates the theater and its performative dimensions; in this way, fire becomes a metaphor for the idea of transformation itself, an idea that figures prominently both in the technique and subject matter of Taymor's art (*Bravo Profiles*). A set/costume designer, mask/puppet maker, choreographer, and director of stage/television/film, Taymor sees herself foremost as one who "transmographies" available materials into new art, one who creates by *playing with fire* (Blumenthal and Taymor 13–17).
2. For detailed discussions on the i.e. mode, see chapters 1–3, and the appendix to this book.
3. See chapter 7, where Hedrick and Reynolds describe in detail the principle of translucency.
4. For detailed discussion on subjective territory, see chapter 1.

5. Hedrick and Reynolds, chapter 7.
6. For another discussion of Artaud's Theater of Cruelty in reference to transversal cinema, see chapter 5 on Polanski's *Macbeth*.
7. Taymor discusses this effect of the mask in her interview for *Bravo Profiles* ("Inside the Creative Mind—Juliet Taymor," dir. Bobbie Birleffi [Bravo Network, November 12, 2000]).
8. To Taymor, this transcultural approach that holds itself accountable for the cultural exchange contrasts greatly with artistic imperialism, which, in Taymor's view, fails to respect the other culture. Taymor gives an anecdote from her travels with a friend in Trunyan to serve as an example of her point of view on the matter. When Taymor and her friend Roland happened upon a religious ritual, Roland plunged himself into the ceremony, giving away his clothing as offerings to the "forest spirits," and making himself the primary focus of the event. Taymor thought his behavior—that of "immersing and losing oneself in another culture, but on *terms that violate* that culture"—to be "disrespectful and dangerous" (Blumenthal and Taymor 17; emphasis added). Although for Roland the experience may have been exhilarating and liberating, from Taymor's perspective, his actions revealed an imperialist motivation that lacked the accountability that ideally would exist in a truly open transcultural exchange, a Kurtz-like response to the *Heart of Darkness* in miniature. This incident became the impetus for Taymor's later work with the Indonesian troupe *Teatr Loh* (meaning both "source" and "oh, my God!"), particularly in her transversal piece on the theme of cultural identities and borders, *Tirai* (meaning "curtain"; Java, Sumatra, and Bali, 1978–1979; New York, 1981). Taymor had treated a similar theme earlier with her first production in Bali, a trilogy titled *Way of Snow* (Java and Bali, 1974–1975; new version New York and Washington, D.C., 1980), which revolved around questions of cultural shifts, identity, and insanity. This early production was multifaceted, incorporating various modes and techniques. In *Tirai*, when Taymor returned to this subject—this time to explore questions of cultural identity and assimilation—she pared down the production, employing only masked and unmasked actors, colliding illusionistic and emblematic modes of representation (see Blumenthal and Taymor 18). In this later production, much like that employed in her 1994 off-Broadway stage production of *Titus Andronicus* and, in cinematic terms, in her film *Titus*, Taymor developed translucency through a more minimal, yet still expansive, theatrical space.
9. Her designs and productions also traverse genre boundaries, forming novel theatrical cross-genre pieces, like her design for a combination Christian passion play and Passover pageant (*The Haggadah*, New York, 1980), her production of *The Lion King*, a musical comedy and puppet show, mass and masque, opera and comedy, and so forth. Nevertheless, Taymor draws from those popular sources, as well as from other experimental theater groups in the United States and abroad, blurring modern distinctions between "high" and mass art. Taymor has incorporated popular forms—including puppetry techniques and conventions of musical theater—in her most daring and decidedly unorthodox productions, like the short-run *Liberty's Taken*

(Ipswich, Massachusetts, 1985), a musical set during the American Revolution. Some of Taymor's avant-garde productions have, like this one, failed to draw large audiences, while others have gained mass appeal. Although certainly unconventional, her expansive, transversal theater has proven ideal for the ever-popular Broadway production of Disney's *The Lion King* (premier New York, 1998), for which Taymor was the first female director to win the Tony award. As Gautam Dasgupta says of *The Lion King*, Taymor "creat[es] a spectacular melding of avant-garde bravura and entertaining popular fare" (qtd. in Schwartz, Skipitares, and Taymor 461). Rather than working only on the margins, Taymor has moved into official culture as well. But in so doing, she has retained her transversal aesthetic, which she has now carried into territories other than that of exclusively avant-garde, experimental theater or "high" art, but also onto television, Broadway, cinema, and often in the margins between these media, grafting qualities of one within the other.

10. Julie Taymor makes this observation in the most recent retrospective of her work, written by Eileen Blumenthal and Julie Taymor, *Playing with Fire* (New York: Harry N. Abrams, 1999), 220.
11. See Shakespeare's *Titus Andronicus*: "I am the sea. Hark how her sighs doth blow. / She is the weeping welkin, I the earth" (3.1.226–27). All excerpts of Shakespeare's play are from Jonathan Bate's New Arden edition, which Julie Taymor used as the basis of her screenplay.
12. See Marcus' speech "O, why should nature build so foul a den, / Unless the gods delight in tragedies?" (4.1.59–60).
13. The "beast in the boudoir" is a phrase coined by Harvey Roy Greenberg in "King Kong: The Beast in the Boudoir—or, 'You Can't Marry That Girl, You're a Gorilla,'" in *Dread of Difference: Gender and the Horror Film*, ed. Barry Keith Grant (Austin: University of Texas Press, 1996), 338–51.
14. For an extensive analysis of Taymor's *Titus* in relation to abjection, see Lisa S. Starks's "Cinema of Cruelty: Powers of Horror in Julie Taymor's *Titus*," in *The Reel Shakespeare*, ed. Lisa S. Starks and Courtney Lehmann (Madison: Fairleigh Dickinson University Press, forthcoming 2002), 121–42.
15. Making a related point about horror films in general, in light of Kristeva's theory of abjection, Barbara Creed transposes the problem of abjection into a cinematic context, identifying the central trope of the horror film as the viewer's encounter with the abject, forcing "the viewing subject's sense of a unified self into crisis, specifically in those moments when the image on the screen becomes too threatening or horrific to watch" (57). See "Horror and the Monstrous Feminine: An Imaginary Abjection," in *The Dread of Difference*, 35–65.
16. Much of the following section on abjection is excerpted from Lisa S. Starks's essay, "Cinema of Cruelty: Powers of Horror in Julie Taymor's *Titus*," in *The Reel Shakespeare*, 121–42.
17. In *Powers of Horror: An Essay on Abjection* (trans. Leon S. Roudiez [New York: Columbia University Press, 1982]), Kristeva provides several examples of the abject and what situations cause it to surface. One is "food

loathing," the disgust of rotten, spoiled food, particularly when it suggests the border in-between, as in her example of the "skin on the surface of milk" (2). Another is the corpse, the quintessence of the abject. In "presenting" rather than representing death, the corpse indicates what one must "thrust aside in order to live," as Kristeva explains (3). Other ambiguous boundaries—the blurred distinction between human and animal, bodily "insides" from "outsides," and the unclean and the clean body—provide further examples of the abject (Kristeva 53). Because the abject is projected onto the maternal body and by extension "the feminine," it is often envisioned in representations of the grotesque female body. Extended beyond the maternal, "the feminine" itself thus "becomes synonymous with a radical evil that is to be suppressed" (Kristeva 70). Although some critics claim that Kristeva's theory merely provides an excuse for misogyny and the oppression of women, we would argue that Kristeva's performative text is descriptive rather than prescriptive in its purpose, laying bare the projections and displacements through which such attitudes may occur both at the level of the subject and society.

18. On the uncanny, see Sigmund Freud, "The Uncanny," in *The Standard Edition of the Complete Psychological Works*, Vol. 27, trans. and ed. James Strachey (London: Hogarth Press, 1955), 219–52. On the distinction between the grotesque and the uncanny, see Wolfgang Kayser, *The Grotesque in Art and Literature*, trans. Ulrich Weisstein (Bloomington: Indiana University Press, 1963); Rosemary Jackson, *Fantasy the Literature of Subversion* (London: Methuen, 1981), 68; and Michael O'Pray, "Surrealism, Fantasy, and the Grotesque: The Cinema of Jan Svankmajer," in *Fantasy and the Cinema*, ed. James Donald (London: BFI, 1989), 253–67, esp. 256. On laughter and the abject, see Julia Kristeva, *Powers of Horror*, 8.

19. We thank Janna Segal for first pointing out the allusions to Kahlo's self-portraits in Taymor's film.

20. The term "ghostings" is derived from Herbert Blau's KRAKEN group, which Taymor was a member of in the 1970s at Oberlin. Blau uses the term "ghostings" to describe his improvisatory approach to Shakespearean performance, which involves "work[ing] our way through every facet and refraction of Shakespeare's text, playing sequences over and over, so there [a]re simultaneously present multiple figurations, or 'ghostings,' not only of character—shadows, doubles, allusions, quotes . . . but of themes, images, phrases, even interteprations, of which in a passing moment there might be several going at once . . ." (262). See Blau's " 'Set Me Where You Stand': Revising the Abyss," *New Literary History* 29.2 (1998): 247–72.

21. Accentuating this prosthetic vision of the human body, Taymor explains that in designing costumes for *Titus*, she used metallics, stone, and animal furs in conjunction with predominantly blue, red, white, and black hues to suggest "the color of veins under pale skin," underscoring the common circuitry that links these human, animal, and machine forms in the cybernetic landscape of the film.

22. For discussion of the phenomenon of becomings, see Gilles Deleuze and Félix Guattari, *A Thousand Plateaus: Capitalism and Schizophrenia*, trans. Brian Massumi (Minneapolis: University of Minnesota Press, 1987): 233–309 and, in direct connection to transversal theory, Bryan Reynolds, *Becoming Criminal: Transversal Performance and Cultural Dissidence in Early Modern England* (Baltimore: Johns Hopkins University Press, 2002), 1–21.
23. Haraway refers to modern warfare as "a cyborg orgy" in "A Cyborg Manifesto: Science, Technology, and Socialist Feminism in the Late Twentieth Century," in *Simians, Cyborgs, and Women: The Reinvention of Nature* (New York: Routledge, 1991), 149–81, 150. She explores the possibilities of "deconstructive relationality" in her coda, "The Actors Are Cyborg, Nature is Coyote, and the Geography is Elsewhere: Postscript to 'Cyborgs at Large,'" in *Technoculture*, ed. Constance Penley and Andrew Ross (Minneapolis: University of Minnesota Press, 1991), 21–26; 23. For an extensive discussion of Taymor's *Titus* in relation to other celluloid incarnations of "cyborg feminism," see Courtney Lehmann, "Crouching Tiger, Hidden Agenda: How Shakespeare and the Renaissance are Taking the Rage Out of Feminism," *Shakespeare Quarterly* 56 (Summer 2002).
24. The opening scenes of Tamora clad in fur are reminiscent of the "cold" image of the female torturer in Leopold von Sacher-Masoch's *Venus in Furs*. For an analysis of Sacher-Masoch's heroines, see Gilles Deleuze's "Coldness and Cruelty" in *Masochism* (New York: Zone Books, 1991), 9–138, esp. 47–55.
25. Says Saturninus of Tamora in Taymor's film, "A goodly lady, trust me; of the hue / That I would choose, were I to choose new" (1.2.265–66).
26. Much of the following section on Young Lucius as Final Boy is adapted from an article in which Lisa S. Starks discusses the Final Boy in terms of abjection and the horror film (see Starks, "Cinema of Cruelty: Powers of Horror in Julie Taymor's *Titus*," in *The Reel Shakespeare*, 121–42, esp. 134–37). In this new context, however, the Final Boy registers differently, in terms of Taymor's transversal aesthetic.
27. Lisa Starks thanks Laurie E. Osborne for first pointing out this connection between Young Lucius and the Final Girl.
28. For a contrast to Taymor's horror-film adaptation, see Christopher Dunne's low-budget, slasher-film version of *Titus Andronicus* (USA 1999).
29. Richard Burt also points out the central role of Young Lucius, although he sees it as functionally different in relation to the film's depiction of violence. See Burt, "Shakespeare and the Holocaust: Julie Taymor's *Titus* is Beautiful, or Shaksploi Meets (the) Camp," *Screening Shakespeare*, ed. Laurie E. Osborne, spec. issue of *Colby Quarterly* 37.1 (2001): 78–106, 94–95.

Works cited

Artaud, Antonin. *The Theater and Its Double*. Ed. and trans. Mary Caroline Richards. New York: Grove Press, 1958.

Bladerunner. Dir. Ridley Scott. Perf. Harrison Ford, Edward James Olmos, and Daryl Hannah. Warner/Ladd/Bladerunner Partnership, 1982.
Blau, Herbert. "'Set Me Where You Stand': Revising the Abyss." *New Literary History* 29.2 (1998): 247–72.
Blumenthal, Eileen and Julie Taymor. *Playing with Fire.* New York: Harry N. Adams, 1999.
Burt, Richard. "Shakespeare and the Holocaust: Julie Taymor's *Titus* is Beautiful, or Shakesploi Meets (the) Camp." *Screening Shakespeare.* Ed. Laurie E. Osborne. Spec. issue of *Colby Quarterly* 37.1 (2001): 78–106.
Clover, Carol. *Men, Women, and Chain Saws: Gender in the Modern Horror Film.* Princeton: Princeton University Press, 1992.
Creed, Barbara. "Horror and the Monstrous-Feminine: An Imaginary Abjection." *The Dread of Difference: Gender and the Horror Film.* Ed. Barry Keith Grant. Austin: University of Texas Press, 1996. 35–67.
——. *The Monstrous-Feminine: Film, Feminism, Psychoanalysis.* London: Routledge, 1993.
Deleuze, Gilles. "Coldness and Cruelty." *Masochism.* New York: Zone Books, 1991. 9–138.
Freud, Sigmund. "The Uncanny." *The Standard Edition of the Complete Psychological Works.* Vol. 27. Trans. and ed. James Strachey. London: Hogarth Press, 1955. 219–52.
Greenberg, Harvey Roy. "King Kong: The Beast in the Boudoir—or, 'You Can't Marry That Girl, You're a Gorilla.'" *Dread of Difference: Gender and the Horror Film.* Ed. Barry Keith Grant. Austin: University of Texas Press, 1996. 338–51.
Grosz, Elizabeth. "Body of Signification." *Abjection, Melancholia and Love: The Works of Julia Kristeva.* Ed. John Fletcher and Andrew Benjamin. London: Routledge, 1989. 80–103.
Haraway, Donna. "The Actors Are Cyborg, Nature is Coyote, and the Geography is Elsewhere: Postscript to 'Cyborgs at Large.'" *Technoculture.* Ed. Constance Penley and Andrew Ross. Minneapolis: University of Minnesota Press, 1991. 21–26.
——. "A Cyborg Manifesto: Science, Technology, and Socialist Feminism in the Late Twentieth Century." *Simians, Cyborgs, and Women: The Reinvention of Nature.* New York: Routledge, 1991. 149–81.
"Inside the Creative Mind—Juliet Taymor." Interview for *Bravo Profiles.* Dir. Bobbie Birleffi. Bravo Network. November 12, 2000.
Jackson, Rosemary. *Fantasy the Literature of Subversion.* London: Methuen, 1981.
"Julie Taymor at Columbia University." Extra feature in *Titus* DVD. Fox Searchlight/Clear Blue Sky Productions, 1999.
Kayser, Wolfgang. *The Grotesque in Art and Literature.* Trans. Ulrich Weisstein. Bloomington: Indiana University Press, 1963.
Kristeva, Julia. *Powers of Horror: An Essay on Abjection.* Trans. Leon S. Roudiez. New York: Columbia University Press, 1982.

Lehmann, Courtney, "Crouching Tiger, Hidden Agenda: How Shakespeare and the Renaissance are Taking the Rage Out of Feminism," *Shakespeare Quarterly* 56 (Summer 2002).

Lurie, Susan. "Pornography and the Dread of Woman." *Take Back the Night*. Ed. Laura Lederer. New York: William Morrow, 1980. 159–73.

O'Pray, Michael. "Surrealism, Fantasy, and the Grotesque: The Cinema of Jan Svankmajer." *Fantasy and the Cinema*. Ed. James Donald. London: BFI, 1989. 253–67.

Reynolds, Bryan. *Becoming Criminal: Transversal Performance and Cultural Dissidence in Early Modern England*. Baltimore: Johns Hopkins University Press, 2002.

Ross, Andrew. "Hacking Away at Counterculture." *Technoculture*. Ed. Constance Penley and Andrew Ross. Minneapolis: University of Minnesota Press, 1991. 107–34.

Schwartz, Bruce D., Theodora Skipitares, and Julie Taymor. "Working with Puppets." *Conversations on Art and Performance*. Ed. Bonnie Marranca and Guatum Dasgupta. Baltimore: Johns Hopkins University Press, 1999. 460–70.

Shakespeare, William. *Titus Andronicus*. Ed. Jonathan Bate. London: Arden Shakespeare, 1995.

Starks, Lisa S. "Cinema of Cruelty: Powers of Horror in Julie Taymor's *Titus*." *The Reel Shakespeare*. Ed. Lisa S. Starks and Courtney Lehmann. Madison: Fairleigh Dickinson University Press, forthcoming 2002. 121–42.

———. *Titus: The Illustrated Screenplay*. New York: Newmarket Press, 2002.

Taymor, Julie. "Directory's Commentary." Extra feature in *Titus* DVD. Fox Searchlight/Clear Blue Sky Productions, 1999.

Titus. Dir. Julie Taymor. Perf. Anthony Hopkins and Jessica Lange. DVD. Fox Searchlight/Clear Blue Sky Productions, 1999.

Titus Andronicus. Dir. Christopher Dunne. Perf. Candy K. Sweet, Robert Reece, and Christopher Dunne. USA, 1999.

Williams, Linda. "When the Woman Looks." *The Dread of Difference: Gender and the Horror Film*. Ed. Barry Keith Grant. Austin: University of Texas Press, 1996. 15–34.

CHAPTER 10

FRIEND OR FO, SHAKESPEARE'S ENDS IS
THE MEANS: REVISING EARLY
MODERN ENGLISH ICONOGRAPHY,
ELISABETTA POINTS TOWARD THE
CRITICAL FUTURE

Bryan Reynolds & Janna Segal

Dario Fo and Franca Rame, Italy's infamous married political satirists, have spent over four decades performing and politicking, using the stage to enact a Gramscian reclaiming of popular culture that simultaneously addresses pertinent sociopolitical issues with the goal of empowering the masses. Taking a nonliterary approach to the theatrical text, Fo and Rame place the immediacy of the issues and the theatrical expectations of their audiences above the posterity of the script, creating what they call "throwaway plays." These are plays written to address current sociopolitical issues that are purposefully left "open" to revision in response to unfolding national and international events and the reactions of their audiences during performances. Their "throwaway theatre"[1] is rooted in popular Italian theatrical traditions, such as the improvisatory, audience-centered *commedia dell'arte*, and is purposefully crafted to provoke the spectators into taking a critical stance on their political and social surroundings. Self-fashioned as a modern-day *guillare* (medieval strolling-player),[2] Fo advertises his and Rame's "illegitimate" theater as defiant of both the authorities in Italy and the common scholarly perspective of the theatrical text as a fixed and unalterable entity that functions as the centerpiece of a performance event.

Positioning himself in opposition to a text-centered theatrical tradition, Fo has insisted that his intention is not to produce literary works with a shelf life, but rather to craft his texts according to their relevance to the present: "My theater is not meant to be passed down in history. I write and perform satire on current, everyday events. These are texts that immediately expend their contents" (qtd. in Scuderi, *Popular Performance* 66). Despite his insistence that his theater is not intended to transcend its own sociohistorical moment, over the past twenty years, he and Rame's work has extended beyond spatial and temporal boundaries, becoming more regularly read and performed in the English language, among others. However, since Fo and Rame's "throwaway plays" are designed to address specific issues, English-speaking translators and adapters have had to find a means of transferring their fluctuating texts not only into another language, but also into political, cultural, historical, and theatrical environments foreign to the original. The results have led to some confrontations between Fo and his fans; Fo has complained that theater practitioners have distorted and *mis*-appropriated his meaning in their attempts at relocating and transposing his texts into new contexts.[3]

The issue of translation and adaptation for the theater has attracted much debate in recent years, and, as Joseph Farrell notes in "Variations on a Theme: Respecting Dario Fo," "it appears that there has been greater interest in the process of transposing a Fo text from its original Italian into other languages than has been accorded the same process with any other writer" (19). The focus on Fo has intensified since the notoriously "illegitimate" playwright won the 1997 Nobel Prize for Literature, an award that "legitimized" his work, transforming the self-proclaimed modern-day *guillare* into a "man of letters." Validated by official culture, Fo has been rewritten in theatrical history as a "serious" political playwright, and his satiric work has been transformed into a canon attractive to producers seeking profit and/or political progress.[4]

While Fo may be the center of debate concerning theatrical translation and adaptation, Shakespeare has occupied a prominent position in this regard for centuries, especially in the sphere of cinematic adaptation, becoming in recent years the most esteemed and successful (dead) screenwriter in film history, and the subject for discussions on such issues as authenticity, accessibility, and "high" and "low" forms of popular media.[5] However heralded as a literary genius and an icon of official culture, Shakespeare has become well known to the masses of the media age through what some might consider a bastardization of his work in mediums lacking the literary refinement, and "high" cultural status, of both the page and the stage. Drawing parallels between Fo and Shakespeare may be seen as a futile attempt to validate the nonliterary work of Fo by comparing it to the literary genius of Shakespeare, but there are a number of important similarities. Both actor-authors' work appears in numerous conflicting editions that frustrate scholars seeking authoritative texts, thereby provoking debate concerning authorship and authenticity. Fo himself has strategically drawn comparisons between his and Shakespeare's theater, venturing tactically into what Donald Hedrick and Bryan Reynolds call "Shakespace," a term they use to describe the numerous "commercial, political, social, and cultural spaces" that Shakespeare has "stimulated, occupied, and affected" throughout history (8).[6] For Fo, Shakespace is a means to substantiate his own work by asserting its relationship to a nonliterary, subversive, European theatrical tradition of which Shakespeare, Fo insists, was a part. In *Manuale Minimo dell'Attore* (*The actor's mini-manual*, published in English as *The Tricks of the Trade*), Fo not only argues that Elizabethan dramatists in general crafted their plays as allegorical critiques of the state machinery of Elizabethan England (123), but describes Shakespeare as an appropriator of *commedia* traditions (63), and as an actor-author whose own troupe of performers used Shakespeare's scripts as springboards for improvisation in performance (87–88).[7] In the *Manuale*, the closest to a poetics of Fo and Rame's work available in print to date, Fo also relies upon *Hamlet* and *Romeo and Juliet* to differentiate his "Theatre of Situation" from character-driven realistic plays that allow audiences to passively indulge in the onstage drama without becoming actively involved in the issues raised by the plot (89–90). For Fo, Shakespeare, like himself, understood that "the situation is the mechanism that grabs the attention of the audience and keeps them on the edge of their seats," and so crafted *Hamlet* as a series of "at least fifteen situations, one after the other," in order to ensure that audiences were involved in the story and compelled to "participate fully in the unfolding of the plot" (89).[8]

Demonstrating his investigative-expansive wherewithal, a combination of transversal thought, courage, and *chutzpah*,[9] Fo goes well beyond referencing Shakespeare when explicating his own theatrical intentions; with transversal ingenuity, he provocatively conjures the image of Shakespeare in his 1984 play, *Quasi per caso una donna: Elisabetta* (*Elizabeth: Almost By Chance a Woman*), appropriating Shakespeare's and Elizabeth I's iconic historiographies and amalgamating them into mechanisms for a critique of traditional gender/class differentiations and the official culture that works to legitimate them. Invoking the icons of what early modern England *means* to spectators today, Fo refashions Shakespeare and the reigning monarch at the onset of Shakespeare's career into propellers for movement through and beyond the subjective territories of modern audiences. Fo has described his theatrical characters as emblematic figures that function, like the Fool in *King Lear*, "to provide a point of dialectical clash which allowed for the presence of contradictory values inside the characters and the story" (*Tricks of the Trade* 108). Through the "dialectical clash" of characters in *Elisabetta*, Fo produces theatricalizations of socioculturally constructed binary structures, and then reveals those structures to be, like the characters in Julie Taymor's 1999 film *Titus* (see chapter 9), fun-house and mirror distortions whose destructive possibilities can be perceived readily.

Fo's open-ended methodology illustrates the potentiality of the investigative-expansive mode of analysis as applied to the theater.[10] Using his characters to develop a critical portrayal of gender and class dichotomies perpetuated by the sociopolitical conductors of state power, Fo offers the audience an opportunity to investigate those binaries. Intensifying his argument against the structures he brings to the stage for scrutiny by expanding the supposed limits of those structures through the process of inversion, Fo offers contradictory codes and values that generate disresolution. Unlike the dissective-cohesive mode, whose goal is to reconstruct the variables into a coherent whole, Fo's investigative-expansive mode of dramaturgy does not seek a cohesive conclusion or holistic, unified meaning, but rather a dynamic, fluctuating collection of variables that reveal the variability and permeability of dichotomous sociocultural demarcations. Splattered across the stage, these structures are shown to be collapsible and, like Fo's texts, always open to revision.

In *Elisabetta*, Fo represents Queen Elizabeth in the last years of her reign, revises Shakespeare as a working-class revolutionary, and interprets *Hamlet* as an incite to rebellion theatricalized for the public at the Globe Theatre. As with many of the other Shakespearean adaptations considered in the pages of this book, such as Roman Polanski's historicalized-horror-genre "period" version of *Macbeth* (1971), Bertolt Brecht's re-politicized adaptation of *Coriolanus*, Robert Wilson's monological re-creation of *Hamlet*, and Taymor's avant-garde cinematic critique of mediatized culture (*Titus*), *Elisabetta* utilizes one of Shakespeare's plays to inspire transversal movement, and, in the process of affecting Shakespace, challenges the construction of Shakespeare in the popular imaginary as a playwright promoting universalisms that transcend spatiotemporal and sociopolitical boundaries. Fo's Shakespeare, simultaneously rooted in Elizabethan England and uprooted to address contemporary political issues, emerges from the past in the service of the present. What distinguishes Fo's transversal use of Shakespeare and his plays from those previously considered, however, is his inclusion of the royal icon of the Elizabethan Age, the Virgin Queen herself, and his exclusion of the period's other major cultural icon, Shakespeare, whose presence, conjured by direct references and passing allusions made throughout the play, is always felt in its absence. Depicted by Fo as the people's playwright of early modern England, Shakespeare appears in his nonappearance as a Fo-like theatrical agitator of official culture, and the parallel between early modern English and postmodern Italian playwright/actor is actualized conceptually by the onstage appearance of Fo as the queen's subversive servant. By referencing and refashioning the iconography of the Renaissance, namely Elizabeth I and Shakespeare, to address twentieth-century issues, Fo creates competing spaces that move audiences backward and forward from early modern England to contemporary Italy to the critical future. These two fictionalized historical figures, the onstage tyrannical monarch and the offstage working-class hero, serve as conduits for both contemporary counter-thought and the ideology of early modern England; they generate a space–time continuum that reminds audiences of the transhistorical legacy of oppressive state apparatuses, the multivariate forms of oppression manifested by official cultures, and the ever-present presence of voices of dissent. Pointing forward as Fo does in his play, we want to suggest in this concluding chapter that scholarship implement Fo's characteristically transversal approach to the text, the performance event, and social history, maintaining a fluid, interactive relationship with the past, present, and emergent future.

An Elizabethan fogery

Quasi pe caso una donna: Elisabetta, as the title suggests, focuses on the first Protestant queen of England, who, like Shakespeare, has recently become a prominent figure in the Hollywood film industry.[11] Unlike newer re-creations of the character that are more "true" to the Elizabeth of lore—the stoic, cold, redhead who must androginize herself to prove her worth in a "man's" position—Fo's queen (originally played by Rame) is an oversexed tyrant distracted from her royal duties by her libido, her obsessional love for Robert Deveruex, Earl of Essex, and her paranoid convictions that Shakespeare's plays are thinly disguised critiques of her reign. Rather than a Virgin Queen who bathes her face in white face-paint and takes a vow of chastity to distract her subjects from her gender and England's break with the Catholic Church,[12] Fo's Elisabetta goes to extremes, with the help of a cosmetic medicine woman, La Donnazza[13] (originally played by Fo), to adapt her body to heightened constructs of idealized femininity through the use of stilts, liposuctioning leeches, and breast-swelling bee stings. Reconstructed by Fo, Queen Elizabeth I becomes a historic site open to revision, a construct herself that can be physically altered to reveal how even women in power are susceptible to the degradation often symptomatic of women's attempts to conform to prescribed gender codes.

In the "Author's Note" preceding the play, Fo describes *Elisabetta* as a series of "forgeries": "Certain sentences attributed to Shakespeare are forgeries; many illusions to historical facts are forgeries; certain characters who appear on stage are downright forgeries, to say nothing of those who are referred to from time to time" (84). The statement could be read as a calculated "disclaimer of authenticity" that protects Fo from being accused of historical inaccuracy, allowing the playwright the freedom to rewrite history without the watchful eye of historians or the burden of reality. Admittedly fictional, Fo is not bound by fact or issues of faithfulness, and since his play is framed within a false context in a fictionalized framework (it is a theatrical piece, after all), there is no reason for anyone to expect him to be anything but unfaithful to the historical or literary data. Doubly false, as a "forgery" and a play, *Elisabetta* is a celebration of the power of the theater to liberate history from the stasis of alleged fact.

Never content unless leaving people guessing, Fo also insists in his "Author's Note" that, while the script is filled with "forgeries," "the body of the text is, I assure you, laden with authenticity" (84). What the playwright defines as "authentic" in the script, however, are the facts as Fo sees them. As the play draws an analogy between early modern England, specific Italian political scandals, such as the 1978 kidnapping and murder of Christian Democrat leader Aldo Moro by the Red Brigades, and contemporary political structures, including capitalism, Fo provides himself with a platform to critique, from his own perspective, the modern sociopolitical climate in Italy and abroad. The "authentic" element of the play becomes, in effect, a subjective theatrical territory, a field of performed fact perhaps completely fictionalized, perhaps not, but drawn from a leftist political perspective and given an authorial air by the voice of a world-renowned political satirist. Written and performed by Fo with the accompaniment of Rame, both of whom have dedicated most of their theatrical careers to critiquing various sociopolitical conductors of state power in Italy—including the "holy" office of the Roman Catholic Church—*Elisabetta* becomes a springboard upon which Fo as Fo behind historical and theatrical masks can perform an analysis of the state machinery's *modus operandi* for oppressing not only the proletariat in general, but women of all classes.

The "truth" as Fo sees it may not be universally accepted, but what is strikingly "authentic" about the play is its treatment of the common association of Elizabeth I with Shakespeare in both official and unofficial spheres in Western culture. Both signifiers of contemporary cultural investment in early modern England, these two historical figures have come to dominate textbooks, films, and popular literature as the quintessential images of their time. Drawn together from the past—a pairing that seems unparalleled, at least in the West—the historical couple has come to symbolize what we identify as the "greatness" of the Elizabethan period: monarchs supported the arts, allowing geniuses like Shakespeare to flourish.[14] Whether accurate or not, the association of Shakespeare with Queen Elizabeth I continues to be fostered in print, on film (*Shakespeare in Love*), and in plays such as Amy Freed's 2001 smash hit *The Beard of Avon*, where Elizabeth I appears not only as Shakespeare's patron, but as his less-talented, overbearing, shrewish collaborator on the original production of *The Taming of the Shrew*.[15]

While the relationship between Shakespeare and Elizabeth I is commonly depicted as a mutually beneficial one—she supported his company of players, at least financially, and his plays entertained her court and demonstrated the greatness of her reign—their relationship is colored more by modern social constructs than hard, historical facts. In their pairing in the popular imagination, Shakespeare serves to humanize the queen, to make her more person than monarch. Presented, as in Freed's play, as the butch in the constructed relationship, Queen Elizabeth's image is softened by her association with the poet/playwright who, as the femme, represents the artistic side of their partnership. Nevertheless, their roles in today's popular imagination are simultaneously reversed: Shakespeare's theatrical and literary output overshadows Elizabeth's reign, and the queen, known today as the English monarch of Shakespeare's era, is validated by her association with the man. Less popularized as the first Protestant queen of England than as Shakespeare's patron, Elizabeth I has taken a second seat, like many women in modern Western cultures, to a man, while at the same time being portrayed as the icy, Virgin Queen who masked her femininity in order to occupy a position reserved for members of the opposite sex.

Unlike the Queen's advisers, who were unsuccessful in their endeavors to resolve Elizabeth's unwed state in the first half of her reign by marrying her off in the hopes of producing a male heir to the throne, modern official cultures have coupled her with a male partner that serves to ease the anxiety aroused by her insistence upon maintaining sole rule over England. Presenting the queen in Shakespeare's image is a means of conforming her reign to contemporary constructs, mapped out and encoded in one's subjective territory, of a woman's role in society. In her manufactured place, Elizabeth I becomes as unthreatening as First Lady Laura Bush, who is also popularized as an avid reader and lover of the arts, and is known through her association with a man (albeit one far less creative than Shakespeare). However historically inaccurate, people have become conditioned to accept Queen Elizabeth I in her more domestic(ated) role as Shakespeare's patron, a part that Fo challenges in his play. By recasting Elizabeth as an unruly queen critical of Shakespeare, and using her image as a site for deconstructing socially prescribed dichotomous conceptions of gender, class, and power, Fo transforms both icons of the Elizabethan Age into conductors for transversal thought. It is such sociostructural subversions and space–time continuum conversions that we envision on the critical horizon. The poststructuralist future of Renaissance studies could inhabit or journey through spatial formations that, like Fo's revisions of Elizabeth and Shakespeare, move tangentially and expansively through spatiotemporal and sociopolitical dimensions, suffusing, exposing, and collapsing diverse variables in the performance of the analytical process itself.

Fomale-on-top?

In *Manuela Minimo dell'Attore*, Fo discusses the "painful" process of producing the text of *Elisabetta*, which included confrontations between he and his wife (53–55). According to Fo, the first draft was completed in 1982, but was rewritten multiple times following Rame's criticisms of, among other things, the role of the Queen, a part Fo had devised specifically for her. After realizing that the title character failed to work in production, Fo claims to have left the project to Rame: "I dropped the whole thing: 'Sort out the character for yourself!'" (55). Apparently Rame followed his command, but Fo admits that he was initially hostile to her attempts to revise his (re)creation: "I stood my ground with a certain irony and detachment: 'This woman thinks she can give lessons to me ... ME ... if I give way on this one, she won't even let me write a Christmas card'" (55). According to Fo, accepting Rame's revisions not only improved the onstage character, but empowered the offstage performer often overshadowed by her husband:[16] "I have to admit that the character began to work much better, and this was not, alas! an isolated experience. Any day at all, she can shoot me down, now that she has developed a taste for it" (55).[17] Fo attaches a message to his anecdotal account of the process of (re)producing the queen for and with Rame, suggesting to female performers, "To obtain space and credibility, perhaps the remedy is to get on with it yourselves, conceivably even, if it is of advantage, in collaboration with men, and to persuade them to address the question and to force them to face up to the problems which confront women" (55). Fo's suggestion of an alternative approach to script development, and his description of his own playwriting process with Rame, points toward what we identify as an "investigative-expansive dramaturgy," a collaborative approach to the theatrical text that fosters and maintains a fluid engagement with the elements within and without the confines of the stage. But while Fo uses the example of *Elisabetta* to suggest that women liberate themselves from a male-dominated theatrical tradition, his comments regarding the playwriting process he and Rame underwent reveal not a liberation from a binary male (active)/female (passive) structure,[18] but an inversion of sex and power roles that, once established, could not be reversed: "Any day now, she can shoot me down, now that she has developed a taste for it!" (55). This inversion–subversion of gender and power roles is replicated in the play in its portrayals of dual "disorderly women" from high and low social classes. Both the Queen, historically a challenge to the gender prescriptions of early modern England, and La Donnazza, the cosmetic medicine woman originally played by Fo, are reminiscent of the "woman-on-top," a *topos* of sexual inversion that Natalie Zemon Davis identifies as dominant in preindustrial European art forms (129).

According to Zemon Davis, the woman-on-top was a disorderly or unruly female figure who, seeking domination over the men around her, inverted gender roles by donning the behavioral codes associated with the opposite sex. The disorderly woman could be used to uphold patriarchal social structures by reaffirming the necessity of male figures to keep women in order; however, as Zemon Davis notes, the sexual inversion could also function to undermine a hierarchical distribution of power: "It was a multivalent image that could operate, first, to widen behavioral options for women within and even outside marriage, and, second, to sanction riot and political disobedience for both men and women in a society that allowed the lower order few formal means of protest" (131). The "multivalent image" of the woman-on-top was prominent in both official and unofficial sociocultural spheres in early modern Europe, but its manifestation varied according to the "high" or "low" cultural forms in which she appeared. In her analysis of sexual inversion in popular festivals, Zemon Davis distinguishes the "two ways" in which popular uses of the figure differed from literary:

> Whereas the purely ritual and/or magical element in sexual inversion was present in literature to only a small degree, it assumed more importance in the popular festivities, along with the carnivalesque functions of mocking and unmasking the truth. Whereas sexual inversion in literary and pictorial play more often involved the female taking on the male role or dressing as a man, the festive inversion more often involved the male taking on the role or garb of the woman, that is, the unruly woman—though this asymmetry may not have existed several centuries earlier. (136)

The two unruly women in *Elisabetta* can be categorized according to this same distinction between "high" and "low"—official and unofficial—cultural formations of the disorderly female figure.

In *Elisabetta*, the first Protestant queen of England, the historical figure who was emblazoned in pages and portraits during and after her reign, is presented in the "literary and pictorial" mode of sexual inversion. Played by a woman, Elisabetta dons a series of traditional male roles and behavioral codes that invert a male (active)/female (passive) binary. Indulging in the masculine position of ruling monarch, Elisabetta tyrannically asserts the authority vested in her royal status in her efforts to verbally and physically dominate her Police Chief, Egerton, her male guards, and her two onstage servants, Donnazza and Martha. Elisabetta also, in keeping with "a rich treatment of [unruly] women who are happily given over to the sway of their bodily senses" (Zemon Davis 134), plays the courter, actively seeking to satisfy her libido by taking a lover, Thomas, in the place of the absent and disgruntled Robert of Essex, "to see whether I was in a fit condition to bear the attentions of a man, in case Robert came" (154). Even the language in which Elisabetta asserts her authority is described as more masculine than a man's; Martha chides the queen for hypocritically taking offense at the Portuguese ambassador's remarks when her "language is much worse" than his vulgar descriptions of the queen (101–02).

While the foul-mouthed, oversexed, male-dominating Elisabetta operates within the "high" art modes of sexual inversion, La Donnazza is typical of the festival forms of the woman-on-top. Played by Fo in both "the role" and the "garb" of a woman from the lower classes,[19] Donnazza both mocks the queen her master and unmasks the monarch's more-than-human veneer. Donnazza first appears on stage "wearing a white mask—a Venetian domino" (108) that, once removed upon Elisabetta's orders, reveals the character as a monstrous figure whose gender is ambiguous even to herself: "I know that I look like a man, an ogre . . . and not even very graceful" (108, 180). Like the "males disguised as grotesque cavorting females" who filled "[t]he ritual and/or magical functions of sexual inversion" in popular festivals (Zemon Davis 137), Donnazza, the "next best thing to a witch" (108), has been sent for by Elisabetta to revert the aging process and transform the queen into a younger, taller, slimmer, chestier version of her former self that will entice the queen's ex-lover, Robert of Essex. Donnazza uses a series of "miraculous poultices" (106) and gruesome folk-remedies—including applying fat-sucking leeches and breast-stinging wasps onto the queen's body[20]— to not only catch Elisabetta's man, but to correct the queen's gender ambiguity. Assigning Robert Cecil's famous quip to a "Portuguese ambassador," the queen learns from an intercepted letter that her sexual identity is confusing her court: " 'They say of her that she's too feminine to be a man . . . and not feminine enough to be a woman . . . ' I'm a hermaphrodite! Elizabeth, Prince Charming!" (101).[21]

Donnazza's efforts to revive the queen's youth and ultra-feminize her form, thereby resolving her ambiguous sexual identity at court, transform the queen into a monstrous caricature that mocks both the Elizabethan monarchy and modern methods of conforming the body to idealized conceptions of how the female gender should look (such as liposuction, breast implants, etc.). Elisabetta emerges from Donnazza's torturous magical procedures degraded and distorted, "like a stuffed owl" with breasts that "undulate" (145). The grotesque procedures and results produce a reduction of the Queen on a sociopolitical level: her exalted status as ruler ordained by God is questioned as her body becomes altered in the hands of her bawdy beautician. Like the medieval fool in Bakhtin's analysis of Carnival, the servant strips her masterly mistress of her power over the people by distorting the iconic image used to promote her status as supreme ruler.[22] But the Queen's physical degradation by Donnazza is two-fold: while the Queen's body is revealed as vulnerable and displayed as such for the mockery of the masses, it is her desire to alter her body for a man that has led to this degradation. Elisabetta's "character flaw" is the correlating "truth" that Donnazza's magical procedures unmask: although she is England's ruling monarch, Elisabetta is vulnerable to the subjective territory generated and maintained by the state machinery of her own realm. This woman-on-top is also a "woman-on-bottom," an "orderly woman" accepting of patriarchal definitions of what it means to be female. In this regard, Fo reminds all women that they may be inadvertently contributing to their own subordination.

Hamlette, the Player Queen, or Shakespeare's play's the thing to catch the conscience of an audience

While disfiguring the queen is Donnazza's primary means of undermining state power, she also utilizes her own body and the body of Shakespeare's *Hamlet* to disrupt the ruling order. Disobeying the hierarchical social structure of Elisabetta's court, Donnazza inverts her subservient social position as she upstages the queen's performance of *Hamlet* with her own *grammelot*[23] version of the play (124–25). Dominating over Elisabetta's performance of *Hamlet* with her own more physicalized, comic version—and thus creating two adaptations-within-a-play of Shakespeare's tragedy, one for the "literary" court, and one for the "festive" masses—Donnazza subverts the power of the monarch's stage presence. Because Donnazza's disorderly performance is given by a woman played by a man playing opposite his wife, this sequence is arguably a moment when Fo, taking center stage away from Rame, is reaffirming a patriarchal structure. However, the theatrical masks in this scene are doubled: Donnazza is competing for the limelight with Elisabetta during the Queen's performance of a reverse-gender version of *Hamlet* that she is giving to prove to Egerton and Martha that Shakespeare's tragedy, thinly disguised as fiction by the characters' sexes, is in actuality "a fiendish attack on my person and my politics" (112). Multiple gender-blurring roles are in operation as Elisabetta uses the power of performance to temporarily become a "Fomale" form of Hamlet—a Hamlette—in an effort to prove to her on- and offstage audience that Shakespeare's character is really a transvestite parody of her intended as "an incitement to rebellion, to revolt!" (128). Donnazza's dual and dueling performance of *Hamlet* emphasizes the gender-layering of the Queen's portrayal as he-as-she simultaneously plays him that supposedly represents her. Gender here is as flexible as Shakespeare's script, whose official cultural status is similarly being challenged as it undergoes treatment on both ends of a binary structure of culture defined by socioeconomic class. At once a literary work whose poetic phrases can be dissected for meaning, and a theatrical piece that can be comically expressed without comprehensible words, *Hamlet* emerges as both "high" and "low" cultural material. The parallel performances of Queen-as-Prince of Denmark-as-Queen and Female Servant-as-Festival-Version-of-Prince of Denmark undermine the privileged sides of gender and sociocultural dichotomies as both characters effectively convince the onstage audience (and with them, the offstage spectators) that *Hamlet* is a sexually inverted critique of the Queen and her court, "a sort of drag act, making fun of the queen" (125, 188), seditiously designed to rouse the masses into mutiny against the ruling body. During Elisabetta's and Donnazza's simultaneous official and unofficial performances of *Hamlet*, Shakespeare's tragic hero is transformed, becoming both a woman-on-top and an inept revolutionary. Presented as an unruly, female-masked-as-male, "exact portrait" of the Queen, Hamlet delivers her "cries of despair . . . [her] curses . . . things [she has] shouted" in her room (91), revealing her private disorderly behavior to the public at the Globe in order to incite rebellion against her rule. As a drag-version of the queen, this Shakespearean Prince(ss) disrupts the ruling order with disruptive imitations of Elisabetta's own unruly actions. However, for Fo, *Hamlet* also functions as a critique of the intellectual's role in a working-class revolution. In a lecture given prior to writing *Elisabetta*, Fo describes how *Hamlet* can be reclaimed for the masses from the "bourgeoisie," who have appropriated Shakespeare and his play, and have transformed Hamlet into a mythological, tragic hero, a "blond, elegant, tall, romantic character who wins the girls' hearts, who struggles, has a few moments of perplexity, and dies in the end in a self-immolation so that all the girls cry in pity for this man's sacrifice" ("Dialogue" 16). Dismissing this officialized interpretation of Shakespeare's character "taught in schools and put on in theatres"—institutions identifiable in transversal terms as sociopolitical conductors of state power—Fo argues that Shakespeare's play is really "a satire about the intellectual of his time who takes advantage of the Cartesian game to give himself alibis and do nothing. He's like the intellectual of today who dreams about revolution but is quite happy to pick up his wages and go home" ("Dialogue" 16).

Fo's reading of Shakespeare's character as an incompetent intellectual "who lets himself get killed because he doesn't know how to struggle" ("Dialogue" 16) is voiced in *Elisabetta* by Donnazza, who attributes the "massacre" in the final act of *Hamlet* to the protagonist's inability to take action: "Everyone knows its Hamlet's fault because he can never make up his mind, he dithers around . . . He could have resolved everything long before" (120, 185). While Donnazza here supports Fo's contention that *Hamlet* is a satiric morality play warning of the disastrous results of passivity, the queen's response to Donnazza's reading provides an alternative to this interpretation. Elisabetta regards Hamlet's indecision as yet another parallel between her and the Danish prince: "And isn't that what I'm accused of too . . . ? [. . .]. Of not eliminating my enemies . . . of not doing anything" (121).[24] Hamlet's indecision thus has the combined effect of implying to the audience that the Queen,

Hamlet's double, is an ineffective ruler who should be overthrown by the public, while also suggesting that the intellectual, Hamlet's other double, must not, like Hamlet and the Queen, remain passive. The two sides of the Hamlet presented in *Elisabetta* are not, however, completely at odds with each other, since both interpretations—Hamlet-as-Queen and Hamlet-as-Incompetent-Intellectual—convey the need to commit to a political movement, which is what Fo describes in a 1984 interview as the critical argument of *Elisabetta*: "It's about the commitment of the intellectual, and the need to participate in world events and take a position" (qtd. in Mitchell 182). Both *Elisabetta* and *Hamlet* (as Fo interprets it) share a central argument concerning the necessity to actively engage in a revolt against a dominant political structure, which in the action of Fo's farce amounts to rebelling against the ruling order of a disorderly queen.

The Hamlet-as-Incompetent-Intellectual may be evident in Donnazza's interpretation of Shakespeare's play, but the Queen's insistence that the character is a sexual inversion of herself is the metaframe of *Elisabetta*. While the Queen's analysis and formal performance of *Hamlet* is given to the onstage audience in the First Act, it is upon her first entrance on stage that the analogy is drawn for the spectator. Emerging with a sword in hand, Elisabetta strikes, à la Hamlet, at a moving figure lurking behind a tapestry that she fears is the vengeful ghost of Mary of Scots (88). Like Hamlet, Elisabetta is haunted throughout by the phantom-figure of a murdered, revenge-seeking relative— although here the ghost, Mary of Scots, was executed at the Queen's command. At the beginning of the Second Act, Elisabetta complains of her ominous dream the night before of Mary, who appeared "Walking around my room as if she owned it ... with no head" (146), while, unbeknownst to the Queen, the Essex *coup d' état* was being planned. The ghost hovers throughout Elisabetta's final monologue, a mosaic of Shakespearean phrases during which Elisabetta speaks to the imagined Mary and visualizes her own grisly death: "That isn't Mary's head / It is my head" (179). This dramatic conclusion to what has until this point been a farcical revisioning of the end of Elizabeth's reign begins with Elisabetta's line, "The grand finale of the last act is beginning. Just like in *Hamlet*" (176). *Elisabetta*, with its queen framed from start to finish as Hamlet, thus concludes "just like" the beginning of the end for the Prince of Denmark.

The parallels between Hamlet and Elisabetta in the opening and closing of the play works cleverly to not only emphasize the Queen's (and Fo's) materialist interpretation of Shakespeare's tragedy as sociopolitical commentary,[25] but also to further blur Elisabetta's onstage gender.

Since she is Hamlet and Hamlet is she, the Queen, once again, is framed as a hermaphroditic figure. This blurring is complicated even further in what we refer to as the Queen's "Fakespearean" final monologue,[26] where she recites lines from, among others, Shakespeare's history plays, *Measure for Measure* and *Julius Caesar*. Speaking strung-together Shakespearean phrases that she believes he appropriated from her— "He steals all his best lines from me" (129)—she speaks as both herself, Shakespeare, and a barrage of Shakespeare's characters, including his (and England's) most famous monarchs. In this final moment, theatrical roles merge with social and political roles, and fact becomes even further smeared with fiction, as the Queen issues lines delivered by Shakespearean kings that Shakespeare fabricated from the pages of history.

The undermining of official social constructs crescendos during the onstage death of Elisabetta's lover, Thomas, whom she adorns in her clothes so that he can "really experience what it means to play the part of a Queen" (158). This line suggests that her position of power is a theatrical role that can be "played" by anyone, regardless of gender or bloodline. Gender and power are here intertwined, as Thomas is dressed in Elisabetta's clothes so that he can "experience what it means" to be in her social position. Thomas finds the "drag act" humiliating, more degrading than serving as the Queen's "little gigolo" (156). Nonetheless, he "makes such a believable girl" (159) that he is mistaken by his own father as the Queen herself, tragically stabbed, and then, after Martha orders the guards to "Take her away" (165), carted off the stage as a woman. The transgendering of Thomas is indicative of the play's overall assessment of sex roles and power positions as socially assigned, performative, and hence alterable. Thomas's onstage murder extends the deconstruction of gender/power/class structures into a critique of the subject position of women in the official territory of both early modern England and modern Western cultures: "to play the part of a Queen" (158) is dangerous and degrading in past and present patriarchal social structures, a theatrical role that can quickly end like a Shakespearean tragedy. The anti-patriarchal sentiment echoes what is now considered first-wave feminist response to the role of women in male-dominated sociopolitical structures: the personal is the political for Elisabetta the Queen, as it is for all people.

Entering Elizaspace, en route to elsewhere

Structuring his farce around the trials and tribulations of the first Protestant Queen of England, Fo, with the *chutzpah* characteristic of the investigative-expansive mode, simultaneously addresses distinct yet controversially interrelated, historical issues. The play functions tangentially as a critique of the degradation of women and an attack on the oppressive effects of a repressive state apparatus Fo imagines was engendered by the title character of his play. Explaining the contemporaneity of his historical farce in a 1984 interview with *La Repubblica*, Fo states, "Hers is the first modern state. She invented the secret service of and modern politics" (qtd. in Mitchell 183). The Queen's "modern state" is defined as "hers," and as such she serves as the symbol for the play's polemic against state power. Yet, she functions on another, parallel plane as a victim of the subjective territory maintained by, in the context of the play, her own court. Fo's rendering of the Queen thus generates multiple meanings, a plurality of codes effecting what Donald Hedrick and Bryan Reynolds identify as the "principle of translucency" (discussed in chapters 7–9): an approach to performance, whether onstage, onscreen, or at a cocktail party, whereby one identity or signifier is imperfectly hidden within another, creating a transversal space wherein spectators are invited to expand and reconfigure the coordinates of their own subjective territory. Elisabetta's translucent effect is highlighted by her historical counterpart: the onstage character carries with her the meanings associated with her "real" precedent, Queen Elizabeth I, that figure whose iconic power peeks through Fo's re-creation. Audiences are asked to decode and deconstruct an onstage figure whose presence provokes an association with a historical past made readily present by the presence of Rame, a performer whose own cultural iconic status—as one of Italy's most famous actresses and as Fo's wife—presents a complication in the decoding process, adding another layer of meaning to the "multivalent image" of Elisabetta, the disorderly Queen emblematic of the dominant sociopolitical order under critical attack.

The translucent effect of Elisabetta was amplified in the 1987 American premiere of *Elisabetta*, where Joe Morton, the actor who played Fo's drag role, performed a prologue that included a letter to Ronald Reagan from the playwright. Ironically addressing the then president as, "Dear Friend and Fellow Actor," the letter included a detailed description of the "possible parallels between the story of Elizabeth in my play, and your own Presidency" (12–13), all of which, Fo insists, were not intentional. The letter, of course, makes it perfectly clear that Fo, framing his farce for an American audience, intended for Elisabetta's gender to be a thin veil over his male, Republican target: "She is a failing leader, losing her memory and her health as her empire collapses around her. We are all lucky that awful things like this only happen in the past" (14–15). In the American production, then, Elisabetta (played by Joan Macintosh) is not only framed as the Prince of Denmark, but also as the American president, which adds another dimension to the interpretation of *Hamlet* that frames the play: if Elisabetta is Hamlet, Reagan, and the Queen of England, then Hamlet, by extension, becomes Elisabetta/Reagan. The theatrical situation questions the former film actor/then U.S. president's masculinity through the confusion of gender and power roles, with Reagan being represented parodically on stage by an unruly woman whose gender-identity is made more ambiguous by his alluded-to presence.

In her analysis of the multivariate speculations concerning Elizabeth's sexuality and conduct that circulated during her forty-five-year reign, Carole Levin notes that the salacious rumors concerning the Queen's sexual behavior testified to the "hostility towards her as an unmarried female ruler whose position transcended the traditional role allotted to women in English Renaissance society" (68). Whether motivated by gender-biased social standards or a desire to bring the Virgin Queen back down to the earth, these rumors generated an alternative version of the carefully crafted image of Elizabeth presented at court. Both the Virgin Queen and the "whore" on the throne operated as conceptualizations of the Tudor monarch that traveled throughout her kingdom, gathering and generating meaning as they continued to pass from present to past. The conceptual and emotional spaces stimulated by Elizabeth were used to promote the hierarchical structure of the English monarchy, as well as to give expression to dissatisfaction, whether based on gender biases, class prejudices, or both, to her rule. We refer to the "spaces" through which the Queen passed, and continues to pass, as an icon as "Elizaspace," an adaptation of Shakespace.[27] Like Shakespace, Elizaspace is a discursive, multidimensional "articulatory space" where critical discourses and coordinates interface, social performances are imbued respectively and relationally with meaning, subsequent communication transpires, and learning is achieved.[28] Also like Shakespace, Elizaspace functions both as a sociopolitical conductor of state power, as evident by the loyalty to the crown produced by the cult of the Virgin Queen during her reign, [29] and as a conduit for transversal movement outside official territory, emerging in the transversal process "as a culturally-imaginable space where [Elizabeth] may function without class in both the socioeconomic and aesthetic senses of the word" (Hedrick and Reynolds 9).

Brilliantly exploiting preexisting relationships between Elizaspace and Shakespace, Fo revises the Renaissance icons and envisions a critical reassessment of sociopolitical structures for an emergent generation of transversal movers and "Shakesers." Although, twenty years after he began work on *Elisabetta*, Fo's carnivalesque rendering of the Virgin Queen also seems, in one sense, a still-life portrait of the feminist polemic of its day, the *mis*appropriated Elizabethan figure at the center of Fo's farce, affecting the principle of translucency, remains an open site for the audience's critical consideration of gender/power and class/culture structures. The Queen, revealed through a series of gruesome beauty treatments as both a tyrannical conductor of state power and a victim of gender roles prescribed by the sociopolitical conductors of her realm's official territory, becomes the apotheosis of the state machinery's inculcating procedures; yet at the same time she functions as "historical" evidence of the effectiveness of the theater—in particular, popular theatrical forms—to unmask the inner workings of state machinery and inspire movement away from the ideological structures it engenders. Through the degradation of its Queen, *Elisabetta* becomes a transhistorical critique of the *modus operandi* of state power, a theatrical text that contains within its body the capacity to inspire transversal movement out of the dichotomous, ideological formations it brings to the stage for investigation. The boundaries of these intertwining dichotomies—in a constant dialectic with each other throughout the play—are expanded through dynamic processes of inversion and conversion: proven to be capable of adoption by opposing forces and of transformation from within a single body, these dichotomies are revealed to be conceptual in form and content. Within the transhistorical world of the play, these social constructs are not transcended, but rather reworked, through their continuous inversion/reversion/conversion. Their shifting structures, like the body of the onstage Queen, begin to lose their shape and collapse, generating a critical space for a future vision, rather than a present revision, of a revolution against a dominant body politic and the intangible boundaries manufactured by its state apparatus and sustained through the imposition of an official, *un*popular culture.

Owing to Elizabeth and Shakespeare's historical association and coupling in the popular imaginary, Elizaspace accompanies a Shakespace in Fo's play, where both the Queen and Shakespeare function "without class." Revamped as carnivalesque figures, Shakespeare and Elizabeth in Fo's revisioning become "classless" conduits for counter-thought whose historicity testifies to what Fo imagines is the transhistorical, dialectical presence of state power and forces of dissidence. Drawing parallels between Elizabethan England and modern political structures, Fo links the state machinery of Elizabeth I's reign to the Christian Democrat Party in Italy to the Republican Party in the United States. Through the "real," imagined, and/or alluded to presence of his revisions of Queen Elizabeth and Shakespeare, Fo takes the viewer on a deconstructive journey of capitalistic state machinery, a theatrical trip through space–time memorialized that defies established spatiotemporal parameters and collapses sociocultural divisions, producing a conceptual terrain where Renaissance icons perpetuated by official Western cultures as "high class" figures become popular, public performers, "classless," classic examples of what Fo imagines is a sociopolitical struggle that has been fought since Elizabeth's formation of the "first modern state" (qtd. in Mitchell 183). Although his historical assertions are debatable, Fo's envisionings of ideologically conflicting Elizaspace and Shakespace, operating from early modern England to the fallout of postmodernism and beyond, point toward intentional deployments of transversal power through sociopolitical performances in the critical future.

Dear Readers,

If you are reading this paragraph, then you have ventured to journey from beginning to end, end to beginning, or from some point on the book's (and/or your own) transversal course. Like the audience of Fo's farce, you have undergone a transhistorical adventure through conceptual configurations of space–time and scholarship. Contained within the pages that you may or may not have perused, the materials—cultural, theoretical, performative, textual, photographic, and so on—are endowed with transversal power whose effect relies upon your engagement. Seeking to stimulate your awareness of the possibilities that lay behind and beyond the cartography of subjective territories, we have invited you to move with, near, or against us through your own navigations, uncovering and discovering new territories in the wake of your reactions to our discourse. Engagement is generally thought of as dialogic; consider the cliché, "It takes two to tango." But, as this collaborative expansion into various parameters of investigation has shown, discourse need not be restricted to delineated conceptions (i.e. three or more also makes for a good mambo; add four or five and you may just have a bacchanalian cha-cha). Our discourse, powered by your imagination and channeled by collective authorship, has welcomed and indeed presumed your participation. If nothing else, we hope you have collected some useful *tchotchkies* on your journey, and thank you for choosing to travel with us.

Sincerely,
The Collaborators

PS:

Before signing off, I, Bryan Reynolds, would like to remind our readers of Dario Fo's provocative criticism of Shakespeare's *Hamlet*: "The play's a satire about an intellectual of his time who takes advantage of the Cartesian game to give himself alibis and do nothing. He's like the intellectual of today who dreams about revolution but is quite happy to pick up his wages and go home" ("Dialogue" 16). As an intellectual and educator who is not happy with merely collecting my wages and going home, developing transversal theory is one of my contributions to the worlds we live in that I believe can be beautiful, resounding with love and compassion, despite the hardships. I believe that the more we experience alternative thoughts and feelings—have experiences alternative to those typical of our lives—the less we will promote unwittingly the sociopolitical conductors that work to oppress us and others. For this reason, I try to inspire transversality, both within myself and in others, as often as possible. While I very much hope that my collaborators and I have convinced you of the value of transversal poetics, if we have not, it would give me great pleasure to know why. So, if you have the time and are interested, please feel free to email me: breynold@uci.edu.

Notes

1. In the December 1974 Italian issue of *Playboy*, Fo referred to his and Rame's theater as "un teatro da bruciare," which is usually translated by critics as "throwaway theatre," but sometimes appears as "theatre for burning," a more literal translation of the Italian (qtd. in Mitchell 95).
2. Explaining his and Rame's 1968 break with the commercial theater in Italy, Fo stated, " 'We were tired of being the *giullari* of the bourgeoisie, for whom by now our criticisms had the same affect as *Alka-Seltzer*, we therefore decided to become the *giullari* of the proletariat' " (qtd. in Tom Behan, *Dario Fo: Revolutionary Theatre* [London: Pluto Press, 2000], 22). For Fo, the medieval *giullari* were theatrical working-class heroes, itinerant performers who defied authorities by satirically giving voice to the concerns of the masses. As Joseph Farrell and Antonio Scuderi note, Fo's conception of the *giullari* has been "disputed by experts who point out that there were purely conventional *giullari* who played for captains and kings" ("Introduction: The Poetics of Dario Fo," *Dario Fo: Stage, Text, and Tradition*, ed. Joseph Farrell and Antonio Scuderi [Carbondale: Southern Illinois University Press, 2000], 1–17, 10).

3. For instance, in reference to Gavin Ricards's adaptation and production of *Morte accidentale di un anarchico* (*Accidental Death of an Anarchist*), Fo, who saw the production after it transferred to the West End in 1980, commented that "the excessive buffoonery" in Richards's version, which made the production "overloaded, verging terribly on the grotesque," was " 'anti-style' . . . not 'style' in some vague sense, but 'style' in the sense of a satirical form of theater that seeks to wound, to disturb people, to hit them where it hurts" (*Theatre Workshops* 67–68).
4. Prior to receiving the Nobel prize, Fo's *Morte accidentale di un anarchico*, written in response to the 1969 Milan bank bombings, had been established as "the all-purpose protest play from Tokyo to London" (Farrell, "Variations on a Theme" 28), and had been produced in commercial, mainstream theaters in Los Angeles (Mark Taper Forum, 1983), the West End (Wyndham's Theatre, 1980–1981; Cottesloe Theatre, Royal National Theatre, 1991), and on Broadway (Belasco Theatre, 1984).
5. For an analysis of Shakespeare in all his multifarious multimedia forms, see, in addition to the other chapters of this book, *Shakespeare Without Class: Misappropriations of Cultural Capital*, ed. Donald Hedrick and Bryan Reynolds (New York: Palgrave, 2000), and *Shakespeare After Mass Media*, ed. Richard Burt (New York: Palgrave, 2002).
6. For further discussions in this collection of "Shakespace," see Bryan Reynolds and Ayanna Thompson's "Inspriteful Ariels: Transversal Tempests," and D.J. Hopkins and Bryan Reynolds's "The Making of Authorships: Transversal Navigation in the Wake of *Hamlet*, Robert Wilson, Wofgang Wiens, and Shakespace." Hopkins and Reynolds's article also appears in *Shakespeare After Mass Media*, ed. Richard Burt (New York: Palgrave, 2002), 265–86.
7. Fo validates his claim that Shakespeare's actors performed "off the cuff" by referencing the "bad" quarto of *Hamlet* (1603), whose authorship he attributes to Richard Burbage (*Tricks of the Trade* 87–88).
8. In his adaptation of *Hamlet*, Robert Wilson similarly conceives of Shakespeare's play as a series of fifteen scenes. For an analysis of Wilson's *Hamlet: A Monologue*, see chapter 2 of this book.
9. For a detailed description of investigative-expansive wherewithal, please see chapter 9.
10. For a detailed discussion of the "investigative-expansive" and "dissective-cohesive" modes of analysis ("i.e. mode" and "d.c. mode"), see the appendix to this book.
11. In addition to her brief appearance in *Shakespeare in Love* (1998), Elizabeth I was the subject of the Oscar-award nominated biopic *Elizabeth: The Virgin Queen* (1998), which revolved around the Queen's tenuous hold of the throne in the early years of her reign.
12. In *The Heart and Stomach of a King: Elizabeth I and the Politics of Sex and Power*, Carole Levin argues that Elizabeth recreated herself in the image of the Virgin Mary to fill the void caused by England's separation from the Catholic Church (Philadelphia: University of Pennsylvania Press, 1994), 26–28. For a detailed biography of Queen Elizabeth, see, among others, Alison Weir's *Elizabeth the Queen* (London: Jonathan Cape, 1998).

13. La Donnazza is translated as Dame Grosslady in Gillian Hanna's translation of the play (*Dario Fo: Plays 2* [London: Methuen, 1997], 83–204), and as Mama Zaza in Ron Jenkins's translation (New York: Samuel French, 1987). All quotations from *Elisabetta* that appear in this essay are taken from Hanna's translation.
14. In his book, *Oxford: Son of Queen Elizabeth I* (London: Oxford Institute Press, 2002), Paul Streitz merges both icons with ingenuity, arguing that Edward de Vere, Seventeenth Earl of Oxford, is not only the author of Shakespeare's plays but also Queen Elizabeth I's son.
15. After its hugely successful 2001 world premiere at South Coast Repertory Theatre, Amy Freed's *The Beard of Avon* went on to enjoy runs in 2002 at The Seattle Repertory Theatre, American Conservatory Theater, Cincinnati Shakespeare Festival, Chicago's Goodman Theatre, and Toronto's CanStage.
16. Much of the recent work on Fo and Rame has compensated for the previous attention solely on Fo by emphasizing Rame's on- and offstage work. See, among other studies, Serena Anderlini-D'Onofrio's "From the *Lady is To Be Disposed Of to An Open Couple*: Franca Rame and Dario Fo's Theater Partnership," *Franca Rame: A Woman on Stage*, ed. Walter Valeri (West Lafayette: Bordighera, 2000), 183–204; Marga Cottino-Jones's "The Transgressive Voice of a Resisting Woman," *Franca Rame: A Woman on Stage*, ed. Walter Valeri (West Lafayette: Bordighera, 2000), 8–58; Pina Piccolo's "Rame, Fo and the Tragic Grotesque: The Politics of Women's Experience," *Franca Rame: A Woman on Stage*, ed. Walter Valeri (West Lafayette: Bordighera, 2000), 115–38; and Sharon Wood's "*Parliamo di donne*: Feminism and Politics in the Theater of Franca Rame," *Dario Fo: Stage, Text, and Tradition*, ed. Joseph Farrell and Antonio Scuderi (Carbondale: Southern Illinois University Press, 2000), 161–80.
17. While Fo's statement implies that *Elisabetta* was the first Fo play Rame collaborated on, his wife had in fact "developed a taste for it" in the mid-1970s, and was the coauthor of her 1977 one-woman show *Tutta casa, lette e chiesa* (*All Home, Bed, and Church*). According to Walter Valeri, the first play Fo and Rame coauthored was the 1976 *La marijuana della mamma è la più bella* (*Mother's Marijuana is the Best*) ["Franca Rame: una donna in scena," in *Franca Rame: A Woman on Stage*, ed. Walter Valeri (West Lafeyette: Bordighera, 2000), 1–4, 3].
18. For an analysis of male (active)/female (passive) as a dominant gender dichotomy in Western culture, see Hélène Cixous' "Sorties: Out and Out: Attacks/Ways Out/Forays," in *The Newly Born Woman* (with Catherine Clément), trans. Betsy Wing (London: I.B. Taurus, 1996). For a discussion on Cixous's perspective, contextualizing it within the history of critical theory, see Bryan Reynolds and Ian Munro, "Hélène Cixous," in *Twentieth Century European Cultural Theorists*, ed. Paul Hanson (Columbia: Bruccoli Clark Layman, 2001), 83–102.
19. Emphasizing the transvestite character's "low" social status and popular roots, Fo has Donnazza speak in "a mixture of North Italian dialects and archaisms," as opposed to standard Italian (Stuart Hood, "Introduction," *Dario Fo Plays: 2* [London: Methuen, 1997], ix–xviii, xv). In her translation

for an English audience, Hanna substitutes Donnazza's speech with "Elizabethan slang (drawn from Nashe, the contemporary chronicler of the Elizabethan underworld), Italian, Cockney rhyming slang, spoonerisms, puns and regional expressions from Scotland, Ireland, Norfolk, Yorkshire and elsewhere, together with words drawn from a dictionary of eighteenth-century slang" (Hood xv). Hanna provides a literal translation of each of the character's lines in an appendix to the text. All of Donnazza's lines cited in this essay are taken from the appendix, but are accompanied by paranthet-icals indicating first where the line(s) appears in the text, and second where it can be found in the appendix provided by Hanna.

20. If intended as an allusion to Elizabeth's White-Anglo-Saxon-Protestant subjects, Donnazza's breast-stinging wasps may be a metaphorical enact-ment of the people's revolt against the Queen.

21. Following her death in 1603, Robert Cecil, in a letter to Sir John Harington, described Elizabeth as "more than a man, and in truth, less than a woman" (qtd. in Levin 147; Marcus, *Puzzling Shakespeare* 53).

22. For Mikhail Bakhtin's analysis of the medieval fool and the sociopolitically subversive features of the carnivalesque, refer to *Rabelais and His World*, trans. Hélène Iswolsky (Cambridge: MIT Press, 1968).

23. *Grammelot* is a verbal and physical performance technique of popular origin that has become a trademark of Fo's work. According to Fo, "*Grammelot* is a term of French origin, coined by Commedia players, and the word itself is devoid of meaning. It refers to a babel of sounds which, nonethe-less, manage to convey the sense of speech. *Grammelot* indicates the onomatopoeic flow of a speech, articulated without rhyme or reason, but capable of transmitting, with the aid of particular gestures, rhythms and sounds, an entire, rounded speech" (*The Tricks of the Trade* [New York: Routledge, 1991], 56). Antonio Scuderi claims that Fo learned the tech-nique from Jacques Lecoq in 1953 while Lecoq was supervising rehearsals of *Il dito nell'occhio* (*A Finger in the Eye*), a revue written by Fo, Giustino Durano, and Franco Parenti that premiered at the Piccolo Teatro ("Updating Antiquity," *Dario Fo: Stage, Text, and Tradition*, ed. Joseph Farrell and Antonio Scuderi [Carbondale: Southern Illinois University Press], 39–64, 59).

24. Alison Weir notes that Elizabeth's chronic indecision infuriated her advi-sors: "A mistress of the subtle art of procrastination, she was marvellously adept at delaying and dissembling, and would usually shelve problems she could not immediately solve" (*Elizabeth the Queen* [London: Jonathan Cape, 1998], 223).

25. Of course, Fo is not alone in his contention that Shakespeare's plays contain commentary on the sociopolitical structures of early modern England. There is a long tradition of this. For discussions of Shakespeare's work in relation to the ideology of early modern England, see, among others, *Political Shakespeare: New Essays in Cultural Materialism*, ed. Jonathan Dollimore and Alan Sinfield (Ithaca: Cornell University Press, 1985); James H. Kavanagh's "Shakespeare in Ideology," *Alternative Shakespeare*, ed. John Drakakis (London: Routledge, 1985), 144–65; Leah S. Marcus's *Puzzling*

Shakespeare: Local Reading and Its Discontents (Berkeley: University of California Press, 1988); and Leonard Tennenhouse's *Power on Display: The Politics of Shakespeare's Genres* (New York: Methuen, 1986). For readings of Shakespearean characters in relation to Elizabeth I, see, among others, Peter Erickson's "The Order of the Garter, the cult of Elizabeth, and class-gender tension in *The Merry Wives of Windsor,*" *Shakespeare Reproduced: The Text in History and Ideology*, ed. Jean E. Howard and Marion F. O'Connor (London: Methuen, 1987), 116–40; Leah S. Marcus's *Puzzling Shakespeare* and "Shakespeare's Comic Heroines, Elizabeth I, and the Political Uses of Androgyny," *Medieval and Renaissance Women: Literary and Historical Perspectives*, ed. Mary Beth Rose (Syracuse: Syracuse University Press, 1986), 135–54; and Leonard Tennenhouse's *Power on Display*.

26. We coined the term "Fakespearean" in collaboration with David Podley.
27. We are grateful to D.J. Hopkins for the slick coinage, "Elizaspace," who, with Bryan Reynolds, is currently writing an article on Elizaspace in the drama and theater of Shakespeare and his contemporaries.
28. For more on "articulatory space," see the introduction to this book.
29. As Levin notes, "By not marrying, by being no-one's mother and everyone's, and by presenting herself as both a virgin to be revered and a sensuous woman to be adored, Elizabeth exerted a strong psychological hold on her subjects" (*The Heart and Stomach of a King: Elizabeth I and the Politics of Sex and Power* [Philadelphia: University of Pennsylvania Press, 1994], 87).

Works cited

Anderlini-D'Onofrio, Serena. "From the Lady is To Be Disposed Of to An Open Couple: Franca Rame and Dario Fo's Theater Partnership." *Franca Rame: A Woman on Stage*. Ed. Walter Valeri. West Lafayette: Bordighera, 2000. 183–204.

Bakhtin, Mikhail. *Rabelais and His World*. Trans. Hélène Iswolsky. Cambridge: MIT Press, 1968.

Behan, Tom. *Dario Fo: Revolutionary Theatre*. London: Pluto Press, 2000.

Burt, Richard. *Shakespeare After Mass Media*. Ed. Richard Burt. New York: St. Martin's Press, 2002.

Cixous, Hélène. "Sorties: Out and Out: Attakcs/Ways Out/Forays." *The Newly Born Woman* (with Catherine Clément). Trans. Betsy Wing. London: I.B. Taurus, 1996.

Cottino-Jones, Marga. "The Transgressive Voice of a Resisting Woman." *Franca Rame: A Woman on Stage*. Ed. Walter Valeri. West Lafayette: Bordighera, 2000. 8–58.

Davis, Natalie Zemon. *Society and Culture in Early Modern France*. Stanford: Stanford University Press, 1975.

Dollimore, Jonathan and Alan Sinfield, eds. *Political Shakespeare: New Essays in Cultural Materialism*. Ithaca: Cornell University Press, 1985.

Elizabeth: The Virgin Queen. Dir. Shekhar Kapur. Perf. Cate Blanchett, Geoffrey Rush, and Joseph Fiennes. Polygram, 1998.

Erickson, Peter. "The Order of the Garter, the Cult of Elizabeth, and Class-Gender Tension in *The Merry Wives of Windsor.*" *Shakespeare Reproduced: The Text in History and Ideology.* Ed. Jean E. Howard and Marion F. O'Connor. London: Methuen, 1987. 116–40.
Farrell, Joseph. "Variations on a Theme: Respecting Dario Fo." *Modern Drama* 61.1 (1998): 19–29.
——— and Antonio Scuderi. "Introduction: The Poetics of Dario Fo." *Dario Fo: Stage, Text, and Tradition.* Ed. Joseph Farrell and Antonio Scuderi. Carbondale: Southern Illinois University Press, 2000. 1–17.
Fo, Dario. "Author's Note." Trans. Gillian Hanna. *Dario Fo Plays: 2.* London: Methuen, 1997. 84–85.
———. "Elizabeth: Almost By Chance a Woman." Trans. Gillian Hanna. *Dario Fo Plays: 2.* London: Methuen, 1997. 83–204.
———. *The Tricks of the Trade (Manuale Minimo dell'Attore).* Trans. Joe Farrell. Ed. Stuart Hood. New York: Routledge, 1991.
———. *Elizabeth: Almost By Chance a Woman.* Trans. Ron Jenkins. New York: Samuel French, 1987.
———. "Prologue." *Elizabeth: Almost By Chance a Woman.* Trans. Ron Jenkins. New York: Samuel French, 1987. 10–15.
———. "Dialogue with an Audience." Trans. Tony Mitchell. *Theatre Quarterly* 9.35 (1979): 11–16.
Fo, Dario and Franca Rame. *Dario Fo and Franca Rame: Theatre Workshops at Riverside Studios, London; April 28th, May 5th, 12th, 13th and 19th 1983.* London: Red Notes, 1983.
Hedrick, Donald and Bryan Reynolds. "Shakespace and Transversal Power." *Shakespeare Without Class: Misappropriations of Cultural Capital.* Ed. Donald Hedrick and Bryan Reynolds. New York: St. Martin's Press, 2000. 3–47.
Hood, Stuart. "Introduction." *Dario Fo Plays: 2.* By Dario Fo. London: Methuen, 1997. ix–xviii.
Kavanagh, James H. "Shakespeare in Ideology." *Alternative Shakespeare.* Ed. John Drakakis. London: Routledge, 1985. 144–65.
Levin, Carole. *The Heart and Stomach of a King: Elizabeth I and the Politics of Sex and Power.* Philadelphia: University of Pennsylvania Press, 1994.
Marcus, Leah S. *Puzzling Shakespeare: Local Reading and Its Discontents.* Berkeley: University of California Press, 1988.
———. "Shakespeare's Comic Heroines, Elizabeth I, and the Political Uses of Androgyny." *Medieval and Renaissance Women: Literary and Historical Perspectives.* Ed. Mary Beth Rose. Syracuse: Syracuse University Press, 1986. 135–54.
Mitchell, Tony. *Dario Fo: People's Court Jester.* Rev. ed. London: Methuen, 1999.
Piccolo, Pina. "Rame, Fo and the Tragic Grotesque: The Politics of Women's Experience." *Franca Rame: A Woman on Stage.* Ed. Walter Valeri. West Lafayette: Bordighera, 2000. 115–38.
Reynolds, Bryan. "Untimely Ripped." *Social Semiotics* 7.2 (1997): 201–18.
Scuderi, Antonio. "Updating Antiquity." *Dario Fo: Stage, Text, and Tradition.* Ed. Joseph Farrell and Antonio Scuderi. Carbondale: Southern Illinois University Press, 2000. 39–64.

Scuderi, Antonio. *Dario Fo and Popular Performance*. New York: Legas, 1998.
Shakespeare in Love. Dir. John Madden, Jr. Perf. Gwenyth Paltrow, Joseph Fiennes, and Dame Judi Dench. Miramax, 1998.
Streitz, Paul. *Oxford: Son of Queen Elizabeth I*. London: Oxford Institute Press, 2002.
Tennenhouse, Leonard. *Power on Display: The Politics of Shakespeare's Genres*. New York: Methuen, 1986.
Valeri, Walter. "Franca Rame: Unna Donna in Scena." *Franca Rame: A Woman on Stage*. Ed. Walter Valeri. West Lafayette: Bordighera, 2000. 1–4.
Weir, Alison. *Elizabeth the Queen*. London: Jonathan Cape, 1998.
Wood, Sharon. "*Parliamo di donne*: Feminism and Politics in the Theater of Franca Rame." Dario Fo: Stage, Text, and Tradition. Ed. Joseph Farrell and Antonio Scuderi. Carbondale: Southern Illinois University Press, 2000. 161–80.

Afterword: Walk Like an Egyptian

Jonathan Gil Harris

> Let Rome in Tiber melt, and the wide arch
> Of the ranged empire fall! Here is my space.
> *Antony and Cleopatra*, 1.1.34–35

I'd like to use this space to talk about space.

Reading the fascinating work of Bryan Reynolds and his fellow transversal "movers and Shakesers," I am struck by the plangency of two related keywords that appear throughout *Performing Transversally*: "territory" and "space." The contributors' express desire to move beyond "the cartography of subjective territories" (379) into "transversal territory"—a location that overlaps with the realm of "Shakespace" in particular, and the confederation of "articulatory spaces" in general—gives some suggestion of how *Performing Transversally* clears new ground in Shakespeare studies by metaphorically constituting "ground" itself as the basis of its theoretical challenge. Yet the spatial tropes that underwrite these and the book's other articulations of transversal theory warrant some critical scrutiny. What and where, exactly, is transversal "territory"? In what sense is transversal "territory" really "outside" subjective "territory"? How far does one have to move, and how much does one have to shake, to get there?

Off the map: space and practice

The tropology of topology is not anything new in Shakespeare studies; indeed, it has become a recurrent if occulted feature of the field's mainstream historicist lexicon. Shakespeare scholars are now thoroughly habituated, if unself-consciously, to the vocabulary of ideological "sites" and discursive "terrains." Even historicism's infamous model of "subversion and containment" is a spatializing hermeneutic: echoing the Cold War strategist George F. Kennan's 1947 article on "containing" the Soviet threat, historicists have often located subversion as a *topos* as much as a practice—Edmund Spenser's Ireland, Thomas Harriot's Virginia.[1] This locative tendency is most evident in Steven Mullaney's *The Place of the Stage: License, Play and Power in Renaissance England* (1988). As the book's subtitle makes clear, Mullaney's is an argument about the politics as well as poetics of early modern space. The drama of the public stage, in his influential account, capitalized on its geographical as well as cultural marginality, inasmuch as the playhouses were situated in the Liberties, a zone of subversive "ambivalence, paradox, and cultural contradiction" beyond the jurisdiction of London (31). This narrative might seem to have something in common with transversal theories of space: the Liberties were for Mullaney territory "outside" the interpellating machinery of the city. Like much historicist scholarship, however, *The Place of the Stage* offers a largely synchronic conception of space. At the beginning of the book, Mullaney helpfully yet tellingly provides a *map* of early modern London and the Liberties. His map undergirds a larger project of cultural cartography that works to reconfigure Renaissance power and subversion within a spatialized grid, whereby authority is placed "here" and subversion "there." In the process, Mullaney's mapping of power eschews temporality: the place of the subversive stage has a decidedly static quality, situated as it is in the freeze-frame of a spatialized historical "moment."[2]

And this is one way in which *Performing Transversally* offers something radically new. "Transversal territory" is not a fixed space that can be located on a map: the ground it occupies moves and shakes. Little wonder that in his introduction, Reynolds includes "terrorist attacks" and "natural catastrophes" in his list of transversalizing spaces. How can a static map represent the convulsions of terrain that characterize an explosion or an earthquake? As Reynolds argues, "transversal territory is a boundless space of metamorphosis" (5). "Shakespace" is "Shake Space"; but it is equally "Shakes Pace," keenly aware of its own velocity or dynamic movement into the future. And that lexical indeterminacy itself conveys something of the impossibility of pinning down transversal "territory." "Shakespace" is continually transformed, over time and even within any one essay—as D.J. Hopkins and Reynolds show in their brilliant chapter on Robert Wilson's (or is it Wolfgang Wiens's? or Shakespeare's? or Lou Reed's? or Hopkins and Reynolds's?) *Hamlet: A Monologue*, a performance piece that self-consciously displaces drama's assumption of an originary, authorized play-text that live performance simply re-presents. Instead, the monologue offers a vision of "Shakespace" as an ongoing, theatrical refashioning without origin—a vision that makes the metamorphic resources of theater available for criticism to re-envisage itself as transversal performance rather than derivative imitation. In this respect, *Hamlet: A Monologue* and, indeed, *Performing Transversally* recall Terence Hawkes's 1985 essay on *Hamlet*, "Telmah," in which he proposes jazz as a template for literary criticism: "For the jazz musician, the 'text' of a melody is a means, not an end. Interpretation in that context is not parasitic but symbiotic in its relationship with its object [. . .] The same unservile principle seems to me to be appropriate to the critic's activity" (330).

If, as Hawkes suggests, some principles of criticism are "unservile," then presumably others are "servile," upholding not only the originary status of the text, but also what Reynolds terms the "state machinery" of subject and object formation (2). Yet even as such machinery seeks to solidify identities and texts within various articulatory spaces, including "Shakespace," this quest cannot help but ultimately fail, given the metamorphic nature of people, things and locations through time. Transversal movements are unavoidable because of the ineluctability of change. So "transversal territory" might be less a place one has to journey to than a location one is always already in, even if one does not know it. Hawkes's adjective provides a useful way of reframing this conundrum: "Servile" and "unservile" describe not only relational positions, but also affective *dis*positions. Perhaps it may be productive to understand transversal space as not just a (universal) protean location, then, but also a (specific) disposition—a disposition that Reynolds and his contributors repeatedly associate throughout *Performing Transversally* with *chutzpah*, a cheeky disregard for the interpellating strictures of "subjective territory" imposed on one, even by oneself.

Hawkes's "unservile" interpretation of *Hamlet* anticipates Wilson's (and Hopkins and Reynolds's) performance-piece in its cheeky reframing—and hence reconstitution—of its object. What starts out as a deconstructive dissection of *Hamlet* abruptly reconfigures itself as a historicist/cultural materialist reading of J. Dover Wilson's reading of W.W. Greg's reading of *Hamlet*. And no sooner has the reader recovered from this parametric disorientation than the essay reconfigures itself again as a manifesto on "jazz" criticism. Hawkes's repeated moving and shaking of his critical kaleidoscope reeks of transversal *chutzpah*. It might be seen also as a prototypical deployment of what Reynolds and James Intriligator have dubbed the investigative-expansive (i.e.) mode of criticism, a mode fully on display in *Performing Transversally*. "Shakespace" trembles throughout this book with the quakes of investigative-expansive reparameterization. We see such tremors in reading(s) of a single play, as in Joseph Fitzpatrick and Reynolds's essay on *Othello*, which is strategically interrupted and reframed by Reynolds's and Janna Segal's intervention into Gilles Deleuze's account of masochism. We see it also in essays that shift the ground from Shakespeare's time to the twentieth century, as in Reynolds's investigations of Shakespeare's *Coriolanus* and Bertolt Brecht's *Coriolan*, or in his shuttling between the different yet uncannily similar "counter-cultures" articulated in Shakespeare's and Roman Polanski's *Macbeth*s.

In its investigative-expansive peregrinations, transversal theory borrows much of its energy as well as its theoretical investments from British cultural materialism. Unlike recent work on so-called material culture, cultural materialism follows Marx in rejecting any static view of materiality as frozen within the cryogenic vault of a historical "moment." Marx criticized Feuerbach for conceiving of materiality "only in the *form* of the *object*"; endorsing Aristotle's definition of "matter" as *dynameos* or potentiality, he instead understood materiality to belong to the diachronic domain of labor and *praxis*, and thus to presume and induce a future.[3] Cultural materialism's theorization of culture as a special instance of diachronic, transformable materiality sharply opposes the new historicism's synchronic "grid" model of power. Its more dynamic understanding of materiality closely parallels transversal theory's reconceptualization of space. Seen under the aegis of *praxis*, transversal space likewise presumes and induces a future. Space, in other words, no longer has three or even two dimensions, as it has tended to in historicist criticism, but (at least) four.

By attending to the diachronic as well as synchronic dimensions of space, *Performing Transversally* responds to the theoretical challenges posed by Michel de Certeau's analysis of space in "Walking in the City"—an essay that Reynolds and Joseph Fitzpatrick have cannily analyzed elsewhere.[4] In a powerful critique of Foucault's account of panoptic power, Certeau distinguishes between "place" and "space." Like Foucault's disciplinary tables, "place" is for Certeau organized into a synchronic grid that becomes legible only from an elevated, god's-eye view. The modern map, with its panoptic perspective, offers the paradigmatic instance of "place." "Space," by contrast, is in Certeau's definition dynamic, disorderly and diachronic. It does not simply form subjects by organizing them into orderly configurations; it is also transformed by these subjects' multiple spatiotemporal movements. Space is not, in other words, a static *being*, but rather a transversal *becoming*—marked and

AFTERWORD / 275

morphed by *praxis*, it is what Certeau terms "a practiced place" (121). As this phrase suggests, Certeau does not define space as a location that one can reach only by leaving place. Rather, Certeau conceives of the relationship between the two dialectically. Place, he argues, is space that has forgotten the practices that have generated it. To use Marx's terminology, if space is metamorphic materiality, constantly worked upon over time by a variety of practices, then place is form, reified space that disavows the processes of labor that have been necessary for its production—especially the walking that is required to map territory. For Certeau, walking is the transformative practice *par excellence* of space: "the long poem of walking manipulates spatial organizations, no matter how panoptic they may be [. . .] it is itself the effect of successive encounters and occasions that constantly alter it and make it the other's blazon: in other words, it is like a peddler, carrying something surprising, transverse or attractive compared with the usual choices" (101).

Cleography

And this brings me to the title of my afterword. "Walk like an Egyptian" is intended to invoke Certeau's "long poem of walking" as a gateway, or come-on, into the province of "Shakespace" that to my mind most insistently tugs in the direction of transversal territory. I refer here, of course, to the Egypt of *Antony and Cleopatra*. This Egypt is far more than a static territory on a map. As numerous readers have noted, the play may lend repeated expression to a divide between the Roman west and the Egyptian east, and in a fashion that recalls the Foucauldian cartography of Said's *Orientalism*: imperial Rome defines itself in geographical and cultural opposition to Egypt, whose carnivalesque "sport" contrasts Rome's Lenten "measure" and "restraint." But this orderly synchronic distinction is shown to be simply a Roman panoptic fantasy of "place" that Egyptian "space" works, with considerable *chutzpah*, to transversally move and shake.

For all the cartographic impulses to which it lends expression, *Antony and Cleopatra* also insists that Egypt cannot be pinned down and made the stable object of panoptic knowledge. Rome's rulers may try to impose order on Egypt's geographical and cultural strangeness, reassuring each other that "the flow o'th'Nile" can be measured "by certain scales i'th'pyramid" (2.7.17–18). But although Egypt may be a client state of the Empire, it repeatedly refuses Roman attempts to map it: like the "o'erflowing Nilus" (1.2.48), like Antony's amorous "dotage," Egypt "o'erflows the measure" (1.1.2). In such moments, Egyptian geography becomes what I term "Cleography." Cleopatra throughout the play functions as a metonymy for, and is often referred to as, "Egypt"; like her, this territory's "o'erflowing" Cleography is characterized by dynamic motion and "infinite variety" (2.2.243). ("Cleography" is related to, but not to be confused with, "Cleospace," a diachronic phenomenon that includes the infinite variety of textual, theatrical, cinematic and other cultural [mis]appropriations of the Egyptian Queen.) Egypt, like its Queen, keeps changing. "Here is my space," Antony says of Egypt; this is not the fixed geography of any mappable Egypt to which he refers, but the Cleographical space of moving and shaking. Crucial to the first scene's image of Egyptian space-in-motion, "stirred by Cleopatra," is the practice of walking. "We'll wander through the streets," Antony says, "and note / The qualities of the people" (1.1.45, 55–56). Here Antony proposes, in effect, to walk like an Egyptian. This walking is not a pseudo-anthropological labor designed to produce orderly, panoptic Roman knowledge of Egyptian subjects, but a mercurial "sport" that transforms both Antony and the space through which he wanders. Like Certeau's walking, Antony's perambulations expose him to "something surprising, transverse or attractive compared with the usual [Roman] choices" (Certeau 101). Wandering thus deflects Antony from the straight and narrow path of panoptic Roman "measure" into a "practiced place" of cultural and erotic novelties.

This might explain what it means for Antony to *walk* like an Egyptian, but what does it mean for him to walk like an *Egyptian*? When the 1980s Los Angeles girl-group The Bangles sang their hit song of the same name, they traded on an immediately recognizable image that has become Western popular culture's favored hieroglyph of the "Egyptian": a figure in profile moving with hands held at right angles to its serpentine arms, one bent arm pointing forward and upward, the other pointing down and backward. But in *Antony and Cleopatra*, to walk like an Egyptian is not to walk like anything as determinate or recognizable as The Bangles's Egyptian. If anything, it is a means by which Antony can *resist* determinacy. "Egyptians," just like Egypt, are impossible to pin down. Hence, when Enobarbus "describes" Cleopatra for the curious Agrippa and Maecenas, he paints a vivid, synaesthetic portrait of the detail surrounding her ("purple [...] sails," "tune of flutes," "silken tackle," "strange invisible perfume," "the feast" [2.2.200, .202, .216, .219, .231]), yet notoriously says nothing about her "own person" other than "it beggared all description" (2.2.204–05).[5] And "describing" the Egyptian crocodile for the curious

Lepidus, Antony likewise conjures up an equally indistinct creature:

> ANTONY   It is shaped, sir, like itself, and it is as broad as it hath breadth.
> It is just so high as it is, and moves with it own organs. It lives by that
> which nourisheth it, and the elements once out of it, it transmigrates.
> LEPIDUS   What colour is it of?
> ANTONY   Of it own colour too.
> LEPIDUS   'Tis a strange serpent. (2.7.41–47)

Lepidus moves transversally out of his subjective territory into new imaginative terrain—though what exactly he sees is hard to guess at, given the sweet tautologous nothings of Antony's description. This nonbeing is distinguished, nonetheless, by movement: Antony's tautologies cannot suppress the irresistibility of motion—the crocodile "moves with it own organs," and "the elements once out of it, it transmigrates." Such dynamism captures what it is to be "Egyptian" in *Antony and Cleopatra*. To walk like an Egyptian is to embrace not a determinate identity, but a metamorphic space of vagrancy. This space is conjured up also by the play's references to Cleopatra as a "gypsy" (1.1.10, 4.12.28), an abbreviated form of "Egyptian" that in early modern England had come to be associated not only with nomadism, but also with criminal impersonation.[6]

Egypt, in other words, is a space in which Antony and other Romans (including Lepidus) can move transversally out of their seemingly fixed subjective territories. This does not mean that their physical and imaginative movements from Rome to Egypt entail trading in one determinate identity for another; it involves rather a shedding of all fixed identities, or at least the pretence of any such identities. In this respect, Egyptian Cleography is a space that materializes Gilles Deleuze and Félix Guattari's visions of *becoming* as opposed to *being*. In *A Thousand Plateaus: Capitalism and Schizophrenia*, Deleuze and Guattari try to think outside the normative subject position of "man" by envisaging the state of "becoming-woman." This is not the same as *being* a woman, for such an ontological endpoint would replicate the (impossible) logic of fixed subjectivities. Rather, becoming-woman entails an openness to non-manliness, and to inhabiting in-between states of non-fixity. Caesar complains that Antony is "not more manlike than Cleopatra, nor the queen of Ptolemy more womanly than he" (1.4.5–6); yet Antony and Cleopatra's sexual role-playing, in which she wears his "sword Philippan" while dressing him up in her "tires and mantles" (2.5.23, .22), produces a fluid space of creative possibility rather than of moral degeneration. Even more insistently, the play addresses what Deleuze and Guattari describe as "becoming-animal." Cleopatra, "the serpent of the Nile," not only possesses that creature's shape-shifting, skin-shedding power, but also kills herself (and defeats Caesar) by nursing an asp at her breast. Snakes are not the only animals that she becomes—or that become her. Under the influence of mandragora, she imaginatively empathizes with Antony's horse: "O happy horse to bear the weight of Antony" (1.5.21). Cleopatra finds erotic pleasure not in retrieving some lost, originally unified self, then, but in communing with, internalizing, and becoming the animal not-self. What the play presents in all these instances are hybrid becomings, transversal movements (like those of the "inspriteful" Ariel of Reynolds and Ayanna Thompson's chapter on *The Tempest*) that "o'erflow the measure" of seemingly fixed identities.

Importantly, these becomings-woman and becomings-animal also lend Cleographical space a diachronic dimension at odds with the static tableaus of Roman cartography. Even if "age cannot wither her" (2.2.242), Cleopatra is by her own admission "wrinkled deep in time" (1.5.29); change is her element, inasmuch as Cleopatra is "whom everything becomes—to chide, to laugh, / To weep" (1.1.51–52). Everything "becomes" Cleopatra: even Romans transversally "become" Cleopatra in gazing on her, as did Pompey the Great, who would "anchor his aspect" gazing on Cleopatra's brow and "die / With looking on his life" (1.5.33–34). But these transformations also "become" her in the sense of "suiting" her, as they would any great actress. Indeed, becoming assumes a theatrical register throughout the play. When the vanquished Antony does his Joni Mitchell routine, looking at (and recognizing himself in) clouds from all sides now, he notably resorts to a theatrical image: he compares their changing forms, which like him refuse to hold any "visible shape," to "Black vesper's pageants" (4.15.14, .8). In their very insubstantiality, Antony recognizes the clouds' and his own transversal materiality; his twilight, actual and metaphorical, shows him that all the world's a (metamorphic) stage.

And this is the second way in which any attempt to subject Egypt to cartographic discipline—a project designed to produce Rome as Egypt's solid opposite—fails. If Egypt is the Cleographical space of becoming, it is equally the repudiated reflection of a Rome whose supposed "measure" and fixity are phantoms. Rome's twin ideals of phallic solidity—masculinity and marble—are demonstrable delusions: Antony loses his invulnerable manliness, just as the "scarce-bearded" Caesar will, and the marble of Roman civilization is already on the road to ruin. Cleographical space extends as much to Rome, then, as to Alexandria. When Cleopatra shudders at the prospect of seeing "some squeaking

Cleopatra boy my greatness / I'th'posture of a whore" (5.2.220–21), she is speaking about only one of the Roman practices of theatricality. Her transformation of the noun "boy" into a verb underlines the play's pervasive eclipsing of "being" by becoming; in Rome as much as Egypt, acting always trumps identity, even if it is not recognized as such. Yet Cleopatra's remark also serves to situate the play's original English audience as much as its Roman characters in the space of Cleography. In a powerful illustration of what D.J. Hopkins, Catherine Ingram and Bryan Reynolds term the acting and framing style of "Nudge, Nudge, Wink, Wink, Know What I Mean? Know What I Mean?" Cleopatra's remark would have allowed audience members in 1608 not only to hear her revulsion at being parodied by Romans walking like Egyptians, but also to acknowledge the performative codes that fashioned femininity on the early modern stage.

Cleopatra's metatheatrical remark provides an instance of the principle of translucency, the wonderfully suggestive hermeneutic model described by Donald Hedrick and Reynolds as the process by which audiences recognize that "one signifier or identity is incompletely concealed within another" (260). Indeed, this principle is arguably at work throughout the play, and *especially* in its characters' apparent assertions of anti-theatrical substantiality. In the last act, Cleopatra seems to gravitate toward a Roman ideal of fixity: "My resolution's placed, and I have nothing / Of woman in me—now from head to foot / I am marble constant" (5.2.238–40). At the same time, however, we can see how such masculinized Roman constancy is itself a theatrical performance, subject to the vicissitudes of "black vesper's pageants." Her Roman resolution requires theatrical props: "Show me, my women, like a queen. / Go fetch my best attires [. . .] bring our crown and all" (5.2.227–28, .232). In escaping capture by Caesar, Cleopatra offers here a signal illustration of Homi Bhabha's theory of colonial mimicry and hybridity, whereby "the insignia of authority becomes a mask, a mockery" (181). Donning such insignia, Cleopatra's transversal performance of imperial Roman solidity mockingly exposes the latter *as* performance, as a pageant proper to the space of Cleography.

"Like a queen": this remark provides an important clue as to how transversal space functions in *Antony and Cleopatra*. Antony *walks* like an Egyptian, and he walks like an *Egyptian*. But he also walks *like* an Egyptian. Simile is both the condition of identity in *Antony and Cleopatra* and what undermines it. Dressed theatrically for battle by Cleopatra, Antony leaves her "like a man of steel" (4.4.33). The clouds in which Antony sees his shapeless reflection are "sometime like a bear or lion" (4.15.3). Cleopatra eulogizes Antony and "his delights" as "dolphin-like" (5.2.88–89). And Caesar, we recall, describes Antony as "not more manlike" than Cleopatra. We might be reminded here of Judith Butler's analysis of Aretha Franklin's "You Make Me Feel Like A Natural Woman": "After all, Aretha sings, you make me feel *like* a natural woman, suggesting that this is a kind of metaphorical substitution, an act of imposture, a kind of sublime and momentary participation in an ontological illusion produced by the mundane operation of heterosexual drag" (27–28). *Antony and Cleopatra* similarly shows identity to be an "ontological illusion"; with the passkey of simile, Cleographical space expands the possibilities for transversal becomings and performances.

I often tease my American students that their much maligned penchant for the word "like" ("Like, Antony is *so* not Superman!" "Cleopatra is, like, *totally* a drama queen!") might actually predispose them to see how identity is a theatrical fiction rhetorically produced through simile. This fiction is most pointed, perhaps, when Antony is playing (like) Antony. The crocodile that Antony notoriously does not describe to Lepidus—"It is shaped, sir, like itself"—provides a template for Antony's own "ontological illusion," inasmuch as it suggests how self-likeness is simultaneously a mode of self-difference. As Cleopatra repeatedly casts Antony in the role of Antony—"Antony / will be himself" (1.1.44–45), "since my lord / Is Antony again, I will be Cleopatra" (3.13.186–87)—we are afforded glimpses of how "Antony" is not a solid core or authorizing origin, but only a sequence of *like*, but not self-identical, repetitions. For those invested in essential identities, this is the source of Antony's tragedy. But I am reminded here also of what Courtney Lehmann, Lisa Starks and Bryan Reynolds have to say about Tamora's "politics of affinity rather than identity" in *Titus* (338). The "likenesses" of *Antony and Cleopatra*, whether with an "other" or with one's "self," similarly suggest a politics of affinity— albeit a politics that entails a play of sameness and difference not just *between* subjects, but also *within* subjects.

If transversality is located anywhere, it is in this Cleographical space of "likeness." What *Antony and Cleopatra* radically suggests is that *all* subjective territories are enabled by transversality; to take on one's name, to internalize one's imago, the ego has to move (as Lacan argues) in a fictional direction signposted by simile: "*I* am like *that*." All successful interpellation thus demands a transversal movement, even if that movement is placed under erasure; the ontological illusion of being depends on, but tends to conceal, the "subjunctive space"—the "as if"—of becoming. If all cognitive territory is in a sense already transversal space, what matters is (once again) not only one's interpellated positionality, perhaps, but also one's *dis*positionality: how better to accept the metamorphic performativity of identity as a means to gaining more unfettered access to, and agency within, a Cleographical space of unbounded transversal possibility?

For Deleuze and Guattari, such access and agency follows from the perception of difference that underwrites "becoming-animal"—as the grounds not for stabilizing the self or for conquering the other, but for self-transformation (243). A transversal criticism, as the contributors to *Performing Transversally* show in their essays, will also seek to transform itself in response to its object. Likewise, in thinking through what it means to walk like an Egyptian, I want to move and shake my own critical kaleidoscope a little. What I have offered so far is a "dissective-cohesive" reading of *Antony and Cleopatra*, albeit one that opens up the transversal possibilities of the play. In investigative-expansive fashion, however, I will now transversalize my discussion by invoking a new frame of analysis. So, à la Cleopatra, I would like to change the scene. Let's sail up the Nile, like a queen, but not to meet Antony; let's sail up the Nile in time to a Cleography of another age: the Egypt of Amitav Ghosh's *In An Antique Land* (1992).

History in the Guise of a Traveler's Tale

In An Antique Land makes no explicit reference to *Antony and Cleopatra*. Yet coming to this book as someone who plays a Shakespearean for his wages, I cannot help but transform it within the terrain of Cleography, even as Ghosh's book has transformed my reading of Shakespeare's play. I read *In An Antique Land*, like *Antony and Cleopatra*, as a parable of transversal Egyptian space confronted with, and potentially divided by, the panoptic interpellations of imperialism. But if *Antony and Cleopatra* dramatizes the cartographic impulse of the Roman Empire, *In An Antique Land* obliquely engages the more recent histories of Western imperial conquest and partition that have worked to efface the transversal spaces, past, present and future, of the Middle East. In the process, Ghosh's book takes up what Bryan Reynolds and Janna Segal, speaking of Dario Fo's dramaturgy, challenge Shakespeare studies to do: "implement Fo's characteristically transversal approach to the text, the performance event, and social history, maintaining a fluid, interactive relationship with the past, present and emergent future" (360).

In An Antique Land is subtitled "History in the Guise of a Traveler's Tale." The traveler's tale, the Oxford English Dictionary (OED) tells us, is possessed of a "mendacious or incredible character." Othello's tales to Desdemona of anthropophagi and men whose heads grow beneath their shoulders is one example; Enobarbus's description of Cleopatra for Agrippa and Maecenas, or Antony's description of the crocodile for Lepidus, are variants of the genre. Like the map, the traveler's tale represents strange territories; but it rejects the cartographic impulse—it is aware of itself as a performance, in which the traveler himself appears as a character. The traveler's tale is thus a movement into transversal space, as opposed to a mapping of panoptic place. The "traveler" in Ghosh's book is at one level Ghosh himself, who as a doctoral student of anthropology in the early 1980s undertook fieldwork in a small village of Egypt. While recounting his interactions with the villagers, however, Ghosh simultaneously tells the story of another pair of travelers to Egypt in the twelfth century: a Jew, Abraham Ben Yiju, and his Indian slave and business associate, Bomma. Ghosh painstakingly pieces together their peregrinations from a paper trail of documents that had until the late nineteenth century been housed in the *geniza* of a synagogue in Fustat, now part of Cairo. Ben Yiju was a devout Jew, the son of a rabbi in North Africa; he was also a merchant, and his business took him to Aden, then to Mangalore in India's Malabar coast, where he lived for twenty years. While in India, he acquired Bomma, who subsequently accompanied Ben Yiju back to Aden and thence to Egypt.

Even as Ben Yiju "identified" as a Jew, he spoke and wrote in Arabic, and was well versed in Sufi traditions of mysticism. He also married a non-Jewish Indian woman, Ashu Nair, while living in Mangalore. *In An Antique Land* thus dramatizes a world of transversal movements across not just the Indian Ocean, but also across the boundaries of subjective territories that we have come to regard as mutually, and even violently, exclusive. In this medieval space, Muslim and Jew, Arab and Indian practiced what Lehmann, Starks and Reynolds term the "politics of affinity"—a transversal encounter that recognizes similarity across difference, and difference in similarity. Ghosh talks, for example, of the web of heterodox religious beliefs that placed medieval Hindu, Muslim and Jewish communities in dialogue. This web allowed the great Egyptian Jewish scholar Maimonides to write a Sufi tract. It also allowed Ben Yiju and Bomma to find affinities across boundaries of culture, religion, class and language: "those inarticulate counter-beliefs [. . .] eventually became a small patch of level ground between them: the matrilineally-descended Tulu and the patriarchal Jew who would otherwise seem to stand on different sides of an unbridgeable chasm" (263).

The case of Sidi Abu Hasira, a medieval "Muslim" saint who was also a Jewish rabbi, likewise suggests a transversal politics of affinity—and in this case, not just between Muslims and Jews, but also *within* Sidi Abu Hasira him "self." In looking for contemporary traces of the transversal spaces that he recognizes in the paper trail of Ben Yiju and Bomma, Ghosh is drawn to Sidi Abu Hasira's shrine in Damanhour, a city near Cairo. He arrives just after a *mowlid*, or festival, for the saint, but he is detained by authorities at the site, who cannot understand why an Indian should be interested in the shrine. As Ghosh is interviewed by a suspicious official, he has a revelation: "But then it struck me, suddenly, that there was nothing I could point to within his world that might give credence to my story—the remains of those small, indistinguishable, intertwined histories, Indian and Egyptian, Muslim and Jewish, Hindu and Muslim, had been partitioned long ago" (339). Ghosh's metaphor of "partition" harks back to his native India, and to the violent cartographical division of Hindus and Muslims that is the legacy of British colonialism. The sundering of India and Pakistan in August 1947 haunts Ghosh's book as a constitutive absence outside his traveler's tale; but this sundering has historical roots that he traces back to the imperial adventures of first the Portuguese, and later the Dutch and the British. In Ghosh's account, the Western imperial project in both India and Egypt produced new divisions of knowledge, new articulatory spaces of "history" and new interpellated—or partitioned—subjective territories. While undergoing interrogation during his detention, Ghosh "began to realize how much success the partitioning of the past had achieved; that I was sitting at that desk now because the mowlid of Sidi Abu Hasira was an anomaly within the categories of knowledge represented by those divisions. I had been caught straddling a border, unaware that the writing of History had predicated its own self-fulfillment" (340).

This historiographical (and cartographical) self-fulfillment may be the legacy of European colonialism, but Ghosh suggests also that it has been aided and abetted in the postcolonial era by the discourses of superpower imperialism. Ghosh tells of another run-in he has, with the Imam of the village in which he does his fieldwork. The Imam accuses Ghosh, a Hindu, of being a primitive cow-worshipper. Their conflict degenerates into a screaming match about which country, Egypt or India, has more advanced "guns and tanks and bombs":

> I was crushed, as I walked away; it seemed to me that the Imam and I had participated in our own final defeat, in the dissolution of the centuries of dialogue that had linked us: we had demonstrated the irreversible triumph

of the language that had usurped all the others in which people once discussed their differences. We had acknowledged that it was no longer possible to speak, as Ben Yiju or his Slave, or any one of the thousands of travellers who had crossed the Indian Ocean in the Middle Ages might have done... Instead, to make ourselves understood, we had both resorted, I, a student of the 'humane' sciences, and he, an old-fashioned village Imam, to the very terms that world leaders and statesmen use at great global conferences, the universal, irresistible metaphysic of modern meaning [...] It was the only language we had been able to discover in common. (236–37)

The language of superpower aggression, Ghosh argues, has become the dominant means of marking difference globally; as such, it is one of the principal impediments to the Cleographical spaces of becoming-other limned in *Antony and Cleopatra* and *In An Antique Land*.

Ghosh's argument with the Imam returns us to the frame of analysis with which Bryan Reynolds begins this book. The events of September 11, 2001 have licensed a potentially cataclysmic build-up of tension between West and East. What would Antony and Cleopatra, or Ben Yiju and Bomma, make of the rhetorical partitioning of subjective territories that Osamaspace (to use Reynolds's coinage), or Axisofevilspace, has sought to effect? Cleographical space, and its politics of affinity, suggests a transversal alternative to the newly universal language of "guns and tanks and bombs" that, Ghosh laments, has "usurped all the others in which people once discussed their differences." In not only espousing but also performing such a politics, *Performing Transversally* is much more than an intervention in Shakespeare studies; it is also a timely moving and shaking of what it means to claim, like Antony, "here is my space."

Notes

1. I have written elsewhere about the overlaps between Greenblatt's and George F. Kennan's discourses of "containment." See Jonathan Gil Harris, "Historicizing Greenblatt's Containment: The Cold War, Functionalism, and the Origins of Social Pathology," *Critical Self-Fashioning: Stephen Greenblatt and the New Historicism*, ed. Jürgen Pieters (New York and Frankfurt: Peter Lang, 1999), 150–73.
2. For a transversal reading of sociocultural spatiality in early modern London that expands on Mullaney's observations about the liminality of the public theater, see Bryan Reynolds, *Becoming Criminal: Transversal Performance and Cultural Dissidence in Early Modern England* (Baltimore: Johns Hopkins University Press, 2002), chapter 4, "Social Spatialization, Criminal Praxis, Transversal Movement," 95–124.

3. See Aristotle, "De Anima," *The Basic Works of Aristotle* Vol. 2, trans. Richard McKeon (New York: Random House, 1941), 555; Karl Marx, "Theses on Feuerbach," *Writings of the Young Karl Marx on Philosophy and Society*, trans. Lloyd D. Easton and Kurt H. Guddat (New York: Doubleday, 1967), 400. I have discussed the "material" of "material culture" elsewhere: see Jonathan Gil Harris, "The New New Historicism's Wunderkammer of Objects," *European Journal of English Studies* 4.3 (2000): 111–23; "Shakespeare's Hair: Staging the Object of Material Culture," *Shakespeare Quarterly* 52.4 (2001): 479–91; and "Atomic Shakespeare," *Shakespeare Studies* 30 (2002): 47–51.
4. See Bryan Reynolds and Joseph Fitzpatrick, "The Transversality of Michel de Certeau: Foucault's Panoptic Discourse and the Cartographic Impulse," *Diacritics* 29:3 (1999): 63–80. My own analysis of Certeau has been very much shaped by Reynolds and Fitzpatrick's.
5. For a fuller analysis of these descriptions of Egyptian nonbeings, see Jonathan Gil Harris, " 'Narcissus in thy face': Roman Desire and the Difference it Fakes in *Antony and Cleopatra*," *Shakespeare Quarterly* 45 (1994): 408–25.
6. Reynolds discusses how early modern English people literally "walked like Egyptians" as a consequence of the vogue for assuming the guise of itinerant gypsies in *Becoming Criminal*, chapter 2, "Becoming Gypsy, Criminal Culture, Becoming Transversal," 23–63.

Works cited

Aristotle. *The Basic Works of Aristotle*. Trans. Richard McKeon. New York: Random House, 1941.
Bhabha, Homi K. "Signs Taken for Wonders: Questions of Ambivalence and Authority under a Tree Outside Delhi, May 1817." *"Race," Writing, and Difference*. Ed. Henry Louis Gates, Jr. Chicago: University of Chicago Press, 1986. 163–84.
Butler, Judith. "Imitation and Gender Insubordination." *Inside/Out: Lesbian Theories, Gay Theories*. Ed. Diana Fuss. New York and London: Routledge, 1991. 13–31.
Certeau, Michel de. *The Practice of Everyday Life*. Trans. Steven Rendall. Berkeley: University of California Press, 1984.
Deleuze, Gilles and Félix Guattari. *A Thousand Plateaus: Capitalism and Schizophreni*. Trans. B. Massumi. Minneapolis: University of Minnesota Press, 1987.
Ghosh, Amitav. *In An Antique Land: History in the Guise of A Traveler's Tale*. New York: Vintage Books, 1992.
Harris, Jonathan Gil. "Atomic Shakespeare." *Shakespeare Studies* 30 (2002): 47–51.
———. "Shakespeare's Hair: Staging the Object of Material Culture." *Shakespeare Quarterly* 52.4 (2001): 479–91.
———. "The New New Historicism's Wunderkammer of Objects." *European Journal of English Studies* 4.3 (2000): 111–23.
———. "Historicizing Greenblatt's Containment: The Cold War, Functionalism, and the Origins of Social Pathology." *Critical Self-Fashioning: Stephen*

Greenblatt and the New Historicism. Ed. Jürgen Pieters. New York and Frankfurt: Peter Lang, 1999. 150–73.

Harris, Jonathan Gil. " 'Narcissus in thy face': Roman Desire and the Difference it Fakes in *Antony and Cleopatra*." *Shakespeare Quarterly* 45 (1994): 408–25.

Hawkes, Terence. "Telmah." *Shakespeare and the Question of Theory*. Ed. Geoffrey Hartman and Patricia Parker. London: Methuen, 1985. 310–32.

Marx, Karl. *Writings of the Young Karl Marx on Philosophy and Society*. Trans. Lloyd D. Easton and Kurt H. Guddat. New York: Doubleday, 1967.

Mullaney, Steven. *The Place of the Stage: License, Play and Power in the Renaissance*. Chicago: University of Chicago Press, 1988.

Reynolds, Bryan. *Becoming Criminal: Transversal Performance and Cultural Dissidence in Early Modern England*. Baltimore: Johns Hopkins University Press, 2002.

Reynolds, Bryan and Joseph Fitzpatrick. "The Transversality of Michel de Certeau: Foucault's Panoptic Discourse and the Cartographic Impulse." *Diacritics* 29.3 (1999): 63–80.

Shakespeare, William. *Antony and Cleopatra*. Ed. Michael Neill. Oxford: Oxford University Press, 1994.

Appendix: Transversal Poetics—I.E. Mode

zooz[1]

The languages of theories

ooz: Talking about something (some kind of "entity" for the lack of a better word) requires that you converse in a particular language. So, for example, let's say that a doctor wants to talk about a "human," she will most likely talk about this entity in the language of medicine—

zoo: —a human is a biological organism comprised of a stomach for digestion, lungs for oxygen supply, a brain to guide its behavior, and so forth.

ooz: Or, an economist might choose to use the terminology of post-Marxist thought. In this case he might analyze and talk about human entities as—

zoo: —production and consumption machines, whose thoughts and actions are governed by the bourgeoisie or dominant class, and whose "fruits of production" are also consumed by the aforementioned dominant class.

ooz: Alternatively, a Lacanian or Freudian literary critic may choose to use the language and tools of psychoanalysis to talk about a human as—

zoo: —the net sum of the actions of the superego, ego, and id—acting both on and over conscious and subconscious information. Such information includes, but is not limited to, memories, dreams, urges, and hallucinations.

ooz: Clearly, descriptions are severely affected by the language used.

zoo: That's true ooz, but really, talking about something goes much deeper than that. It's not just the language. These descriptions are not different simply in terms of the words being used. It's not like I say—

ooz: "the sky is blue,"

zoo: and you say,

ooz: "la ciel est bleue"

zoo: or

ooz: "blocha ra chroka."

zoo: No, its much larger than words; it's even the concepts and tools that are brought to bear on the entities. Is there some way to talk (more broadly) about the diverse worlds that each kind of analysis brings to the entities or questions under examination?

ooz: It is our aim here to share with you, the reader(s), one way that we have found to speak, in a general manner, about analyses and theories. To do this, we ask that you accompany us on an analytical journey into what we call "transversal space." We want to begin our travel with a statement, meant in the broadest terms. What makes two theories different is that they operate with and within different parameterizations.

zoo: For example, in the descriptions of humans we went through a minute ago, we would say that the doctor was operating within some kind of biological/medical-dissective parameterization, the economist was in a sociopolitical non-Keynesian-dissective parameterization, and the psychologist was working in a socio-wanna-be-medical-trying-to-be-scientific parameterization. The nature of different parameterizations is something we discuss in detail elsewhere.[2] However, to aid us in our present journey, we will briefly describe several aspects of parameterizations.

Parameterizations

ooz: First of all, as just emphasized, any theory or discussion will be about something or some things (be they body parts, humans, literary texts, and/or political systems). We move toward a spatiotemporal metaphor insofar as these "things" (entities and concepts) often exist in space and time; and therefore we talk about these things—the specific subjects of a particular investigation—as the chosen "area of exploration." It's that stuff which interests us. It's that region of spatiotemporal conceptual space into which we will venture and inside of which we will stroll, probe, navigate, and chart. As we journey along we will construct a matrix illustrating different aspects of parameterizations, the first column of which is shown in table A.1. The lists in each column are not ordered or correlated, but chosen to show the unlimited range of options.

Table A.1

Area of Exploration

Body Parts
Humans
Literary Texts
Political Systems
Molecules
Shakespeare[3]

zoo: Next, after the area of exploration is parameterized, the stuff within this area will be approached with some "critical mode" (see Table A.2). The critical mode is something that is always influenced by the sociopolitical environment in which the analysis takes place. Because our culture is saturated with the scientific method, many critical modes have some aspect of this broad method infused within their analysis. But then we have more specific critical modes, such as "deconstructionist," "new historicist," "behaviorist," or "medical" modes.

Table A.2

Area of Exploration	*Critical Mode*
Body Parts	Scientific Method
Humans	Deconstructionist
Literary Texts	New Historicist
Political Systems	Behaviorist
Molecules	Medical
Shakespeare	Astrology[4]

ooz: Of course, because the items in each column of our table are not in any particular order, almost any critical mode can be applied to an area of exploration—the theorist's mode of choice will be guided by primarily pragmatic considerations. Each different critical mode, when applied to a chosen area of exploration, will privilege some fundamental parts—some elementary units. For example, when a medical critical mode is applied to humans, such elementary units as body organs might be discussed. Or when a behaviorist mode is applied to the same area of exploration, such elementary units as stimuli and responses might be the focus. We chose to keep our classificatory terminology as relative and general as possible; thus we refer to these elementary units as "elems" (see Table A.3). A final example might help make this clearer: physicists, using largely the scientific method, might posit elems of molecules, or perhaps electrons, or even quarks or strings.

Table A.3

Area of Exploration	Critical Mode	Elems
Body Parts	Scientific Method	Characters
Humans	Deconstructionist	Electrons
Literary Texts	New Historicist	Body Organs
Political Systems	Behaviorist	Stimuli/Responses
Molecules	Medical	Ecosystems
Shakespace	Astrology	Concepts

zoo: Furthermore, a theory or analysis will usually also determine or declare certain "laws" that operate on the elems. These laws will specify how the elems act or interact and how such actions might lead to more complex entities. We call such complex entities "assemblages." For example, a literary analysis might decide that the elems of texts are the characters of the text and that the laws by which they interact are simply folk-psychological modes of interaction. By following these laws, the character elems form the larger assemblages known as situations and plots. Table A.4 shows some possibilities for each aspect of a parameterization.

Table A.4

Area of Exploration	Critical Mode	Elems	Laws	Assemblages
Body Parts	Scientific Method	Characters	Moral Codes	Governments
Humans	Deconstructionist	Electrons	Grammar	Plots
Literary Texts	New Historicist	Body Organs	Thermodynamics	Languages
Political Systems	Behaviorist	Stimuli/Responses	Folk Psychology	Humans
Molecules	Medical	Ecosystems	Law of Demand	Automobiles
Shakespace	Astrology	Concepts	Karma	zooz[5]

ooz: The schema that you see in table A.4 is, zooz believes, a general way to delineate any theory. Specifically, when you conduct or participate in a theoretical analysis of any kind, you necessarily function within a parameterization that will channel your analysis.

zoo: To be sure, I don't mean to imply that all theoreticians are cognizant of the parameterization(s) in which they operate. Unfortunately, this is often not the case: the theoretician operates largely unaware of their own parameterization and the constraints it imposes. Consequently, many analyses are likely to end up in what zooz calls a "progressive quagmire." According to zooz,

> Progressive quagmires are research states, indeed states of being (if you will), that are manifest when the analytical tools which were believed to fuel progress prove unable to resist the analysis' momentum and thus are incapable of generating new directionality and expansion because the analysis is pushed along a rigid course. And yet the researcher, having experienced past "successes" with them, and being urged on by social, cultural, political, and paradigmatic conventions, is reluctant to part with these investigative tools and, by extension, the (de)limiting assumptions underlying them.

Parameterization spaces

ooz: We launched this journey by talking about theories and then taking a "step back" to see a bigger picture. By doing this we have gained an understanding of parameterizations and the ways in which they form and inform analyses. Now we would like to take yet another step back and consider the nature of parameterizations.

zoo: Moving again into a spatiotemporal metaphor, we can think of each aspect of a particular parameterization (area of exploration, critical mode, and so on) as defining a sort of "dimension." For example, a psychologist might operate on a "dimension of human behavior."

ooz: Here zooz wants you thinking of a big open field filled with "human behaviors."

zoo: A chemist might be operating in a space that is formed by the intersection of the "molecules dimension" and the "scientific method dimension."

ooz: Imagine a glass cube filled with people in white lab coats using magnifying glasses to look at tiny dust motes.

zoo: All of these images are fine until we, as humans, have to imagine the intersection of many of these dimensions at once. Our three-dimensional "mind's eye"—

ooz: (probably embodied in parietal- and frontal-lobe structures operating on visual cortex over space and time)

zoo: —just can't handle it. Or can it?

ooz: I think so. Let's step back further from this madness for a minute. Any parameterization can be described as some "subspace" that has been carved out of an infinitely expansive and infinitely smooth higher-dimensional space.

zoo: Yes, but it's important to realize that a subspace is really just one of an infinite number of possible slices taken from a higher-dimensional space. It is like we have a big room that is full of millions and millions of little glass cubes each representing a different parameterization subspace.

ooz: There is something amiss here, right? Even in the image of the glass-cube filled room it seems obvious that each theorist, operating within their own parameterization cube, is only able to move into and observe a small piece of the larger space.

zoo: Precisely, ooz. (Isn't it great that we tend to agree on all things?) The theorists are entangled within and confined by the static parameterizations of their analysis—they are unable to transverse the space between their analytic parameterization and the parameterizations of other analysts working nearby.

ooz: And this is a problem because there is much to be gained by being both willing and able to move into neighboring spaces of parameterization. There is power in this agency, a power in the sharing, awareness, sociality, and transformation that such mobility enables and encourages. We refer to this blurring of boundaries and slipping through the smooth "transversal space" between and among parameterizations as "transversal movement." We call the forces that inspire and realize such movements "transversal power." We will explain these terms as we progress.

Transversing parameterization spaces

zoo: But how can such a space be entered? In other words, in the realm of analysis, how do we engage in transversal movement?

ooz: Well, zoo, we must recall what got us into our little glass cubes in the first place.

zoo: As zooz mentions elsewhere, it is the static and narrowing nature implicit in the concept of parameterization that encumbers us and encloses us in these limited analytic and experiential spaces.[6]

ooz: It's as if thinkers thought that critical approaches (such as the scientific method and the new historicism) would set them free, in turn allowing them to examine the whole space. But in endowing themselves with the ability to explore, they are at the same time making alternative subspaces inaccessible, as well as constraining the area they can explore. This is because their particular critical approach is so narrowly tailored to their particular area of exploration that it fails to yield useful insights when applied to other areas of exploration. Metaphorically, then, we can talk about their circumscribed analysis as movements in a "two-dimensional subspace" that has been sliced from a larger, richer, interconnected "three-dimensional space."

zoo: How then can an analysis become transversally empowered?

ooz: The answer is simple: just open up the boundaries of parameterization. But don't just mindlessly and irresponsibly open the boundaries. Instead, engage with and take control of the boundaries; treat them not as limits but as gateways through which an investigation can pass in all directions—lineally, laterally, synchronically, diachronically.

zoo: Yes! Become transversally empowered and navigate your parameterizations as you explore something.

ooz: You must become, as zooz writes elsewhere,—

zoo: —like a vintner on grapes, dancing upon the critical mass in question, obliterating any pretense of two dimensionality or totalization, complicating the subject of investigation, and causing constant eruptions out of all dissective-cohesive two-dimensional spaces.[7]

ooz: Here, dissective-cohesive refers to the spaces created by critical modes, like the scientific method, that breakdown the subject matter under investigation into constituent parts in hopes of ultimately piecing it back together as a comprehensible totality. However, because all things share an interconnectedness that gives them multidimensionality, such reduction is only ever a fantasy. It is a fantasy that arises because investigators inhabiting two-dimensional conceptual spaces see only shadows cast from spaces with more dimensions, and believe that such reduction is possible. It is the impossibility of achieving such a goal that suggests two-dimensional space.

zoo: That is, a dissective-cohesive critical mode disallows what we call an "investigative-expansive mode of analysis," which always takes into consideration the contingency and relatedness of all things, and remains ever ready to reparameterize in response to glitches and new discoveries throughout the course of the analysis. Thus, whereas the dissective-cohesive researcher finds herself spinning her wheels in a progressive quagmire, the investigative-expansive researcher moves into and out of transversal space.[8]

ooz: But wait, isn't this whole approach fairly similar to what is often referred to as an "interdisciplinary" analysis?

zoo: Ah, we do see how one might be tempted to misinterpret transversal analysis as a form of interdisciplinary analysis. However, the radical nonlinearity of transversal analysis makes it a different kind of beast altogether, even though it necessarily enjoys much interdisciplinarity.

ooz: Yes, I see what you mean. Interdisciplinary analysis does tend toward the linear. The difference is metaphorically like the difference between taking different routes to get home versus taking different routes, and possibly different vehicles, on a trip without a specific destination in mind.

zoo: Exactly. Although interdisciplinarians (like transversalians) do utilize the tools and methods of many different fields (consider the new historicists, for instance), they tend to have a fixed goal set for each investigation to which they are aiming. Transversal analysis, on the other hand, roams freely over/through/around/and out of realms of analysis.

ooz: Transversal analysis values the journey—the transversal movements—more than any final goal.

zoo: In fact, there is no final goal (just as there is no final Truth), there is only the journey and all the gems that are collected along the way.

ooz: Transversal movements occur, then, when any aspect of a parameterization is blurred, crossed, or shattered. Transversal power is obtained through such movements, possibly making subsequent transversal movements possible.

zoo: These movements are often provoked when new observations are made that challenge the current parameterization (consider the parametric impact of early observations made with the telescope).

ooz: However, new observations are not always necessary for transversal movements to occur. Ideally, this process happens when a researcher or theorist makes a conscious effort to move transversally.

zoo: However, there are many cases where transversal eruptions can force even the most rigidly parameterized analysis into transversal space. Such eruptions are similar to those paradigm-shifting moments referred to by Thomas Kuhn as a "scientific revolution."

ooz: We are thinking of analyses like classical physics before Einstein came along—or the literary analysis of the new criticism before the new historicism of Greenblatt and his clan—or even the dimensionally constrained music of Europe before Bach or Charlie Parker.

zoo: As this last example suggests, transversal power is not just something that a theory can have. In fact, transversal power is manifested and perhaps even conducted whenever parameterizations are being bent, grown, or surfed. The parameterization in question might be related to a theory (like we have been talking about); but, for instance, the actions of a human also take place within a particular parameterization.

ooz: Of course, when our area of exploration is human actions, the parameterization in question does not specify a part of, say, "theory-space" (like our glass cubes from before). Instead the parameterization delimits and usually limits what we refer to as "subjective territory."

zoo: Subjective territory, as some of us have written elsewhere,

> is delineated by conceptual and emotional boundaries that are normally defined by the prevailing science, morality, and ideology.... It is the existential and experiencial realm in and from which a given subject of a given hierarchical society perceives and relates to the universe and his or her place in it. (Reynolds, "Devil's House" 146–47)

ooz: Even though we, as subjects, are in a different kind of territory (subjective rather than theoretic), transversal power operates in much the same way. Whenever the bounds of a territory are liberated, transversal power is present and transversal movement has begun.

zoo: In subjective territory, it is through movements such as overcoming shyness, transgressing gender roles, acting in theater, kinds of interdisciplinary analysis, and political, cultural, and intellectual iconoclasm that transversal power rears its head. The individual making these movements can become a conductor and move out of his or her circumscribing subjective territory, and into what we call "transversal territory."

ooz: In zooz's words,

> Transversal territory is entered through the transgression of the conceptual boundaries and, usually by extension, the emotional boundaries of subjective territory. (Reynolds, "Devil's House" 149)

zoo: Transversal territory is entered when subjective territory is either departed, expanded, or subversively intersected. Transversal power, as a mechanism for experiential alternatives, leads to and emerges from the dynamism, enunciation, and amplification of transition states. Transition states can be manifest and transversal power can be released in such diverse ways as heating water, intoxication by hallucinogenic drugs, military draft resistance, or heresy.

ooz: How about a specific historical example?

zoo: Okay. Consider the early modern English public theaters that operated from 1576 until 1642, when they were closed down by the state. As one of us argues elsewhere, these theaters were channels for transversal power because they popularized for the first time in English history the idea that one's social identity could be changed because it only exists insofar as it is convincingly performed. In other words, one's social identity is always already changeable and must be performed to be. Like the actors on the stage, who functioned within the stage illusion, it became popular in early modern England for people of one class or gender to pretend to be—to act and pass as—someone of another class or gender. For this reason, the state passed laws against people cross-class and cross-gender dressing on the street.[9]

ooz: Generally speaking, we think transversal power is a fundamental force of change, for groups as well as for individuals.

zoo: It promotes revolution, as well as reconfiguration.

ooz: It is that through which movement and exploration typically occurs.

zoo: Now that you have an idea of what transversal power is—

ooz: —the power of metamorphosis that both fuels and is fueled by the transversing of parameterizations—

zoo: —and some examples of how it can emerge through theories and actions, we will take you to the next stage of this analytic journey. This journey, or perhaps it is better thought of as a "trip," will be running on transversal power.

ooz: Please unfasten your seatbelts.

zoo: And we will examine some ways by which transversal power is both existent and created in the world around us.

ooz: We would like to take you on an excursion . . .

From molecules, to Jesus Christ, to The Grateful Dead, and beyond

zoo: We only have several pages to cover several millenia, so this will have to be brief.

ooz: Because transversal theory insists on the arbitrariness of any investigative sequence, we could begin our analysis anywhere.

zoo: How about molecules as our initial elems? They have worked well as elementary units for biochemists.

ooz: Okay, then molecules it is. Let's imagine that these little chaps are just sort of hanging out in the ocean, minding their own business—

zoo: (pardon the anthropomorphism)

ooz: —but, over time, these elems will form (through what some might describe as laws of thermodynamics, entropy, and evolution) large-scale assemblages, such as single-celled beings.

zoo: It may not be clear yet, but when these elems become these particular kinds of assemblages, transversal power crystallizes. These assemblages (and their actions) are outside the area of exploration, and thus beyond the parameterizations of traditional chemistry and thermodynamics.

ooz: To follow these new assemblages, we must transversally glide into parameterizations that talk about biological entities. No longer will molecules be the elems of our analysis. Now we have to have new elems. These elems can be thought of, in some very misleading sense, as higher-level elems. They might be things like "males and females" or "edible and inedible" or "cats and spiders."

zoo: In this particular case, transversal power was actualized when entities emerged that fell outside the bounds of our current analytic realm. Consequently, we were forced to reparameterize. Over billions of years, and perhaps innumerable transversal movements and reparameterizations, we could arrive at humans.

ooz: Once we have chosen humans as the elems of our analysis, something new appears.

zoo: Because humans have will and intentionality, they can intentionally choose to engage in transversal movements. They can decide to move outside the parameterizations that have enclosed them in their subjective territories. But there is a problem here: often humans are unaware that they are living within a subjective space—or, what is more often the case, they are unaware that any realms exist outside of their subjectified parameterizations.

ooz: It is here, at the level of human experience, that conductors of transversal power become essential. A conductor can be, among other things, an assemblage that empowers other assemblages; it can wake them up to their parameterizations, and inform them of the strength that can be found in transversal movements.

zoo: Sometimes these conductors are individuals. As an example, consider the man known as Jesus of Nazareth.

ooz: This assemblage/conductor lived in a world where subjective territory was fastidiously maintained by the traditions, laws, and edicts of Judaism.

zoo: This individual spread transversal power by trying to shake the individuals of his day out of their parameterizations. He did this with words ("turn the other cheek") and actions (by, for example, raising the dead). He advocated another way of life, a way that seemed alien to the subjectified populace, a way that could lead them out of their subjectification and into new territory.

ooz: He advocated a reconceptualization of the individual and his or her behavior. The individual, the elem of this analysis, could, Jesus said, become a part of a larger assemblage: the Kingdom of Heaven.

zoo: Insofar as he inspired transversal movement, Jesus was an individual who functioned as a conductor of transversal power. But, it is not only individuals who are capable of conducting transversal power and producing transversal movement. Transversal power is potentially present and potent at all levels of analysis, and at all scales.

ooz: A final example should make this clear. Consider the larger-scale sociopolitical assemblage known as The Grateful Dead.

zoo: Active for almost forty years, The Grateful Dead are more than just a rock-n-roll band, more than just a group of people hanging out and playing music (incidentally, the assemblage is continuing in various forms without their founder Jerry Garcia, who died in 1995). Because of The Grateful Dead's massive popular and countercultural appeal, they are an assemblage that has influence over other assemblages at many levels, including the physical, aesthetic, philosophical, and ideological.

ooz: Through their phenomenal presence and power they work to create environments where transversal power is continually unleashed and conducted.

zoo: Recreational drug use, nomadism, and iconoclasm all characterize The Grateful Dead movement.

ooz: Jesus of Nazareth and The Grateful Dead, as zooz puts it, are vehicles and models for experimentation and adventurousness. They are conductors of transversal power, and calls for transversal movement.

Coda

The futility of reduction was once eloquently represented by Ludwig Wittgenstein:

"In order to find the real artichoke, we divested it of its leaves."[10]

Ludwig, you took the task to heart; there is no artichoke without the leaves. But if the leaves are left behind:

> Andouille Fire Roasted Artichoke & Red Pepper Salad
> Crab & Shitake Ragout in Pan Seared Artichokes
> Artichoke Ravioli in Red Wine Truffle Sauce
> Curried Artichokes with Mango Yogurt Dip
> Fried Artichokes with Spicy Tamarind Chutney
> Rosemary Dripped Artichoke with Gorgonzola Sauce

Although it may be the heart that matters most,
No order can peel the real.
Transversal table etiquette is shared by zooz:
"Forget the real. Keep on plucking and eating, and trying new sauces."

Notes

1. zooz is a pseudonym for the assemblage of "elems" more commonly known as James Intriligator and Bryan Reynolds, or vice versa. For a definition of "elems," see "Transversal Poetics: I.E. Mode," 3–4.
2. See Stephen Kosslyn and James Intriligator, "Is Cognitive Neuropsychology Plausible? The Perils of Sitting on a One-Legged Stool," *Journal of Cognitive Neuroscience* 4.1 (1992): 96–106. Italian translation by Maurizio Riccucci under the title, "La neuropsicologia cognitiva è plausibile? Ovvero i rischi di sedere su uno sgabello a una gamba sola," *Systemi Intelligenti* 6.2 (1994): 181–205.
3. Donald Hedrick and Bryan Reynolds define "Shakespace" in their introductory essay to *Shakespeare Without Class: Misappropriations of Cultural Capital*, ed. Donald Hedrick and Bryan Reynolds (New York: St. Martin's Press, 2000), 3–47: "The historical spaces through which Shakespeare has passed as an icon have been diverse and numerous. Shakespeare has stimulated, occupied, and affected countless commercial, political, social, and cultural spaces. These Shakespearean or Shakespeare-influenced spaces, however conventional, alternative or transversal, we refer to as 'Shakespace,' a term that accounts for these particular spaces and the time-speed, the 'pace', at which they move from generation to generation and era to era" (10).
4. According to astrology, zooz's zodiac sign would be Pisces–Cancer or, conversely, Cancer–Pisces, making zooz an "assemblage" of character traits

300 \ APPENDIX

determined by the stars. For a definition of "assemblages," refer to "Transversal Power: I.E. Mode," 4.
5. Please see note 1.
6. See Donald Hedrick and Bryan Reynolds, "Shakespace and Transversal Power," *Shakespeare Without Class: Misappropriation of Cultural Capital* (New York: St. Martin's Press, 2000), 3–47.
7. For a detailed discussion of the necessity of breaking out of two-dimensional spaces in the field of cognitive neuroscience, see Stephen Kosslyn and James Intriligator, "Is Cognitive Neuropsychology Plausible? The Perils of Sitting on a One-Legged Stool," *Journal of Cognitive Neuroscience* 4.1 (1992): 96–106.
8. For examples and explanations of critical applications of the investigative-expansive mode, see: Bryan Reynolds, *Becoming Criminal: Transversal Performance and Cultural Dissidence in Early Modern England* (Baltimore: Johns Hopkins University Press, 2002); Bryan Reynolds and Joseph Fitzpatrick, "The Transversality of Michel de Certeau: Foucault's Panoptic Discourse and the Cartographic Impulse," *Diacritics* 29.3 (2000): 63–80; and Bryan Reynolds and Joseph Fitzpatrick, "Venetian Ideology or Transversal Power?: Iago's Motives and the Means by which Othello Falls," *Critical Essays on Othello*, ed. Philip Kolin (New York: Routledge, 2002), 203–19, and in *Performing Transversally: Reimagining Shakespeare for the New Millennium* (New York: St. Martin's Press, 2002).
9. See Bryan Reynolds, *Becoming Criminal*, and Bryan Reynolds, "The Devil's House, 'or worse': Transversal Power and Antitheatrical Discourse in Early Modern," *Theatre Journal* 49.2 (1997): 143–67.
10. Ludwig Wittgenstein, *Philosophical Investigations* (Oxford: Basil Blackwell, 1986), 66E: 164.

Works cited

Hedrick, Donald and Bryan Reynolds. "Shakespace and Transversal Power." *Shakespeare Without Class: Misappropriations of Cultural Capital*. Ed. Donald Hedrick and Bryan Reynolds. New York: St. Martin's Press, 2000. 3–47.

———. "La neuropsicologia cognitiva è plausibile? Ovvero i rischi di sedere su uno sgabello a una gamba sola." Trans. Maurizio Riccucci. *Systemi Intelligenti* 6.2 (1994): 181–205.

Kosslyn, Stephen and James Intriligator. "Is Cognitive Neuropsychology Plausible? The Perils of Sitting on a One-Legged Stool." *Journal of Cognitive Neuroscience* 4.1 (1992): 96–106.

Reynolds, Bryan. *Becoming Criminal: Transversal Performance and Cultural Dissidence in Early Modern England*. Baltimore: Johns Hopkins University Press, 2002.

———. "The Devil's House, 'or worse': Transversal Power and Antitheatrical Discourse in Early Modern England." *Theatre Journal* 49.2 (1997): 143–67.

———. "Venetian Ideology or Transversal Power?: Iago's Motives and the Means by which Othello Falls." *Othello: New Critical Essays*. Ed. Philip Kolin. New York: Routledge, 2002. 203–19.

Reynolds, Bryan and Joseph Fitzpatrick, "The Transversality of Michel de Certeau: Foucault's Panoptic Discourse and the Cartographic Impulse." *Diacritics* 29.3 (2000): 63–80.
Wittgenstein, Ludwig. *Philosophical Investigations*. Oxford: Basil Blackwell, 1986. 66E: 164.

Notes on Collaborators

JOSEPH FITZPATRICK is a doctoral candidate in English at Duke University. His dissertation addresses metaphors of organicism and mechanization in British, American, and Russian modernist texts. He is also currently working on a study of montage techniques in Joyce, Dos Passos, and Kataev. With Bryan Reynolds, he has published articles on transversality and Michel de Certeau, Michel Foucault, and Shakespeare (fitzpat@duke.edu).

DONALD HEDRICK is Professor of English at Kansas State University and founding director of the Program in Cultural Studies. He has been a visiting professor at Cornell University, Amherst College, and the University of California, Irvine, and held the O'Connor Chair of Literature at Colgate University. With Bryan Reynolds he is coeditor of *Shakespeare Without Class: Misappropriations of Cultural Capital* (Palgrave 2000). His publications include articles on architecture, Renaissance drama, film, and cultural theory. His most recent article, "Advantage, Affect, History, and *Henry V*," is forthcoming in the "Imagining History" special issue of *PMLA*. Professor Hedrick is currently working on projects on movie trailers, and on valuation and entertainment in the early modern period (hedrick@ksu.edu).

D.J. HOPKINS is a doctoral candidate in the UCI/UCSD Joint Ph.D. Program in Theatre and Drama. His publications have appeared in *Modern Drama, Theatre Topics, TheatreForum, Theatre Survey, Theatre InSight*, and in the collection *Shakespeare After Mass Media* edited by Richard Burt. Hopkins also holds an M.F.A in dramaturgy from UCSD (dhopkins@ucsd.edu).

CATHERINE INGMAN is an actor, currently completing her M.F.A. at the University of Washington (2003). She has also trained and performed at the Actors Theatre of Louisville, the Guthrie Theater, and in New York. She earned her B.A. in English at Harvard University, where her primary studies were early modern English drama and performance theory (cathingman@yahoo.com).

DR. JAMES INTRILIGATOR has been wandering transversally through conceptio-space–time for over thirty-five years. Spatiotemporally, he has meandered through Los Angeles, San Diego, London, Boston, Lyon, and Culver City. Conceptiotemporally, he has strolled through acting, writing, computer science, mathematics, art, philosophy, psychology, cognitive neuroscience, business, and consulting. Highlights include degrees from the University of California (B.A. in Philosophy and Psychology) and Harvard University (Ph.D. in Cognitive Neuroscience), marriage to his wife (Susanne), the birth of his son (Eli), and the upcoming birth of his daughter (Ella). Tangible artifacts he has left behind include over fifty articles/presentations appearing in leading scientific journals/conferences, several U.S. patents, and a permanent interactive museum exhibit in Switzerland (jamesi@post.harvard.edu).

COURTNEY LEHMANN is Assistant Professor of English and Film Studies at the University of Pacific, and she is the Director of the Pacific Humanities Center. She is an award-winning teacher and the author of *Shakespeare Remains: Theater to Film, Early Modern to Postmodern* (Cornell University Press, 2002), as well as coeditor, with Lisa S. Starks, of two volumes of Shakespeare and screen criticism: *Spectacular Shakespeare: Critical Theory and Popular Cinema* (Fairleigh Dickinson, University Press, 2001) and *The Reel Shakespeare: Alternative Cinema and Theory* (Fairleigh Dickinson, 2002). Her most recent publication, "Crouching Tiger, Hidden Agenda: How Shakespeare and the Renaissance are Taking the Rage out of Feminism," is an analysis of Shakespeare, film, and feminism in *Shakespeare Quarterly* (Summer 2002) (clehmann@uop.edu).

BRYAN REYNOLDS is Associate Professor and Head of Doctoral Studies in Drama at the University of California, Irvine. He is the author of *Becoming Criminal: Transversal Performance and Cultural Dissidence in Early Modern England* (Johns Hopkins University Press, 2002), coeditor, with Donald Hedrick, of *Shakespeare Without Class: Misappropriations of Cultural Capital* (St. Martins/Palgrave, 2000), and has published articles on the work of Shakespeare, Cixous, Certeau, Dekker, Deleuze, Foucault, Guattari, Middleton, Polanski, Rousseau, and Robert Wilson. He is also a playwright, a director of theater, and a screenwriter (breynolds@uci.edu).

JANNA SEGAL is Associate Instructor and doctoral student in the UCI/ UCSD joint Ph.D. program in Drama and Theatre. She completed her B.A. in Theatre at the University of California at Santa Cruz, and her M.A. in Theatre Literature, History, Criticism, and Theory at California State University, Northridge. Her essay, written with Bryan

Reynolds, "The Reckoning of Moll Cutpurse: A Transversal Enterprise," is forthcoming in Craig Dionne and Steve Mentz eds, *Rogues in Early Modern English Literary Culture* (University of Michigan Press, 2003) (segalj@uci.edu).

LISA S. STARKS is Assistant Professor of English at University of South Florida, St. Petersburg. She has published works on Marlowe, Shakespeare, cinema, and psychoanalysis, with articles in *Marlowe, History, and Sexuality* (ed. Paul Whitfield White, 1998) and *Shakespeare and Appropriation* (ed. Christy Desmet and Robert Sawyer, 1999). Starks's articles have also been published in *Literature and Psychology, Early Modern Literary Studies*, and *Post Script: Essays in Film and the Humanities*. Starks has also guest-edited two special issues of *Post Script* on Shakespeare and film, and has coedited, with Courtney Lehmann, two book collections on the subject: *Spectacular Shakespeare: Critical Theory and Popular Cinema* (Fairleigh Dickinson University Press, 2002) and *The Reel Shakespeare: Alternative Cinema and Theory* (Fairleigh Dickinson University Press, 2002). Her most recent article, "'Remember Me': Psychoanalysis, Cinema, and the Crisis of Modernity," is in a special issue of *Shakespeare Quarterly* on Shakespeare and cinema (Summer 2002) (starks@stpt.usf.edu).

AYANNA THOMPSON is Assistant Professor of English Literature at the University of New Mexico, where she teaches courses on Renaissance and Restoration drama. Interested in the production of racial identity and the theatrical violence that often accompanied these early modern constructions, she has published articles on Shakespeare and contemporary film adaptations of his works. She is currently completing her first monograph, *Staging Torture: Constructing Race and Empire in Early Modern Theatre* (thompson@unm.edu).

ZOOZ: for over ten years zooz has been gliding through transversal spaces. Having origins in the spatiotemporally limited interactions between, amongst, and around Bryan Reynolds and James Intriligator, zooz has now managed to transcend space–time (as all transversal conductors must). Although the core physically coincident nexus of energy exchanges are sporadic, zooz continues to extend into, across, beyond, and out of contemporary, past, future, and ever-changing conceptual-active spaces. zooz has an article forthcoming in *GESTOS: Teoría y Práctica del Teatro Hispánico*, and is currently exploring such spaces as interpersonal companionship, interpersonal consciousness, and extrapersonal culture.

Additional contributors

JONATHAN GIL HARRIS is Professor of English at George Washington University, where he teaches Shakespeare, Renaissance Literature, and critical theory. He is the author of *Foreign Bodies and the Body Politic: Discourses of Social Pathology in Early Modern England* (Cambridge, 1998) and coeditor, with Natasha Korda, of *Staged Properties in Early Modern England* (Cambridge, 2002). He has just finished a new book entitled *Etiologies of the Economy: Dramas of Mercantilism and Disease in Shakespeare's England* (harrisj@ithaca.edu).

JANELLE REINELT is Associate Dean of the Claire Trevor School of the Arts at the University of California, Irvine. She is a former editor of *Theatre Journal*, and the vice president of the International Federation for Theatre Research. Her books include *After Brecht: British Epic Theatre, Critical Theory and Performance*, with Joseph Roach, *Crucibles of Crisis: Performance and Social Change*, and, with Elaine Aston, *A Cambridge Companion to Modern British Women Playwrights* (jreinelt@uci.edu).

Index

abjection, 14–15, 218, 222, 225, 226–7, 237, 239nn. 14, 15, 16, 239–40n. 17
 see also Kristeva, Julia
absent spaces, 7, 8
Adelman, Janet, 54–6, 78n. 8, 89, 90, 95–6
affective presence, 8, 27n. 8, 191, 195, 209n. 8
Almereyda, Michael
 Hamlet (dir. Almereyda), 139, 141, 156–9, 161
 Nadja (dir. Almeredya), 157
Al Qaeda, 12, 17, 21
 see also Osamaspace
Althusser, Louis, 23n. 1, 23–4n. 2, 24–6n. 4, 56, 58, 64, 67, 76, 78n. 13, 79n. 17
Anderlini-D'Onofrio, Serena, 266n. 16
Anderson, Laurie, 29, 34
antitheatricalists, 159
Antony and Cleopatra,
 see Shakespeare, William
Ariel, *see* Rodó, José Enrique
Aristotle, 15, 274, 285n. 3
Artaud, Antonin, 112, 118, 121–4, 220, 227, 238n. 6
articulatory spaces, 2, 7–8, 9, 271, 273, 283
As You Like It, *see* Shakespeare, William
 see also Gosnell, Raja
Atkins, Susan, 111, 122, 130

Auslander, Philip, 50n. 1, 160, 164n. 5
Austin, J.L., 166n. 14
author-function, xvii, 45–9, 51n. 9, 141
authorship, 12–23, 45–9

Bach, Johann Sebastian, 294
Bakhtin, Mikhail, 255, 267n. 22
Bangles, The, 276
Barrymore, Drew, 140, 150
Barthes, Roland, xiv, 40
Bartkowiak, Andrzej
 Romeo Must Die (dir. Andrzej Bartkowiak), 140
Barton, John, 144, 211n. 29
Bate, Jonathan, 211n. 31, 239n. 11
Bateson, Gregory, 147
The Beard of Avon, *see* Freed, Amy
becoming(s), 158–9, 175, 191, 233, 241n. 22, 274–5, 277–9, 280–1
becoming(s)-animal, 230, 277–8, 281
becoming(s)-other(s), 7, 175, 185, 230
becoming(s)-woman, 236, 277–8
Behan, Tom, 264n. 2
Beier, A.L., 166n. 21
Belhassen, S., 210n. 24
Belsey, Catherine, 79n. 19
Berghahn, V.R., 106n. 4
Bevington, David, 72–3, 74, 79n. 19, 79–80n. 21, 80n. 24, 80–1n. 25

308 \ INDEX

Bhabha, Homi, 279
bin Laden, Osama, 12–13, 17, 19, 21
 see also Osamaspace
Bladerunner (dir. Ridley Scott), *see* Scott, Ridley
The Blair Witch Project (dir. Eduardo Sanchez), *see* Sanchez, Eduardo
Nelson, Tim Blake
 O (dir. Tim Blake Nelson), 141, 163n. 2, 165n. 8
Blau, Herbert, 219–20, 146, 240n. 20
Bloom, Allan, 55–6, 67, 77n. 3, 78n. 13
Blumenthal, Eileen, 218, 219–20, 221, 239n. 10
Boose, Lynda A., 140–1, 156–7
Bourdieu, Pierre, 24–6n. 4
Boyd, Gregory, 49
Bradley, A.C., 57, 62, 65, 77n. 5, 101
Braithwaite, Edward, 210n. 23
Branagh, Kenneth, xvii
 As You Like It, 163n. 2
 Hamlet (dir. Kenneth Brannagh), 157
 Love's Labor's Lost, 163n. 2
 Macbeth, 163n. 2
Brecht, Bertolt, xvii, 248, 274, 106n. 7
"Big Time, Lost," 85, 105n. 1
Coriolan, 85–105, 106n. 6
"The Solution" ("Die Lösung"), 86–7, 106n. 8
Verfremdung, 162, 166–7n. 23
Bristol, Michael D., 78n. 16, 89
Brockbank, Philip, 89, 93, 94, 100–1, 106–7n. 14, 107n. 16
Brotten, Jerry, 208n. 2
Brown, Paul, 190, 208n. 3
Buchel, Charles, 202, 205
Bugliosi, Vincent, 117

Burt, Richard, 30, 139, 140–1, 141–3, 156–7, 225, 226, 265nn. 5, 6
Burton, D.E., 210n. 21
Bush, George Jr.; President Bush, 17
Bush, Laura, 252
Butler, Judith, 166nn. 14, 15, 280
Butler, Martin, 105n. 3

Carlson, Marvin, 146, 147
Carmody, Jim, 163–4n. 3
Cartelli, Thomas, 210–11n. 25
Castiglione, Baldessare, 171
 Castiglione model, 177–8, 181, 185–6
Cattaneo, Peter
 The Full Monty (dir. Peter Cattaneo), 157
Cecil, Robert, 113, 255, 267n. 21
Césaire, Aimé, 196–7, 210n. 21
 A Tempest (*Une Tempête*), 196–200, 205–6
Chamberlain, John; Lord Chamberlain, 97
Christ, Jesus, 13, 296, 298
 see also JesusChristspace
chutzpah, 217, 247, 259–60, 273, 275
 see also investigative-expansive wherewithal
C^3I, 222–5, 228–9, 232
 see also Haraway, Donna
Cixous, Hélène, 266n. 18
Clare, Janet, 105n. 3
Clark, John Pepper, 210n. 23
Clark, Stuart, 132n. 6
Cleography, 275–81
Cleospace, 276
Clover, Carol, 234–5
Cohen, Alain J.J., 160–1
Cohen, Derek, 211n. 30
Coleridge, Samuel Taylor, 36, 66, 89, 106–7n. 14

commodity-fetish, 118
Contempt (Le Mépris), see Godard, Jean-Luc
Coriolan, see Brecht, Bertolt
Coriolanus, see Shakespeare, William
 see also Brecht, Bertolt
Cottino-Jones, Marga, 266n. 16
Creed, Barbara, 228, 239n. 15
Cruise, Tom, 161
crystalline artifact, 131
crystalline effect, 121, 127, 131
crystalline narration, 112, 126, 131
cultural materialism, 274
 see also subversion/containment paradigm
Cummings, Peter, 68

Danson, Lawrence, 79n. 19
Dasgupta, Guatam, 221, 238–9n. 9
Davenant, William
 The Tempest, or The Enchanted Island, 191–6, 198, 205–6, 210n. 15
David, Jacques-Louis, 228
Davis, Natalie Zemon, 253–4, 255
Deats, Sara Munson, 68, 72–3, 79n. 19, 79–80n. 21
de Certeau, Michel, 274–5, 276
deconstruction(ist), 19–20, 22, 289–90
Degas, Edgar, 229
Dekker, Thomas, 27n. 6, 133n. 15
Deleuze, Gilles, xvi, xvii, 24–6n. 4, 112, 274
 action-image, 118–19
 any-space-whatever, 119–21
 Cinema 2: The Time-Image, 118–29, 131–2
 "Coldness and Cruelty," 68–76, 241n. 24
 crystal-image, 120–2, 126, 131–2
 dream-image, 120
 movement-image, 118–19
 recollection-image, 120

A Thousand Plateaus: Capitalism and Schizophrenia, 64–5, 66, 241n. 22, 277–8, 281
 time-image, 118–20
Deneuve, Catherine, 117
Derrida, Jacques, 20, 25, 27n. 11, 166n. 14
Deutscher, Isaac, 97
Deveruex, Robert; Earl of Essex, 186, 249
Didion, Joan, 116, 131
dissective-cohesive mode (d.c. mode), 6–7, 32, 54–5, 58, 62–3, 190, 248, 265n. 10, 293
 see also transversal poetics
Dohrn, Bernadine, 133n. 17
Dollimore, Jonathan, xiii, 78n. 16, 89, 104, 267–8n. 25
dramaturgy, 43–9; 163–4n. 3
 actor-dramaturg, 138–9
 audience-dramaturg, 162
 investigative-expansive dramaturgy, xvi, 248, 253, 281
 see also Fo, Dario; Nudge, Nudge, Wink, Wink, Know What I Mean, Know What I Mean?; Pavis, Patrice; Wiens, Wolfgang
Dryden, John
 The Indian Emperor, or the Conquest of Mexico by the Spaniards, 193
 The Tempest, or The Enchanted Island, 191–6, 198, 205–6, 210n. 15
Dunne, Christopher
 Titus Andronicus (dir. Christopher Dunne), 241n. 28
Dunne, Dominick, 17
Dusinberre, Juliet, 144

Eagleton, Terry, 38
Eastwood, Clint, 116

Ebert, Roger, 114
Eco, Umberto, 146
Einstein, Albert, 294
Eliot, T.S., 98
Elisabetta: Almost By Chance a Woman (Quasi per caso una donna: Elisabetta), see Fo, Dario
Elizabeth (dir. Shekhar Kapur), see Kapur, Shekhar
Elizabeth I, Queen, 87, 96, 186, 261, 265n. 12, 267nn. 21, 24, 268n. 29
   as icon, with Shakespeare, 248–9, 251–2, 266n. 15
   rumors about, 261
   witchcraze, 112–13, 132nn. 3, 6
   see also Elizaspace; Fo, Dario
Elizabethan Act of 1563, 112–13, 132n. 6
Elizaspace, 259–63, 268n. 27
Ellis-Fermor, Una, 176
empathy, xv, 4–5, 13, 57
Engels, Friedrich
   Communist Manifesto, 87–8, 90–1
Erickson, Peter, 267–8n. 25
Eyes Wide Shut (dir. Stanley Kubrick), see Kubrick, Stanley

Fakespearean, 259, 268n. 26
Farnham, Willard, 89, 106–7n. 14
Farrell, Joseph, 246, 264n. 2, 265n. 4
Feuerbach, Ludwig, 274, 285n. 3
Fian, John, 113
Fiedler, Leslie, 114
Fiennes, Joseph, 142, 165n. 10
Final Boy, 234–7
Final Girl, 234–6
   see also Clover, Carol
Finkelpearl, Philip J., 105n. 3
Fitzpatrick, Joseph, 11, 24n. 3, 27n. 7, 50nn. 3, 5, 77n. 1, 80n. 22, 133n. 14, 164n. 4, 186n. 1, 208–9n. 5, 274, 285n. 4, 300n. 8

Fo, Dario, xvi, 245–63, 264nn. 1, 2, 265nn. 3, 4, 266nn. 16, 17, 267–8n. 25, 281
   Elisabetta: Almost By Chance a Woman (Quasi per caso una donna: Elisabetta), 247–63, 266nn. 13, 17, 266–7n. 19, 267n. 20
   grammelot, 256, 267n. 23
   guillareli, 245, 246, 264n. 2
   and *Hamlet,* 256–9, 265n. 7
   and Shakespeare, 246–9
   Theatre of Situation, 247
   throwaway theatre, 245–6, 264n. 1
Folger, Abigail, 111, 117
Forbidden Planet, 211–12n. 32
Foucault, Michel, 115, 274
   History of Sexuality, 24–6n. 4
   "What is an Author," 46–7
Franklin, Aretha, 280
Freed, Amy
   The Beard of Avon, 251, 266n. 15
Freud, Sigmund, 9–10, 13, 24–6n. 4, 46–7
   "The Uncanny," 240n. 18
   see also Freudspace
Freudspace, 9–10, 12–13
Frey, Charles, 209n. 10
Frykowski, Voityck, 111, 117
The Full Monty (dir. Peter Cattaneo), see Cattaneo, Peter
Fuller, Graham, 137, 140
Fuseli, Henry, 203
future-absent-spaces, 8
future-present-spaces, 7, 8

Garber, Marjorie, xiii, 153
Garcia, Jerry, 298
   see Grateful Dead, The
Ghosh, Amitav
   In An Antique Land, 281–4
Gildon, Charles, 89
Gilmore, John, 131

Gitlin, Todd, 133n. 17
Godard, Jean-Luc
    Contempt (Le Mépris), 139,
        160–1
Goffman, Erving, 147–8, 174
Goldenthal, Elliot, 221
Gosnell, Raja
    Never Been Kissed (dir. Raja
        Gosnell), 139, 140–1, 150,
        156
Grass, Günter, 89, 106–7n. 14
Grateful Dead, The, 296, 298
Greenberg, Harvey Roy, 239n. 13
Greenblatt, Stephen, xviii, 24–6n. 4,
    57, 78n. 12, 115, 208n. 1,
    209nn. 7, 10, 284n. 1, 294
Greg, W.W., 41, 273
Griffiths, Trevor, 209n. 7
Gronbeck-Tedesco, John, 68, 80n.
    23
Guattari, Félix, xvii, 26, 76
    Molecular Revolution, 24n. 3,
        50n. 3
    *A Thousand Plateaus: Capitalism
        and Schizophrenia*, 64–5, 66,
        241n. 22, 277–8, 281
Guevara, Ernesto Che, 200, 201
Guffey, George, 209n. 9
Gunpowder Plot, 113, 124
Gurr, Andrew, 174, 211n. 30

Hall, Kim, 194, 209–10n. 13
Hamlet, *see* Shakespeare, William
    see also Almereyda, Michael;
        Fo, Dario; Wilson, Robert
Hamlet (dir. Michael Almereyda),
    see Almereyda, Michael
Hamlet (dir. Kenneth Branagh),
    see Branagh, Kenneth
Hamlet: A Monologue, *see* Wilson,
    Robert
Hankey, Julie, 78n. 11
Hanna, Gillian, 266n. 13, 266–7n.
    19

Haraway, Donna, 222–3, 224, 225,
    231, 232, 234, 241n. 23
Haring-Smith, Tori, 144, 153–4
Harris, Jonathan Gil, xvii, 284n. 1,
    285nn. 3, 5
Hawke, Ethan, 141, 157, 161
Hawkes, Terence, 273
Hazlitt, William, 36
Hedrick, Donald, 9, 27n. 9, 30, 50n.
    2, 95, 107n. 17, 165n. 9, 178,
    187nn. 2, 3, 208n. 1, 217–18,
    247, 260, 261, 265n. 5, 279,
    299n. 3, 300n. 6
Hefner, Hugh, 112
Heinemann, Margot, 98, 105,
    106n. 6
Helter Skelter, *see* Manson, Charles
Henry V, *see* Shakespeare, William
Henry VI, Parts 1–3, *see* Shakespeare,
    William
Henslowe, Philip, 142
Hill, Christopher, 106n. 11
Hill, Errol, 202, 211n. 28
Hitler, Adolf, 13, 97, 98, 105
Hoffman, Michael
    A Midsummer Night's Dream (dir.
        Michael Hoffman), 157–8
Hogarth, William, 202
Holderness, Graham, 184
Holinshed, Raphael, 113–14, 133nn.
    11, 12
Honig, Edward, 89, 106–7n. 14
Hood, Stuart, 266–7n. 19
Hopkins, D.J., xvii, 29, 48, 163–4n.
    3, 174, 265n. 6, 268n. 27, 272,
    273, 279
Horowitz, Jeffrey, 220
Howard, Jean, xiii, 176, 177
Huffman, Clifford Chalmers, 89,
    106–7n. 14
Hulme, Peter, 209n. 10
Hunter, G.K., 53–6, 76–7, 77n. 6
Hüsges, H., 105
Hussein, Saddam, 21

Ide, Richard, 89, 101, 106–7n. 14
ideograph, 219–20
see also Blau, Herbert
In An Antique Land, see Ghosh, Amitav
incoherent discourse, 190
The Indian Emperor, or the Conquest of Mexico by the Spaniards, see Dryden, John informatics of domination, 224, 231, 232
see also Haraway, Donna
Ingman, Catherine, 174
Intriligator, James, xv, 6, 27n. 7, 32, 50n. 6, 77–8n. 7, 215, 273, 299nn. 1, 2, 300n. 7
investigative-expansive mode (i.e. mode), xv–xvii, 6–9, 11, 22, 27n. 7, 32–3, 55, 77–8n. 7, 215–17, 237n. 2, 273–4, 287–99, 300n. 8
 area of exploration, 288–9
 assemblages, 290
 critical mode, 289
 elems, 289–90
 investigative-expansive dramaturgy, see dramaturgy
 investigative-expansive thinker (i.e. thinker), 7–8
 investigative-expansive wherewithal, 215–17, 218–19, 221–2, 235, 247, 265n. 10
 laws, 290
 progressive quagmire, 291

Jackson, Rosemary, 240n. 18
Jacobean Act of 1604, 113, 132n. 6
Jaffa, Harry, 55–6, 67, 77n. 3, 78n. 13
Jagendorf, Zvi, 89, 107n. 15
James I, King, 96–7, 98, 100, 124
 The Basilicon Doron, 96
 Daemonologie, 113
 The Trew Law of Free Monarchies, 96
 witchcraze, 113–14, 132nn. 3, 6
Jameson, Frederick, 24–6n. 4, 164n. 5
Jenkins, Ron, 266n. 13
JesusChristspace, 17
Johnson, Samuel, 100–1, 106–7n. 14
Jorgenson, P.A., 107n. 18
Julius Caesar, see Shakespeare, William
Junger, Gil
 10 Things I Hate About You (dir. Gil Junger), 139, 141, 144, 150–6, 163

Kael, Pauline, 117
Kahlo, Frida, 229, 232
Kállay, Géza, 68, 79n. 19
Kapur, Shekhar
 Elizabeth (dir. Shekhar Kapur), 165n. 10, 265n. 11
Kasabian, Linda, 111
Kastan, David Scott, 208n. 2
Kavanagh, James H., 267–8n. 25
Kayser, Wolfgang, 240n. 18
Keegan, John, 98
Kenner, Ron, 131
Kermode, Frank, 208n. 1
Kernan, Alvin B., 105n. 3, 211n. 31
Kidman, Nicole, 161
King Lear, see Shakespeare, William
Kishlansky, Mark, 89
Kline, Kevin, 158
Kosslyn, Stephen, 299n. 2, 300n. 7
Krenwinkel, Katie, 111, 122, 130
Kristeva, Julia, 222, 227, 228, 239nn. 14, 15, 16, 239–40n. 17
Kubiak, Anthony, 123, 133n. 18
Kubrik, Stanley
 Eyes Wide Shut (dir. Stanley Kubrick), 161–2
Kuhn, Thomas, 294

Lacan, Jacques, 24–6n. 4, 280
Laclau, Ernesto
  Hegemony and Socialist Strategy,
    24–6n. 4
  *New Reflections on the Revolution of
    Our Time*, 24–6n. 4
Lamb, Charles, 36
Lamming, George, 210n. 23
Lang, Fritz, 160–1
The Last Action Hero (dir. John
  McTiernan), *see* McTiernan, John
Leavis, A.R., 62
Ledger, Heath, 151, 155
Lee, Canada, 206
Lehman, Courtney, 14–15, 241n.
  23, 280, 282
Levin, Carole, 261, 265n. 12, 267n.
  21, 268n. 29
Londré, Felicia Hardison, 143–4
Luhrman, Baz
  William Shakespeare's Romeo+Juliet
    (dir. Baz Luhrman), 165n. 7
Lurie, Susan, 225

Macbeth, *see* Shakespeare, William
  see also Polanski, Roman
Macbeth (dir. Roman Polanski), *see*
  Polanski, Roman
MacCallum, M.W., 89, 106–7n. 14
Macfarlane, Alan, 132n. 3
Madden, John
  Shakespeare in Love (dir. John
    Madden), 10, 13, 141–3,
    165n. 10, 251, 265n. 11
Maguire, Laurie E., 41
Manson, Charles, 11, 113, 115, 121,
  122, 123, 125–6, 127, 129–31,
  133nn. 13, 17
  Helter Skelter, 111, 121
  Manson murders, 111, 114–18,
    121, 122, 123, 126, 127,
    129–31
  see Polanski, Roman; witchcraft

Marcus, Leah, 41–3, 45, 144, 267n.
  21, 267–8n. 25
Marlowe, Christopher, 142–3
Mary, Queen of Scots, 258
Marx, Karl, 9–10, 13, 46–7, 274,
  275
  Communist Manifesto, 87–8,
    90–1
  *A Contribution to a Critique of
    Political Economy*, 104
  The German Ideology, 104
  "Thesis on Feuerbach," 274,
    285n. 3
  see also Marxspace
Marxspace, 9–10, 12–13
masochism, 68–77, 80n. 22
  see also Deleuze, Gilles
Maus, Katherine Eisaman, 210nn.
  16, 17
McKenzie, Jon, xiii
McMullan, John, 166n. 21
McTiernan, John
  The Last Action Hero (dir. John
    McTiernan), 141
Measure for Measure, *see* Shakespeare,
  William
The Merchant of Venice, *see*
  Shakespeare, William
The Merry Wives of Windsor, *see*
  Shakespeare, William
Middleton, Thomas, 27n. 6, 133n.
  15
A Midsummer Night's Dream, *see*
  Shakespeare, William
A Midsummer Night's Dream (dir.
  Michael Hoffman), *see* Hoffman,
  Michael Miller, F., 204
Mitchell, Tony, 258, 260, 263,
  264n. 1
*The Mock Tempest: Or, the Enchanted
  Castle*, 192
Mohammed, 12–13
Montrose, Louis, 78n. 16

314 \ INDEX

Monty Python's Flying Circus, 79n. 18, 163n. 1, 174
Moro, Aldo, 250
Morrisette, Billy
  Scotland, P.A. (dir. Billy Morrisette), 163n. 2, 164n. 6
Morton, Joe, 260
The Most Cruell and Bloody Murther Committed by an Innkeeper's Wife, 124–5
Mouffe, Chantal
  Hegemony and Socialist Strategy, 24–6n. 4
Muir, Kenneth, 113–14, 132nn. 8, 10, 133nn. 11, 12
Mullaney, Steven, 166n. 21, 272, 284n. 2
Müller, Heiner, 37
Munro, Ian, 266n. 18

Nadja (dir. Almeredya), see Almereyda, Michael
Never Been Kissed (dir. Raja Gosnell), see Gosnell, Raja
Newes From Scotland, 132n. 4
new historicism, xiii, 208n. 1, 274, 292, 274
  see also subversion/containment paradigm
Newman, Karen, 77n. 2
Nixon, Rob, 210n. 19, 210–11n. 25
Novak, Maximilian, 209n. 9
Novy, Marianne L., 144
Nudge, Nudge, Wink, Wink, Know What I Mean, Know What I Mean?, 137–9, 159–63, 163n. 1, 174, 279
Nudge-Wink-Know, 161–3

O (dir. Tim Blake Nelson), see Nelson, Tim Blake
official territory, 11, 13, 31
O'Pray, Michael, 240n. 18

Orlin, Lena Cowen, 72, 73, 79n. 19, 79–80n. 21
Osamaspace, 12–21, 284
Othello, see Shakespeare, William
Othello (dir. Oliver Parker), see Parker, Oliver

Paltrow, Gwyneth, 143, 165n. 10
Parent, Stephen, 111
Parker, Charlie, 294
Parker, Oliver
  Othello (dir. Oliver Parker), 78n. 15
Patterson, Annabel, 89, 106n. 10, 132n. 9
Pavis, Patrice, 138, 139, 163–4n. 3
Pechter, Edward, 79n. 19
Pericles, see Shakespeare, William
Pettet, E.C., 87, 88, 89, 106nn. 10, 11, 12, 106–7n. 14
Philips, James Emerson, 89, 106–7n. 14
Piccolo, Pina, 266n. 16
Podley, David, 268n. 26
Polanski, Roman
  cinematic Theater of Cruelty, 118, 121–4, 126, 131, 238n. 6
  Macbeth (dir. Roman Polanski), xvi, 11, 111–32, 238n. 6, 248, 274
  Rosemary's Baby, 111, 114
  see also Deleuze, Gilles; witchcraft
politics of affinity, 232, 280, 282–4
Pollard, A.W., 40–1
Porter, Joseph, 177
post-cinema(tic), xvi, 140, 156–63
Poulantzas, Nicos
  State, Power, Socialism, 23n. 1
  Political Power and Social Classes, 23–4n. 2
present-spaces, 7, 8
principle of translucency, xiii, 171–2, 217–18, 260, 279
  see also translucent effect
projective transversality, 173–6

INDEX / 315

Rabkin, Norman, 89, 106–7n. 14, 187n. 2
Rackin, Phyllis, xiii, 176, 181, 184
radical accountability, 226–7
Raleigh, Sir Walter, 96
Rame, Franca, xvi, 245–6, 249, 250, 252–3, 256, 260, 264n. 2, 266nn. 16, 17
see also Fo, Dario
Reagan, Ronald, 260–1
Red Brigades, 250
Reed, Lou, xiv, 29, 34, 48, 50n. 1, 51n. 9, 272
Rendra, W.S., 237n. 1
Retamar, Roberto Fernández
"Caliban: Notes Toward a Discussion of Culture in Our America," 199–201, 205–6
Reynolds, Bryan, xiii–xvii, 29, 30, 32, 50n. 2, 68, 76, 77–8n. 7, 79n. 18, 80n. 22, 106n. 8, 107n. 20, 174, 184, 215–17, 217–18, 260, 264, 268n. 27, 271, 272, 273, 274, 278, 279, 280, 281, 282, 284, 299nn. 1, 2
Becoming Criminal: Transversal Performance and Cultural Dissidence in Early Modern England, 23n. 1, 23–4n. 2, 24–6n. 4, 27n. 8, 50n. 3, 77nn. 1, 133n. 14, 164n. 4, 186n. 1, 187n. 4, 208–9n. 5, 209n. 8, 233, 241n. 22, 284n. 2, 285n. 6, 300nn. 8, 9 with Joseph Fitzpatrick, "The Transversality of Michel de Certeau: Foucault's Panoptic Discourse and the Cartographic Impulse," 24n. 3, 27n. 7, 50nn. 3, 5, 77n. 1, 133n. 14, 164n. 4, 186n. 1, 208–9n. 5, 284n. 4, 300n. 8

"The Devil's House, 'or worse':
Transversal Power and Antitheatrical Discourse in Early Modern England," 23n. 1, 23–4n. 2, 31, 50nn. 3, 4, 57–8, 66–7, 77n. 1, 78n. 16, 101, 107n. 20, 133n. 14, 158–9, 162, 164n. 4, 166n. 21, 186n. 1, 187n. 4, 208–9n. 5, 295, 300n. 9
with Ayanna Thompson, "Inspriteful Ariels: Transversal Tempests," 265n. 6
with Donald Hedrick, "Shakespace and Transversal Power," 27n. 9, 50n. 2, 165n. 9, 208n. 1, 247, 261, 299n. 3, 300n. 6
with Donald Hedrick, *Shakespeare Without Class: Misappropriations of Cultural Capital*, 265n. 5
with D.J. Hopkins, "The Making of Authorships: Transversal Navigation in the Wake of *Hamlet*, Robert Wilson, Wolfgang Wiens, and Shakespace," 163–4n. 3, 265n. 6
with Ian Munro, "Hélène Cixous," 266n. 18
with Janna Segal, "The Reckoning of Moll Cutpurse: A Transveral Enterprise," 24–6n. 4, 27n. 6, 133n. 15
with Joseph Fitzpatrick, "Venetian Ideology or Transversal Power? Iago's Motives and the Means by which Othello Falls," 300n. 8
Richard III, *see* Shakespeare, William
Ridley, M.R., 65, 77n. 5
Rix, Lucy, 210n. 20
Roach, Joseph, 210n. 15

Robbins, Rossell Hope, 132nn. 3, 6
Rodó, José Enrique
    Ariel, 200–1
Romeo and Juliet, *see* Shakespeare,
    William
Romeo Must Die (dir. Andrzej
    Bartkowiak), *see* Bartkowiak,
    Andrzej
Rommen, Ann-Christin, 44
Rose, Mary Beth, 72, 73, 79n. 19,
    79–80n. 21
Ross, Andrew, 231
Rossiter, A.P., 89, 101, 106–7n. 14
Rothwell, Kenneth, 114
Rouse, John, 37, 47, 107n. 21, 162,
    166–7n. 23
Rubin, Jerry, 133n. 17

sadism, 68–77, 80n. 22
sadomasochism, 68–77, 80n. 22
Said, Andrew, 275
Sanchez, Eduardo
    The Blair Witch Project (dir.
    Eduardo Sanchez), 157
Sarnecki, Judith Holland, 210n. 22
Sartre, Jean-Paul, 24
Scarry, Elaine, 123, 133n. 19
Schechner, Richard, 146, 147, 166n. 13
Schlegel, Augustus William, 55–6,
    58, 76, 77n. 6, 78n. 9
Schwartz, Bruce, 219, 221, 222,
    238–9n. 9
Schwarzenegger, Arnold, 141
Scofield, Martin, 89, 107n. 15
Scot, Reginald, 113, 132n. 5
Scotland, P.A. (dir. Billy Morrisette),
    see Morrisette, Billy
Scott, Ridley
    Bladerunner (dir. Ridley Scot), 231
Scuderi, Antonio, 246, 264n. 2,
    267n. 23
Sebring, Jay, 111

Segal, Janna, xvii, 11, 24–6n. 4, 27n.
    6, 68, 76, 79n. 18, 80n. 22,
    133n. 15, 240n. 19, 274, 281
Shakespace, xvi, 9–11, 30, 31, 247,
    261, 272, 299n. 3
Shakespeare in Love (dir. John
    Madden), *see* Madden, John
Shakespeare-relatedness, 140
Shakespeare, William
    Antony and Cleopatra, 184, 271,
        275–82, 284
    As You Like It, 141, 150
    Coriolanus, 9, 85–105, 274
    Hamlet, 40–3, 45, 143, 146, 161
    Henry V, 40, 171–97
    Henry VI, Parts 1–3, 88
    Julius Caesar, 259
    King Lear, 41, 248
    Macbeth, 11, 113–14, 115–16,
        117, 122, 123, 124–5, 126,
        127, 129–30, 132nn. 8, 1,
        133n. 12, 274
    Measure for Measure, 259
    The Merchant of Venice, 165–6n. 12
    The Merry Wives of Windsor,
        40, 41
    A Midsummer Night's Dream, 143
    Othello, xvi, 11, 53–77, 77nn. 4,
        5, 78nn. 13, 14, 80n. 24,
        80–1n. 25, 274, 282
    Pericles, 40
    Richard III, 41
    Romeo and Juliet, 13, 40, 141,
        142, 143, 165n. 7, 247
    Sonnet 56, 152, 166n. 17
    The Taming of the Shrew, 40–1,
        139, 141, 143–9, 150,
        153–4, 155, 156, 165nn. 11,
        12, 166n. 16, 174, 251
    Titus Andronicus, 165–6n. 12,
        215, 228, 232, 234, 239nn.
        11–12, 241n. 25; as icon,
        with Elizabeth I, 248–9,

251–2, 266n. 14; as
  theatrical character, see
  Madden, John; Freed, Amy
  see also Shakespace; Fo, Dario
Shakespearean loser, 141–3, 166n. 20
  see also Burt, Richard
Shapiro, Michael, 144
Siegel, Lee, 161
Simmons, J.L., 89, 106–7n. 14
Sinfield, Alan, xiii, 56, 62–3, 78n.
  16, 267–8n. 25
Skipitares, Theodora, 219, 221, 222,
  238–9n. 9
Skura, Meredith Ann, 209n. 7
sociopolitical conductors, 2, 3–4, 13,
  23n. 1
Socrates, 12
Sorge, Thomas, 89, 91–2
Stalin, Joseph, 96, 97–8
Stanislavski, Konstantin, 146
Starks, Lisa S., 239nn. 14, 16, 241n.
  27, 280, 282
state machinery, xv, 2, 3, 23–4n. 2,
  31, 99, 273
state power, 23–4n. 2, 31, 99
Stern, Jane and Michael, 133n. 17
Stiles, Julia, 141, 151, 155, 157,
  166n. 19
Stone, Oliver, 114
Stow, John, 87
Streitz, Paul, 266n. 14
subject
  identity-formation, see subjective
    territory; transversality
  of investigation, see
    investigative-expansive mode
    (i.e. mode)
subjectification
  see subjective territory
subjective territory, xv, 3–4, 24–6n.
  4, 31–2, 57, 99–100, 102,
  119–20, 144–5, 148–9,
  217–18, 295

subjunctive space, 4–5, 120
subjunctivity, 120
subversion/containment paradigm,
  62–3, 78n. 16, 272, 284n. 1
Summers, Montague, 209n. 9
Survin, Darko, 97, 98, 106n. 6,
  107n. 19

Takakuwa, Yoko, 79n. 19
The Taming of the Shrew, see
  Shakespeare, William
  see also Junger, Gil
Tate, Sharon, 111, 116, 117–18,
  121, 124
Taylor, Gary, 41
Taymor, Julie
  Fool's Fire, 222, 237n. 1
  The Haggadah, 238–9n. 9
  Juan Darien: A Carnival Mass,
    221, 222
  King Stag, 221
  Liberty's Taken, 238–9n. 9
  The Lion King, 230, 238–9n. 9
  Tirai, 238n. 8
  Titus (dir. Julie Taymor), xv,
    14–15, 215–37, 238n. 8,
    239n. 11, 240n. 21, 241nn.
    24, 25, 26, 29, 248, 280
  transversal aesthetic, 218–22,
    238n. 8, 238–9n. 9
  Way of Snow, 222, 238n. 8
Teatr Loh, 238n. 8
A Tempest (Une Tempête), see Cèsaire,
  Aimê
The Tempest, see Shakespeare, William
  see also Cèsaire, Aimê; Davenant,
    William; Dryden, John;
    Retamar, Roberto Fernández;
    Rodó, José Enrique
The Tempest, or The Enchanted Island,
  see Davenant, William; Dryden,
  John
The Tempest, An Opera, 192

318 \ INDEX

Tennenhouse, Leonard, 181, 267–8n. 25
terrorism, 123
   authorship, 16–21
   September 11, 2001, 12–13, 16–21
   theatrical, 123–4
   see also Manson, Charles; Polanski, Roman
theater of accountability, 225
Theater of Cruelty, *see* Artaud, Antonin
   cinematic Theater of Cruelty, 118, 121–4, 126, 131, 220
   see also Polanski, Roman; Taymor, Julie
10 Things I Hate About You, *see* Junger, Gil
Thiong'o, Ngugi wa, 210n. 23
Thompson, Ann, 208n. 1
Thompson, Ayanna, 9, 265n. 6, 278
Titus (dir. Julie Taymor), *see* Taymor, Julie
Titus Andronicus, *see* Shakespeare, William
   see also Taymor, Julie
Titus Andronicus (dir. Christopher Dunne), *see* Dunne, Christopher tragedy, 12–18
Traister, Rebecca, 165n. 8
translucent effect, xiv, 176, 195, 222, 260–1
translucent reading, 180
transversality
   of early modern England's public theatre, 50n. 3, 158–9, 162, 233, 284n. 2, 296
   transversal aesthetics, 219, 233
   transversal authorship, 45–9
   transversal dislocations, 171, 224
   transversalians, 294
   transversal intercourse, 191
   transversal methodology, *see* investigative-expansive mode (i.e. mode)
   transversal movement, xv, 4–5, 31–2, 57–8, 101, 102, 120, 292, 294–5
   transversal poetics, 7, 287–99
   transversal power, 5, 31–2, 66–7, 292, 294–5
   transversal territory, xv, 4–5, 31–2, 58, 66–7, 101, 119–20, 272, 295
   transversal theory, 2–8
   transversal thinker(s), 22
   see also Shakespace; state machinery; state power; subjective territory
Tricomi, Albert H., 96
trip-image, 126–7
Turner, Henry Ashby Jr., 86, 106n. 5
Turner, Victor, 146, 166n. 13
Tut, King, 12
Tynan, Kenneth, 112, 127–9

Valeri, Walter, 266n. 17
Vaughan, Virginia and Alden, 209n. 7
Venus, Nina, 105n. 1, 106n. 8
Volker, Klaus, 106n. 7

Wallace, David, 210n. 23
Warren, Michael, 41
Watson, Charles, 111, 130
Wayne, John, 116
Weir, Alison, 265n. 12, 267n. 24
Weller, Barry, 145
Wells, Stanley, 41, 210n. 14
Wiens, Wolfgang, 33, 34, 35, 36, 39, 43–9, 50n. 7, 272
Wilcox, Laurie, 175, 176
Will, George, 208n. 1
Willet, John, 106nn. 6, 7
William Shakespeare's Romeo+Juliet (dir. Baz Luhrman), *see* Luhrman, Baz

Williams, Gwyn, 145
Williams, Linda, 226
Willis, Deborah, 208n. 2
Wilson, John Dover, 93
Wilson, Robert
  authorship, 45–9, 50n. 7
  Hamlet: A Monologue, xiv, 29–49,
    50n. 7, 51n. 8, 265n. 8, 272–3
  textual medium, 43–5
  transversal performance, 35–6
witchcraft, 111–32
  Manson family accused of, 113,
    116–17, 122, 130–1
  nineteen-sixties counterculture,
    114, 115, 116–18, 130
  Polanski accused of, 117–18
  in Polanski's *Macbeth*, 114–16,
    122–32
  in Shakespeare's *Macbeth*, 115–16,
    124–6
witches, *see* witchcraft
Wittgenstein, Ludwig, 299
Wizinski, Sy, 115, 116
Wood, Sharon, 266n. 16
Worthen, William B., 33, 37–9, 46,
  140, 166n. 14

Zeeveld, Gordon W., 89, 106–7n.
  14
Zimmerman, Paul, 114
Žižek, Slavoj, 24–6n. 4
zooz, 299nn. 1, 4

GPSR Compliance
The European Union's (EU) General Product Safety Regulation (GPSR) is a set of rules that requires consumer products to be safe and our obligations to ensure this.

If you have any concerns about our products, you can contact us on

ProductSafety@springernature.com

In case Publisher is established outside the EU, the EU authorized representative is:

Springer Nature Customer Service Center GmbH
Europaplatz 3
69115 Heidelberg, Germany

www.ingramcontent.com/pod-product-compliance
Lightning Source LLC
LaVergne TN
LVHW041620060526
838200LV00040B/1358